Breaking Tolerance to Anti-Cancer Cell-Mediated Immunotherapy

T CELL METABOLISM AND CANCER IMMUNOTHERAPY

VOLUME 8

Breaking Tolerance to Anti-Cancer Cell-Mediated Immunotherapy

Series Editor: Benjamin Bonavida, PhD

Volume 1: Autophagy in Immune Response: Impact on Cancer Immunotherapy
Edited by Salem Chouaib

Volume 2: Immunotherapy in Resistant Cancer: From the Lab Bench Work to Its Clinical Perspectives
Edited by Jorge Morales-Monto and Mariana Segovia-Mendoza

Volume 3: Immunotherapeutic Strategies for the Treatment of Glioma
Edited by Michael Lim and Christopher Jackson

Volume 4: NK cells in cancer immunotherapy: Successes and Challenges
Edited by Anahid Jewett and Yuman Fong

Volume 5: Immune Landscape of Pancreatic Cancer Development and Drug Resistance
Edited by Batoul Farran and Ganji Purnachandra Nagaraju

Volume 6: Principles of Immunotherapy in Breast and Gastrointestinal Cancers
Edited by Michele Ghidini

Volume 7: γδ T Cell Cancer Immunotherapy
Edited by Marta Barisa

Volume 8: T Cell Metabolism and Cancer Immunotherapy
Edited by Jianxun Song

Upcoming Volumes in Series:

Breaking Tolerance to Pancreatic Cancer Unresponsive to Immunotherapy
Edited by Kasuya Hideki and Itzel Bustos Villalobos

Gammadelta T Cells in Cancer Immunotherapy
Edited by Ilan Bank

Breaking Tolerance to Anti-Cancer Cell-Mediated Immunotherapy

T CELL METABOLISM AND CANCER IMMUNOTHERAPY

VOLUME 8

Edited by

Jianxun Song
Department of Microbial Pathogenesis and Immunology, Texas A&M University Health Science Center, Bryan, TX, United States

ELSEVIER

ACADEMIC PRESS
An imprint of Elsevier

Academic Press is an imprint of Elsevier
125 London Wall, London EC2Y 5AS, United Kingdom
525 B Street, Suite 1650, San Diego, CA 92101, United States
50 Hampshire Street, 5th Floor, Cambridge, MA 02139, United States

ISBN: 978-0-443-13827-0

For information on all Academic Press publications
visit our website at https://www.elsevier.com/books-and-journals

Publisher: Stacy Masucci
Senior Acquisition Editor: Linda Versteeg-Buschman
Editorial Project Manager: Samantha Allard
Production Project Manager: Selvaraj, Raviraj
Cover Designer: Vicky Pearson Esser

Typeset by STRAIVE, India

Cover image insert

Immunotherapy as a human immune system therapy concept as a biomedical or biomedicine oncology treatment using the natural cancer-fighting properties of the body as a 3D render.

Aims and scope of series "Breaking Tolerance to Anti-Cancer Cell-Mediated Immunotherapy"

The role of the immune system in the eradication of cancers has been investigated for several decades with controversial findings. The controversy was the result of a poor understanding of the underlying mechanisms that govern responsiveness and unresponsiveness. Hence, significant advances have been made with respect to the regulation of the host immune response against cancer, and several immunotherapeutics have been recently introduced and used clinically. These include both antibody and cell-mediated immunity targeting the cancer cells. Such immunotherapies led to significant clinical responses in various cancer types that were unresponsive to conventional therapies. However, only a subset of cancer patients responds to such immunotherapeutics, and there is also a responding subset that develops resistance to further treatment. Various studies have examined potential underlying mechanisms involved in resistance and identified a variety of gene products that play pivotal roles in maintaining the resistant phenotype of the cancer cells to cell-mediated immunotherapy.

The main objective of the proposed series "Breaking Cancer Resistance to Cell-mediated Immunotherapies" is the development of individual volumes that are focused on the application of particular sensitizing agents that, when used in combination with cell-mediated immunotherapy, result in the reversal of resistance.

A variety of different classes of immunosensitizing agents have been reported. Each individual volume will focus on one class of immunosensitizing agents and their effects on the reversal of cell-mediated immune resistance in different cancers. The emphasis will be on biochemical, molecular, and genetic mechanisms by which the sensitizing agents mediate their effects individually and/or in combination with immunotherapy. Each editor will compile nonoverlapping review chapters on the therapeutic role of specific sensitizing agents used in combination with conventional immunotherapy and the reversal of resistance. There will also be an emphasis on the discrimination of responses obtained in various cancer types.

The scope of the series is to provide updated information to scientists and clinicians that is valuable in their quest to gather information, carry out new investigations, and develop novel immunosensitizing agents that are both more potent and also that might be active whereby the existing ones were not active.

Benjamin Bonavida, PhD (Series Editor)

About the Series Editor

Dr. Benjamin Bonavida, PhD (Series Editor), is currently Distinguished Research Professor at the University of California, Los Angeles (UCLA). His research career, thus far, has focused on basic immunochemistry and cancer immunobiology. His research investigations have ranged from the mechanisms of cell-mediated killing and sensitization of resistant tumor cells to chemo-/immunotherapies, characterization of resistant factors in cancer cells, cell-signaling pathways mediated by therapeutic anticancer antibodies, and characterization of a dysregulated NF-κB/Snail/YY1/RKIP/PTEN loop in many cancers that regulate cell survival, proliferation, invasion, metastasis, and resistance. He has also investigated the role of nitric oxide in cancer and its potential antitumor activity. Many of the above studies are centered on the clinically challenging features of cancer patients' failure to respond to both conventional and targeted therapies. The development and activity of various chemosensitizing agents and their modes of action to reverse chemo- and immunoresistance are highlighted in many refereed publications.

Acknowledgments: The Series Editor acknowledges the Department of Microbiology, Immunology, and Molecular Genetics and the UCLA David Geffen School of Medicine for their continuous support. The Series Editor also acknowledges the assistance of Mr. Rafael Teixeira, Acquisitions Editor for Elsevier/Academic Press, and the excellent assistance and support of Ms. Samantha Allard, Editorial Project Manager for Elsevier/Academic Press, for their continuous cooperation throughout the development of this book.

Aims and scope of the volume

Aims: This volume aims to provide a comprehensive examination of immunotherapy approaches, particularly focusing on the adoptive cell transfer (ACT) of cytotoxic T lymphocytes (CTLs). The book intends to explore the mechanisms behind the resistance of pathogenic cells to immune-therapeutic modalities, which limits the overall success of immunotherapy. By delving into the role of immunomodulators in improving the local environment, such as the tumor microenvironment (TME), the volume seeks to enhance the understanding of how these factors can be manipulated to selectively recognize and destroy pathogenic cells. The primary audience includes researchers, clinicians, and students in the fields of immunology, oncology, and cellular biology.

Scope: This book covers a broad range of topics within the field of immunotherapy and immunomodulation, including but not limited to:

- Detailed analyses of adoptive cell transfer (ACT) techniques involving CTLs.
- Mechanisms of resistance in pathogenic cells to current immunotherapeutic modalities.
- Role of immunomodulators in altering the TME to enhance immunotherapy effectiveness.
- Advances in targeting T cell metabolism as a strategy for improving immune regulation and therapeutic outcomes.
- Identification of new therapeutic targets and approaches to enhance the efficacy of immunotherapy by focusing on T cell metabolism.
- Case studies and experimental findings that illustrate the practical applications of these approaches in clinical settings.

The volume emphasizes both the theoretical underpinnings and practical applications of these advanced immunotherapeutic strategies, aiming to bridge significant knowledge gaps in T cell metabolism, immunomodulation, and overall immunotherapy. By providing insights into novel therapeutic opportunities, this book seeks to contribute to the development of more effective immunotherapeutic interventions. Topics outside the scope include unrelated metabolic pathways.

This volume is expected to serve as a valuable resource for advancing the understanding and application of immunotherapy, ultimately leading to improved outcomes in the treatment of various diseases.

Volume Editor Biography

Jianxun Jim Song, PhD
Professor, Department of Microbial Pathogenesis and Immunology at Texas A&M University Health Science Center, Bryan, TX

Dr. Song is renowned for his pioneering research on the immune system, often likened to the body's own set of superheroes. His work focuses on understanding how these immune cells recognize and combat pathogens, such as viruses and bacteria, and how they can be enhanced to better defend the body. His recent advancements in immunotherapy—particularly in cancer treatment—showcase innovative approaches to harnessing the body's natural defenses to fight disease.

At the Texas A&M University School of Medicine, Dr. Song's research represents a pinnacle of scientific inquiry and translational impact in the field of Cancer Immunology. His groundbreaking work spans four pivotal projects that explore the intricacies of T-cell immunity, metabolic reprogramming, and immunomodulatory targets. These projects have significant implications for enhancing cancer immunotherapy and addressing autoimmune disorders, making Dr. Song's contributions essential to both the scientific community and clinical practice.

Preface

The landscape of modern medicine is being transformed by the advent of immunotherapy, a powerful approach that harnesses the body's immune system to fight diseases, including various cancers and autoimmune disorders. Within this revolutionary field, the adoptive cell transfer (ACT) of cytotoxic T lymphocytes (CTLs) and regulatory T cells (Tregs) has shown remarkable promise. These methods leverage the body's inherent defense mechanisms to target and eradicate pathogenic cells. Despite the significant advancements, one of the major hurdles in immunotherapy is the resistance posed by pathogenic cells, which substantially limits the success of these treatments. Overcoming this resistance is critical for improving patient outcomes and advancing the efficacy of immunotherapy.

This volume is dedicated to addressing these challenges by exploring the role of immunomodulators in enhancing the local immune environment, such as the tumor microenvironment (TME). By improving these environments, immunomodulators can bolster the immune system's ability to selectively recognize and destroy pathogenic cells, thereby enhancing the effectiveness of immunotherapeutic interventions. The insights gathered in this book aim to significantly advance our understanding and application of immunotherapy, particularly in the context of cancer treatment.

The content of this volume is comprehensive, covering a wide array of topics within the realm of immunotherapy and immunomodulation. It includes detailed analyses of the mechanisms by which T cells function, how they can be metabolically reprogrammed to improve their efficacy, and the identification of immunomodulatory targets that can be manipulated to enhance therapeutic outcomes. By delving into these areas, this book not only reveals critical determinants of immune regulation but also uncovers unique therapeutic opportunities to improve immunotherapy through targeting T cell metabolism.

Our understanding of T cell metabolism and its impact on immune regulation has evolved significantly over the past few years. T cells, which play a crucial role in the immune response, rely on specific metabolic pathways to function effectively. By understanding these pathways, researchers can develop strategies to enhance T cell function, making them more effective in recognizing and destroying pathogenic cells. This volume aims to bridge the gap between fundamental research and clinical application, providing readers with a comprehensive overview of the latest advancements in T cell metabolism and its implications for immunotherapy.

In addition to theoretical discussions, this volume includes practical applications and case studies that illustrate the real-world impact of these advancements. The case studies provide valuable insights into how these theoretical concepts are being translated into clinical practice, highlighting the potential for improved patient outcomes. These examples underscore the importance of interdisciplinary collaboration in advancing the field of immunotherapy, as the integration of research findings from various disciplines is crucial for developing effective treatments.

The contributors to this volume are leading experts in the field, whose collective research and insights have made this book an invaluable resource for researchers, clinicians, and students. Their dedication to advancing our understanding of immunotherapy and immunomodulation has resulted in a compilation of cutting-edge research and innovative approaches that have the potential to transform patient care. I am deeply grateful to these contributors for their hard work and commitment to excellence.

As we continue to explore the complexities of the immune system and its role in disease, it is my hope that this volume will serve as a catalyst for further innovation and discovery. By providing a comprehensive overview of the current state of the field and highlighting areas of future research, this book aims to inspire the next generation of scientists and clinicians to push the boundaries of what is possible in immunotherapy. The ultimate goal is to improve patient outcomes and provide new hope for those battling cancer and other diseases.

In conclusion, this volume represents a significant step forward in our understanding of immunotherapy and its potential to revolutionize patient care. By addressing the challenges of pathogenic cell resistance and exploring the role of immunomodulators in enhancing the immune response, we can pave the way for more effective and targeted treatments. I invite you to delve into the rich content of this book and join us in the quest to unlock the full potential of the immune system in the fight against disease.

Dr. Jianxun Jim Song
Professor of Microbial Pathogenesis and Immunology
Texas A&M University Health Science Center

Contents

7. The combination of local treatment and immunotherapy

Bo Zhu, Jianguo Sun, Qingzhu Jia, Zhi Xu, Yongdong Feng, Luping Zhang, Lingyou Sun, Yaxian Qi, Bin Wang, Yu Liu, Lei Wang, Zheng Jin, Yuxuan Wei, and Shimeng Liang

8. T cell metabolism and cancer immunotherapy

Qianqian Li and Zhuangzhuang Zhang

9. Exploring the interplay of kinases and transcription factors in immunometabolism

Darby J. Ballard, Liqing Wang, Jugal Kishore Das, Anil Kumar, Yijie Ren, and Jianxun Song

10. Targeting engineered bacteria to improve cancer immunotherapy

Aiswarya Dash, Sonakshi Bahuguna, and Jugal Kishore Das

11. *Salmonella*: A bacterial warrior in cancer immunotherapy

Wei Kong, Lingchen Fu, and Yixin Shi

12. Oncolytic virus for cancer immunotherapy

Penghui Yang and Hao Yang

13. Advances in immunotherapy for colorectal cancer

Shuai Lu, Benqiang Rao, and Penghui Yang

Contributors

Sonakshi Bahuguna Department of Life Science, School of Natural Sciences, Shiv Nadar University, Uttar Pradesh, India

Haotian Bai Department of Molecular Microbiology and Immunology, Johns Hopkins Bloomberg School of Public Health, Baltimore, MD; Department of Microbiology and Immunology, Lineberger Comprehensive Cancer Center, Chapel Hill, NC, United States

Xue-Feng Bai Department of Pathology, College of Medicine and Comprehensive Cancer Center, Ohio State University, Columbus, OH, United States

Darby J. Ballard Department of Microbial Pathogenesis and Immunology, Texas A&M University Health Science Center, Bryan, TX, United States

Gerold Bepler Department of Oncology, Karmanos Cancer Institute, Wayne State University School of Medicine, Detroit, MI, United States

Jugal Kishore Das Department of Microbial Pathogenesis and Immunology, Texas A&M University Health Science Center, Bryan, TX, United States; Department of Life Science, School of Natural Sciences, Shiv Nadar University, Uttar Pradesh, India

Aiswarya Dash Department of Medicinal Chemistry and Biophysics Kemihuset (K), Campus, Umeå University, Umeå, Sweden

Yongdong Feng Xinqiao Hospital, Army Medical University, Chongqing, China

Lingchen Fu Biodesign Center for Applied Structural Discovery, Arizona State University, Tempe, AZ, United States

Yukai He Georgia Cancer Center; Department of Medicine, Medical College of Georgia, Augusta University, Augusta, GA, United States

Shao-Hsi Hung Department of Melanoma Medical Oncology, The University of Texas MD Anderson Cancer Center, Houston, TX, United States

Mohamed Shawky Hussein Georgia Cancer Center, Medical College of Georgia, Augusta, GA, United States

Qingzhu Jia Xinqiao Hospital, Army Medical University, Chongqing, China

Zheng Jin Xinqiao Hospital, Army Medical University, Chongqing, China

Wei Kong Biodesign Center for Applied Structural Discovery, Arizona State University, Tempe, AZ, United States

Anil Kumar Department of Microbial Pathogenesis and Immunology, Texas A&M University Health Science Center, Bryan, TX, United States

Qianqian Li Guangdong-Hong Kong-Macao University Joint Laboratory of Interventional Medicine, The Fifth Affiliated Hospital, Sun Yat-sen University, Zhuhai, China

Yating Li Department of Melanoma Medical Oncology, The University of Texas MD Anderson Cancer Center, Houston, TX, United States

Shimeng Liang Xinqiao Hospital, Army Medical University, Chongqing, China

Cho-Hao Lin Department of Pathology, College of Medicine and Comprehensive Cancer Center, Ohio State University, Columbus, OH, United States

Jin-Qing Liu Department of Pathology, College of Medicine and Comprehensive Cancer Center, Ohio State University, Columbus, OH, United States

Yu Liu Xinqiao Hospital, Army Medical University, Chongqing, China

Shuai Lu Department of General Surgery, Beijing Anzhen Hospital Affiliated to Capital Medical University, Beijing, China

Agnes Malysa Cancer Biology Graduate Program; Department of Oncology, Karmanos Cancer Institute, Wayne State University School of Medicine, Detroit, MI, United States

Rui Mao Georgia Cancer Center, Medical College of Georgia, Augusta, GA, United States

Yaxian Qi Xinqiao Hospital, Army Medical University, Chongqing, China

Benqiang Rao Department of Gastrointestinal Surgery/Clinical Nutrition, Beijing Shijitan Hospital Affiliated to Capital Medical University, Beijing, China

Yijie Ren Department of Microbial Pathogenesis and Immunology, Texas A&M University Health Science Center, Bryan, TX, United States

Yixin Shi School of Life Sciences, Arizona State University, Tempe, AZ, United States

Shailbala Singh Department of Melanoma Medical Oncology, The University of Texas MD Anderson Cancer Center, Houston, TX, United States

Jianxun Song Department of Microbial Pathogenesis and Immunology, Texas A&M University Health Science Center, Bryan, TX, United States

Jianxun Jim Song Department of Microbial Pathogenesis & Immunology, School of Medicine, University of Texas A&M, Bryan, TX, United States

Jianguo Sun Xinqiao Hospital, Army Medical University, Chongqing, China

Lingyou Sun Xinqiao Hospital, Army Medical University, Chongqing, China

Yisong Y. Wan Department of Microbiology and Immunology, Lineberger Comprehensive Cancer Center, Chapel Hill, NC, United States

Bin Wang Xinqiao Hospital, Army Medical University, Chongqing, China

Lei Wang Xinqiao Hospital, Army Medical University, Chongqing, China

Liqing Wang Department of Microbial Pathogenesis and Immunology, Texas A&M University Health Science Center, Bryan; Department of Biochemistry and Biophysics, Texas A&M University, College Station, TX, United States

Yuxuan Wei Xinqiao Hospital, Army Medical University, Chongqing, China

Zhi Xu Xinqiao Hospital, Army Medical University, Chongqing, China

Hao Yang Faculty of Hepatopancreatobiliary Surgery, The First Medical Center of Chinese PLA General Hospital, Beijing; Department of General Surgery, Affiliated Taian City Central Hospital of Qingdao University, Taian, China

Penghui Yang Faculty of Hepatopancreatobiliary Surgery, The First Medical Center of Chinese PLA General Hospital, Beijing, China

Cassian Yee Department of Melanoma Medical Oncology, The University of Texas MD Anderson Cancer Center, Houston, TX, United States

Luping Zhang Xinqiao Hospital, Army Medical University, Chongqing, China

Xiaohong Mary Zhang Department of Oncology, Karmanos Cancer Institute, Wayne State University School of Medicine, Detroit, MI, United States

Zhuangzhuang Zhang Guangdong-Hong Kong-Macao University Joint Laboratory of Interventional Medicine, The Fifth Affiliated Hospital, Sun Yat-sen University, Zhuhai, China

Bolin Zhao Department of Pathology, College of Medicine and Comprehensive Cancer Center, Ohio State University, Columbus, OH, United States

Bo Zhu Xinqiao Hospital, Army Medical University, Chongqing, China

CHAPTER

1

Modulating T cell metabolism to improve T cell therapy

Yating Li[*], *Shao-Hsi Hung*[*], *Shailbala Singh*, *and Cassian Yee*

Department of Melanoma Medical Oncology, The University of Texas MD Anderson Cancer Center, Houston, TX, United States

Abstract

Cancer immunotherapy has transformed conventional approaches to cancer treatment, leveraging the body's immune system to target cancer cells. While adoptive cell therapy and immune checkpoint inhibitors have yielded substantial successes in treating hematological malignancies and immunogenic cancer types, a considerable proportion of patients with solid tumors do not respond to these treatments. The effectiveness of cancer immunotherapy relies on various factors, encompassing both endogenous and synthetic immunity. Moreover, patients exhibit improved outcomes when there is an augmented population of stem-like memory or precursor characteristics in endogenous or adoptively transferred T cells. In this chapter, we will discuss the basics of T cell metabolism and metabolic features in the tumor microenvironment and focus on strategies to overcome T cell metabolic constraints for an enhanced antitumor immune response.

Abbreviations

2-HG 2-hydroxyglutarate
α-KG alpha-ketoglutarate
A2AR A2A adenosine receptor
ADP adenosine diphosphate
AHR aryl hydrocarbon receptor
AKT protein kinase B
AMP adenosine monophosphate
AQP9 aquaporin 9
Arginase-1 enzyme that hydrolyzes arginine
ATP adenosine triphosphate
cAMP cyclic adenosine monophosphate
CAR-T chimeric antigen receptor T cell
CB-839 glutaminase inhibitor

[*]These authors contributed equally.

T Cell Metabolism and Cancer Immunotherapy
https://doi.org/10.1016/B978-0-443-13827-0.00008-1

CD	cluster of differentiation
CD28	cluster of differentiation 28
CD8 T cells	cytotoxic T cells
CD80	cluster of differentiation 80
CD86	cluster of differentiation 86
DNA	deoxyribonucleic acid
DO	indoleamine 2,3-dioxygenase
DON	6-diazo-5-oxo-l-norleucine (glutamine antagonist)
ERK	extracellular signal-regulated kinase
EZH2	enhancer of zeste homolog 2
FA	fatty acid
FAO	fatty acid oxidation
GAPDH	glyceraldehyde 3-phosphate dehydrogenase
GLS1	glutaminase
GLUT	glucose transporter
GLUT1	glucose transporter 1
HDACi	histone deacetylase inhibitor
HIF-1α	hypoxia-inducible factor 1-alpha
IACS-6274	glutaminase inhibitor
ICB	immune checkpoint blockade
IDH	isocitrate dehydrogenase
IDO	indoleamine 2,3-dioxygenase
IFNγ	interferon gamma
IL	interleukin
IL-2	interleukin-2
IL-12	interleukin-12
JAK	janus kinase
KEAP1	kelch-like ECH-associated protein 1
LDH	lactate dehydrogenase
MAPK	mitogen-activated protein kinase
MCT	monocarboxylate transporter
MDSC	myeloid-derived suppressor cell
MEK	mitogen-activated protein kinase
mTOR	mammalian target of rapamycin
NAD+	nicotinamide adenine dinucleotide
NADH	nicotinamide adenine dinucleotide (reduced form)
NFAT	nuclear factor of activated T cells
NK cells	natural killer cells
NMA	nonmyeloablative
OXPHOS	oxidative phosphorylation
PD-1	programmed cell death protein 1
PD-L1	programmed death-ligand 1
PEG	polyethylene glycol
PGC1α	peroxisome proliferator-activated receptor gamma coactivator 1-alpha
PGDH	phosphoglycerate dehydrogenase
PI3K	phosphoinositide 3-kinase
PKC	protein kinase C
PP2	protein phosphatase 2
RNA	ribonucleic acid
ROS	reactive oxygen species
siRNA	small interfering RNA
SLC1A5	solute carrier family 1 member 5 (glutamine transporter)
SNAT1	sodium-coupled neutral amino acid transporter 1

SNAT2	sodium-coupled neutral amino acid transporter 2
STAT5	signal transducer and activator of transcription 5
TAG	triglyceride
TCR	T cell receptor
TDO	tryptophan 2,3-dioxygenase
TGFβ	transforming growth factor beta
TIL	tumor-infiltrating lymphocyte
TIM3	T cell immunoglobulin and mucin domain containing-3
TLR	toll-like receptor
TME	tumor microenvironment
Tregs	regulatory T cells
TSC	tuberous sclerosis complex
VEGF	vascular endothelial growth factor
ZAP70	zeta-chain-associated protein kinase 70

Conflict of interest

No potential conflicts of interest were disclosed.

The basics of T cell metabolism

Brief overview of T cell metabolism

Naïve T cells adapt oxidative phosphorylation (OXPHOS) to support their basal metabolic consumption for long-term survival. Upon antigen stimulation, either from an infection or a tumor, these naïve cells undergo differentiation, clonal expansion, and acquisition of effector functions, which present significant energetic requirements. Activated T cells rapidly switch their metabolic program to rely on aerobic glycolysis for high energy demand and synthesis of biomass [1–3]. In addition, the pentose phosphate pathway and glutaminolysis pathway [4] are upregulated to further support clonal expansion. Upon pathogen clearance, the majority of effector T cells contract, while a subset of memory cells persists to provide long-term protection. These memory cells adapt a metabolically quiescent state and rewire to fatty acid oxidation (FAO) and OXPHOS to sustain self-renewal [5–7]. Metabolic reprogramming during different stages of T cells indicates specific requirements for energy sources. It represents more than a consequence of T cell activation but rather a parameter that dictates cell fate, proliferation, and function.

Nutrient sources

Glucose

Glucose, fatty acids, and glutamine are three major sources of energy fuel for T cells [1]. Glucose is transported into T cells via glucose transporter 1 (GLUT1) and subsequently converted to pyruvate via the glycolytic pathway [8,9]. Pyruvate, as a critical metabolic intermediate, can be broken down to acetyl-CoA and enter the tricarboxylic acid (TCA) cycle in mitochondria, where OXPHOS is engaged to fully oxidize carbon compounds and produce 32 ATP molecules. Alternatively, pyruvate is converted to lactate in the cytoplasm by lactate dehydrogenase (LDH) even in the presence of sufficient oxygen. This process of aerobic glycolysis is also known as Warburg Effect [10], a hallmark of cancer cell metabolism.

The conversion from pyruvate to lactate is highly inefficient in terms of ATP production. T cells switch to this metabolic program upon activation as it rapidly generates a vast number of metabolic intermediates as building blocks for macromolecule synthesis to meet the high demand of biomass during cell proliferation, differentiation, and effector function.

Lipid metabolism

In addition to glucose, lipids are another important source of energy and can often function as signaling mediators that affect T cell function. Extracellular lipids, such as fatty acids and cholesterols, are taken up by T cells through fatty acid (FA) receptors, which include the G protein-coupled receptors, CD36, fatty acid transport proteins, and lipoprotein receptors. Lipids in extracellular environment have multiple effects on T cell function depending on the type of lipids, length of the FA, degree of saturation, and level of oxidation. For example, short-chain fatty acids (SCFAs) promote memory differentiation [11], while long-chain fatty acids (LCFAs) [12] or cholesterol [13] impair the effector function of CD8 T cells. The pool of intracellular lipid content is maintained by balancing lipid de novo synthesis, FAO, and lipid storage. Upon T cell activation, critical enzymes involved in fatty acid synthesis are upregulated to generate sufficient lipid content for cell proliferation. Acetyl-CoA, which is catabolized in the glycolytic pathway, is carbonylated to malonyl-CoA, entering fatty acid synthesis (FAS) pathway for the generation of various fatty acids [14]. Alternatively, it can be converted to mevalonate to synthesize cholesterol [15]. In memory T cells, however, lipid catabolism [6] is engaged by FAO in mitochondria, which generates acetyl-CoA to fuel the TCA cycle for energy production. Interestingly, substrates for FAO in central memory CD8 T cells are not FAs taken up from their environment, but rather the lipids de novo synthesized from extracellular glucose [16]. In contrast, regulatory T cells (Treg) heavily rely on extracellular FAs for their metabolic expenditures [17]. They express high levels of fatty acid-binding proteins (FABPs) for FA transport as well as enzymes for FAO. Depleting extracellular FAs or inhibiting FAO impairs inducible Treg differentiation and function [18,19].

Amino acid metabolism

Another essential nutrient source for T cells is glutamine. T cells upregulate glutamine transporter and glutaminolysis-related enzymes upon activation [20,21]. Alpha-ketoglutarate that is generated from glutaminolysis can directly fuel TCA cycle. Glutamine also maintains redox homeostasis by converting to glutathione, which balances reactive oxygen species (ROS) produced by electron transport chain (ETC) in mitochondria. In addition, glutamine provides precursors for nucleotide and amino acid synthesis to support T cell proliferation. However, blocking glutaminolysis has differential effects on distinct T cell subsets [22]. In a study using either pharmacological inhibitor or genetic deletion to block glutaminase—the first step of glutamine metabolism, Rathmell and others found the cytotoxic T lymphocytes and Th1 cells displayed improved effector function while differentiation of Th17 cells was impaired, suggesting cell type-specific requirement for glutamine metabolism [23].

Mitochondria in T cell metabolism

Despite relying on aerobic glycolysis during the effector phase, activated T cells upregulate OXPHOS activity and later rewire to FAO for memory cell differentiation.

The electron transport chain, which transfers electrons via redox reactions to couple OXPHOS, is located at the mitochondria inner membrane, rendering mitochondria the central venue for oxidative metabolism. Mitochondria also function as a critical module to integrate various signaling events, including calcium signaling upon TCR stimulation [24]. In addition, the positioning of mitochondria is another important parameter affecting T cell function. They accumulate in cellular compartments with high demand of ATP, such as immune synapse for T cell stimulation [25] and polarized region for migration [26]. Mitochondrial biogenesis and dynamics also affect T cell activation and differentiation [27–30]. Upon stimulation, T cells rapidly increase mitochondrial mass and mitochondrial DNA to support proliferation. Mitochondria fission in effector T cells promotes cristae expansion for reduced ETC efficiency and increased aerobic glycolysis, while they fuse in memory T cells with tightened cristae so that ETC protein complexes are in close proximity for efficient oxidative metabolism. Furthermore, mitochondrial ROS, as a byproduct of ETC, is indispensable for T cell signaling [31]. Even though high ROS level is detrimental to T cells, transient increase of ROS triggered by TCR stimulation is necessary to activate critical transcription factors NF-kB, AP-1, and NFAT for T cell stimulation [32,33]. Overall, the physiological activities of mitochondria are critical for multiple aspects of T cell function.

Metabolic features of regulatory T cells

Tregs present a unique subset of T cells that suppress the immune system to maintain tolerance and prevent tissue damage. Treg cells can originate from thymus (tTreg) or differentiate from naïve CD4 T cells in the periphery (pTreg) [34,35]. They can be distinguished by the expression of Helios and neuropilin-1 [36,37]. Yet both rely on the master transcription factor Foxp3, which dictates their metabolic programs to be highly similar. Tregs employ a distinct metabolic program than other types of T cells. Instead of relying on glycolysis, activated Tregs mainly utilize FAO for energy production [17,18]. They express high levels of CD36 for lipid transportation and proteins involved in lipid metabolism [38]. Supplementing Tregs with exogenous SCFA enhances proliferation and suppressive function [39]. On the other hand, inhibiting lipid uptake or FAO impairs Tregs [40,41]. Though glycolysis is considered to be required for Treg migration, the metabolic reprogramming away from glycolysis is driven by Foxp3, which represses expression of genes in the glycolysis pathway and promotes oxidative metabolism [42]. Due to high dependency on oxidative metabolism, mitochondrial ROS production is much higher in Tregs than effector T cells. However, the elevated ROS level is not detrimental to Tregs due to high intracellular level of antioxidant glutathione [43]. In turn, ROS maintains Foxp3 expression via transcription factor BACH2 [44,45]. Another essential fuel for effector T cells, amino acids, is not required for Treg metabolism. Preventing uptake of glutamine by genetically deleting amino acid transporter decreases effector T cell differentiation while sparing Treg generation [23]. In line with this, depleting exogenous glutamine redirects CD4 T cell differentiation to Tregs rather than effector phenotype [22]. Overall, the distinct metabolic profile of Treg may present survival benefits that stand out in a harsh tumor microenvironment, which will be discussed later.

TME affects T cell metabolism

Tumor microenvironment is composed of both cancer cells and various noncancerous cells, including immune cells, stromal cells, fibroblasts, endothelial cells, and pericytes, shaping a highly heterogenous niche [46]. Each cell type has distinct metabolic programs, displaying unique preferences for nutrients. For example, cancer cells are the top consumer of glutamine [47] while myeloid-derived suppressor cells (MDSCs) have the highest uptake of glucose [48–50]. The uneven distribution of nutrients further contributes to metabolic heterogeneity within the TME. Most solid tumors have insufficient angiogenesis and irregularly branched vasculature, rendering limited vascular exchange. Nutrients and oxygen become limited, forcing different types of cells to compete for fuel, while the waste removal is constrained, leading to buildup of certain toxic metabolic end products. Hypoxia, as a hallmark of cancer, compromises cells that require oxidative metabolism. It promotes effector T cell exhaustion while enhancing Treg and MDSC differentiation [41,51]. Necrotic regions, often found within the TME, are niches with severe and prolonged nutrient deprivation, which present as a selection pressure for cancer cells [52]. Cancer cells evolve to cope with the metabolic stress, whereas T cells are less metabolically plastic for survival. Overall, intrinsic cancer cell metabolism, anatomical location, and metabolic interaction of cancer cells with neighboring cells contribute to the metabolically harsh TME. This metabolic stress renders tumor-infiltrating T cells exhausted and forces them to adopt a metabolic program distinct from resting or acutely activated T cells, profoundly impacting T cell function as well as the overall antitumor immunity.

Nutrient deprivation

Fast-proliferating cancer cells require abundant nutrients to support their large demand for biomass synthesis and energy for essential functions. Oncogenic pathways or loss of tumor suppressors often rewire cancer cell metabolism to "Warburg effect" [53], where upregulation of glycolysis, even in the presence of sufficient oxygen, leads to glucose deprivation. In addition, cancer cells upregulate lipolysis and glutaminolysis, further depleting the availability of common carbon nutrients, such as lipids, cholesterol, and glutamine.

Similar to cancer cells, activated CD8 T cells engage in aerobic glycolysis, competing with cancer cells for glucose. Glucose deprivation in TME limits infiltration, proliferation, and effector function of CD8 T cells [54]. Mechanistically, phosphoenolpyruvate, a glycolytic metabolite, facilitates TCR-dependent Ca + and NFAT signaling in CD4 T cells, which is decreased in a glucose-poor condition [55]. The low glucose availability also suppresses the production of cytokines and effector molecules, such as IFNγ, both transcriptionally and translationally [56,57]. Lastly, glucose deprivation inhibits EZH2-mediated epigenetic programming of effector CD8 T cells, resulting in a subverted antitumor immunity [58].

Besides glucose, amino acid levels such as glutamine, arginine, tryptophan, and alanine are altered in TME. Several types of cancer upregulate glutamine transporter SLC1A5 [59,60] and genes involved in glutaminolysis, resulting in potential competition of glutamine between cancer cells and effector T cells. Arginine is another example of how an altered amino acid level impairs T cell function. Some cancer cells rely on arginine for macromolecule synthesis, lowering the extracellular level of arginine in TME and access for T cells. Lack of intracellular arginine reduced T cell proliferation and effector function [61]. Moreover, MDSCs upregulate Arginase-1 to consume arginine, further reducing the extracellular concentration to suppress T cell function [62]. Meanwhile, Nitric oxide (NO) synthase expressed by MDSC breaks down arginine to generate NO, which induces T cell apoptosis and inhibits T cell activation and proliferation [63–65].

Immunosuppressive metabolites

Tumor metabolites rapidly accumulate within the TME due to lack of sufficient vasculature for waste clearance. Certain metabolites are highly immunosuppressive, dampening the antitumor immunity and contributing to cancer cell immune evasion.

Lactic acid and acidosis

Cancer cells engage aerobic glycolysis, breaking down glucose to pyruvate and then converting to lactate by -lactate dehydrogenase (LDH) [66]. An increasing amount of lactate is transported out to the extracellular milieu via monocarboxylate transporter (MCT) [67]. As a proton pump, MCT exports proton and lactate simultaneously, generating lactic acid, which results in a highly acidic TME [68]. The high concentration of lactic acid in TME prevents the export of intracellular lactate within T cells, thereby inhibiting T cell glycolysis, reducing NFAT signaling, and impairing T cell function [69,70]. Moreover, regulatory T cells and MDSCs are able to adapt to the low pH environment [71]. Lactate upregulates PD-1 in Tregs [72] while inducing arginase-1 expression in MDSCs [73,74], both critical for their suppressive functions. Clinically, high serum LDH level is associated with poor prognosis and reduced overall survival in various solid tumors. In human cervical cancers and melanoma, high lactate concentration in primary lesions strongly correlates with the incidence of metastasis as well as tumor recurrence [75,76].

Kynurenine

Kynurenine is generated from tryptophan catabolism by enzymes indoleamine-2,3-dioxygenase (IDO) or tryptophan-2,3-dioxygenase (TDO). Both cancer cells and MDSCs express high levels of IDO to deprive tryptophan in TME, leaving an excessive amount of kynurenine. Kynurenine induces T cell apoptosis [77,78], inhibits antigen-specific T cell proliferation [79], and suppresses T cell cytotoxicity [80]. In addition, kynurenine functions as a potent ligand for aryl hydrocarbon receptor (AHR) [81,82]. Activation of AHR on CD8 T cells upregulates PD-1 expression [83], while in CD4 T cells it induces Treg differentiation [84,85].

Adenosine

Adenosine is a well-established immunosuppressive metabolite generated by hydrolyzing ATP. Cancer cells, MDSCs, as well as Tregs express CD39 and CD73, which convert ATP to ADP/AMP and AMP to adenosine, respectively [86,87]. There are four G-protein-coupled adenosine receptors, namely A1, A2A, A2B, and A3, with A2AR being the dominant receptor expressed on T cells [88]. Signaling through A2A receptor, adenosine inhibits IL-2 and IFNγ production and upregulates PD-1 expression on effector T cells [89,90], while promoting Treg differentiation by enhancing Foxp3 expression [91].

D-2HG

D-2-hydroxyglutarate (2-HG) is an oncometabolite generated from α-KG by the mutated isocitrate dehydrogenase (IDH), often found in glioma [92,93]. It promotes tumorigenesis, tumor growth, and invasiveness by epigenetic remodeling in a cancer cell-intrinsic manner. D2-HG also exerts a direct suppressive effect on CD8 T cells by inhibiting calcium signaling and NFAT signaling [94]. A recent study reported LDH being a molecular target of D-2HG [95]. Inhibiting LDH in activated CD8 T cells rewires the metabolic program toward OXPHOS, decreasing cytotoxicity and impairing proliferation.

The source of immunosuppressive metabolites in TME is not limited to cancer cells. Meanwhile, they can target multiple cell types simultaneously and affect not only cellular metabolism but also intracellular signaling, robustly impairing the antitumor immunity through multiple mechanisms.

Modulating T cell metabolism

Preconditioning and supplementing with cytokines

For an effective T cell expansion following stimulation, T cells require three separate signals, encompassing antigen recognition, co-stimulation, and cytokine-mediated stimulation. Distinct T cell expansion and differentiation are driven by the signals from various cytokines received by T cells. In the manufacture process of tumor-infiltrating lymphocyte products, high-dose of IL-2 has been extensively employed, enabling the expansion of a minimal number of cells from tumor biopsies into billions [96]. IL-2 signaling is critical to T cell activation, survival, and differentiation [97]. However, upon receiving high doses of IL-2 in vitro, T cells possess high aerobic glycolysis, propelling effector T cells toward terminal differentiation [98,99]. These T cells exhibit reduced persistence and compromised antitumor capacity [100]. However, T cells derived from stem-like or central memory characteristics demonstrate heightened antitumor activity [101]. Thus, a variety of cytokines has been integrated into the preparation and manufacturing process of adoptive cell therapy in order to ensure and improve T cell fitness upon infusion. Cytokines from the γ-chain (γc) family, including interleukin-7 (IL-7), IL-15, and IL-21, are commonly employed for preconditioning T cells.

IL-7 is essential for maintaining naïve and memory T cell homeostasis and survival [102,103]. Following IL-7 signaling, memory CD8 T cells elevate the expression of the glycerol channel aquaporin 9 (AQP9), leading to increased uptake of glycerol and upregulation of triglyceride (TAG) synthesis. The accumulation of lipids, in turn, sustains memory T cell function [104]. Conversely, IL-15 signaling fosters the utilization of fatty acids as an energy source through FAO, bolstering spare respiratory capacity and promoting mitochondrial biogenesis [7]. Thus, supplementing IL-7 and IL-15 following TCR stimulation enhances the differentiation of CD8 T lymphocytes into memory stem-like characteristics [105]. Moreover, preconditioning CAR-T cells with IL-7 and IL-15 ex vivo enhances their responses when combined with immune checkpoint blockades in vivo animal models [106,107].

IL-21 plays a crucial role in regulating the differentiation of CD8 T cells while preserving metabolic states akin to naïve T cells through upregulation of FAO [108,109]. IL-21 signaling maintains metabolic quiescence and enhances mitochondrial fitness in activated

CD8 T cells, thereby augmenting CD8 T cell stemness and antitumor capacity [99]. Furthermore, employing IL-21 in the preparation of adoptive cell transfer products can enhance T cell expansion and promote the generation of an increased population of memory stem-like T cells [110,111]. These approaches can be extended to include the combination of histone deacetylase inhibitor (HDACi) with IL21 to reprogram effector T cells into memory T cells [112]. Additionally, another study has indicated that administering a fusion protein of anti-PD-1 antibody with IL-21 induces the differentiation of naïve T cells into memory stem-like T cells, leading to a notably enhanced antitumor effect in vivo [113].

Supplementation of nutrients and metabolic preconditioning

T cells dynamically adjust their metabolism in response to the environmental cues, necessitating specific nutrients to effectively respond to various stimuli. The metabolic competition within the tumor microenvironment often results in T cell dysfunction, as crucial nutrients like glucose are depleted by tumor cells or other immune populations [54,56]. Therefore, recent studies emphasize the importance of supplementing nutrients for T cells to adapt and thrive within the hostile tumor microenvironment or preconditioning T cells under metabolic stress to undergo metabolic rewiring before encountering the nutrient-depleted tumor microenvironment.

Cancer cells outcompete methionine for T cells, resulting in abnormal epigenetic patterns in T cells due to the lack of the donor epigenetic methylation. The alteration in the epigenetic landscape diminishes STAT5 activity and attenuates the antitumor capacity of T cells. Supplementing methionine to mouse models or patients can increase T cell polyfunctional cytokine release and T cell survival through rescuing STAT5 activity [114]. Besides, the accessibility of L-arginine regulates T cell functionality [115]. Increasing intracellular L-arginine levels directs activated T cells' metabolism toward OXPHOS rather than glycolysis, thus restraining T cell differentiation. Supplementing L-arginine in the diet of tumor-bearing mice can also bolster antitumor responses, enabling T cells to sustain a central memory-like state and enhance T cell survival [61]. Additionally, the lack of glutamine in the tumor microenvironment can hinder the activation of naïve T cells. Upon activation, T cells increase the expression of the amino acid transporter ASCT2, enhancing glutamine uptake and activating mTORC1-mediated metabolic pathways [21]. To improve glutamine availability for T cells, a glutamine antagonist known as 6-diazo-5-oxo-l-norleucine (DON) has been utilized in a mouse model of colorectal tumors. Mice treated with DON exhibit enhanced tumor control, with tumor-infiltrating lymphocytes maintaining energy homeostasis through metabolic reprograming and displaying a durable memory-like phenotype [116]. Moreover, inosine can serve as an alternative carbon source for activated T cells in a glucose-deprived environment. Inosine supplementation enhances the secretion of polyfunctional cytokines in T cells and extends mouse survival [117]. The increased metabolism of inosine in CAR-T cells triggers the expression of stemness-associated genes and facilitates metabolic reprogramming by enhancing mitochondrial capacity [118]. Instead of fueling glucose-restricted T cells with essential nutrients, substituting acetate can induce epigenetic reprogramming, enhancing chromatin accessibility in glucose-deprived T cells. This heightened histone acetylation boosts the expression of effector genes and pathways related to immune response. Acetate supplementation leads to increased IFNγ release in tumor-infiltrating T lymphocytes [119].

Rather than providing supplementary essential nutrients to alter T cell metabolism, subjecting T cells to metabolic stress prior to their exposure to nutrient deprivation can enhance their metabolic resilience when they encounter a nutrient-deficient tumor microenvironment. During T cell priming or ex vivo expansion, naïve T cells with limited access to glucose not only display a more memory-like pattern but also enhance T cell homing ability and antitumor functionality [120]. Transient glucose restriction in activated T cells drives metabolic remodeling aimed at preserving mitochondrial activity and restoring energetic balance. After re-exposure to glucose, T cells previously subjected to transient glucose restriction demonstrated increased anabolic metabolism, further facilitating tumor clearance in murine model [121]. Glucose is a crucial nutrient required to support T cell-mediated immunity. The phenomenon of increased T cell functionality in vivo when glucose access is limited in vitro can be attributed to induced metabolic reprogramming and improved mitochondrial activity. T cells adjust their metabolism and efficiently adapt to nutrient-deprived tumor microenvironments, particularly following metabolic stress in advance.

Modulating mitochondrial dynamics

In the immunosuppressive tumor microenvironment, T cells encounter metabolic obstacles that impair their metabolic fitness and their ability to fight against tumors [122]. The mitochondria play a crucial role in regulating energy demand and supply, thereby influencing T cell development. Mitochondrial dynamics not only influences T cell fate but also governs T cell metabolic processes [27]. Research has focused on enhancing mitochondrial fitness to bolster T cell functionality within the tumor microenvironment.

Upon infiltration into the tumor microenvironment, T cells consistently exhibit diminished mitochondrial function due to loss of PPAR-gamma coactivator 1α (PGC1α). PGC1α governs mitochondrial biogenesis and is suppressed by prolonged Akt signaling in T cells infiltrating tumors. Overexpression of PGC1α leads to metabolic reprogramming of tumor-specific T cells, increasing secretion of type I cytokines, and further enhancing the intratumoral antitumor immune response [123]. Moreover, bezafibrate, an activator of PGC-1α/PPAR complexes, enhances mitochondrial function and augments FAO in T cells. Bezafibrate additionally promotes the proliferation and differentiation of naïve T cells into effector T cells, leading to improved tumor control and displaying a synergistic effect when combined with immune checkpoint blockade [124].

Due to activation of T cell receptor and PD-1 signaling pathways, tumor-infiltrating lymphocytes demonstrate reduced mitochondrial mass and compromised mitochondria biogenesis. Consequently, this induces a state of terminal exhaustion in CD8 T cells via epigenetic reprogramming. The addition of nicotinamide ribose significantly diminishes the buildup of depolarized mitochondria, thereby enhancing T cell mitochondrial fitness and subsequently improving the antitumor response [125]. Another analogous study demonstrated that persistent antigenic stimulation hindered ADP-coupled mitochondrial oxidative phosphorylation, resulting in additional suppression of nucleotide triphosphate synthesis. This suppression is attributed to the ROS-mediated increase in the NADH/NAD+ ratio in chronically stimulated T cells. Consequently, this inhibits T cell proliferation and upregulates genes associated with T cell exhaustion. Introducing antioxidants can revive the proliferation and self-renewal capabilities of T cells that have been chronically stimulated [126].

Studies have demonstrated that metformin can boost FAO and facilitate the generation of long-lived memory T cells in cases of impaired AMP-activated kinase activation [5]. Besides, metformin enhances T cell functionality by preventing mitochondrial dysfunction under conditions of hypoxia enriched with ROS, thereby improving the antitumor response [127]. Furthermore, the use of Mdivi-1, a mitochondrial fission inhibitor, can promote mitochondrial fusion, augment the memory T cell population, and enhance metabolic fitness in tumor-infiltrating lymphocytes [27,30]. These studies underscore the significance of modulating mitochondrial dynamics to enhance T cell metabolic reprogramming within the nutrient-deprived tumor microenvironment.

Inhibiting PI3K/AKT/mTOR and MAPK pathways

Upon activation of T cells, two main downstream signaling pathways become activated to enable metabolic reprogramming and foster the differentiation of effector T cells: the PI3K/AKT/mTOR pathway and the MEK/ERK pathway. Inhibiting these signals during T cell activation increases the likelihood of T cells exhibiting a pattern more resembling memory cells, thus improving their longevity and functionality.

Pharmacological inhibition of PI3Kδ signaling, specifically in CAR-T cells, not only increases the presence of less differentiated T cells but also enhances T cell persistence [128]. Additionally, a similar study also showed that blocking either PI3Kγ or PI3Kδ yields a higher proportion of central memory T cells and improves T cell function against tumor [129–131]. When it comes to ex vivo expansion of CAR-T cells, tonic signaling prompted by the CD3ζ ITAMs often decreases naïve and stem-like phenotypes. By incorporating PI3K inhibition into the manufacturing process, CAR-T cells maintain a less differentiated state and restore CAR-T cells' memory population [132,133].

While expanding tumor-infiltrating lymphocytes ex vivo, inhibiting Akt prompts these cells to adopt a transcriptional, metabolic, and functional phenotype that closely resembles memory cells. Tumor-infiltrating lymphocytes treated with an Akt inhibitor show an increase in spare respiratory capacity, equipping them with additional energy reserves to cope with stress. Consequently, this results in enhanced T cell persistence and survival in a murine model [134]. The same effect was discovered in ex vivo expansion of CD19-targeting CAR-T cells. Akt inhibition generates more memory-like patterns and demonstrates superior tumor control in xenograft murine model [135]. A parallel study also indicated that inhibiting Akt in activated T cells ex vivo enhances the secretion of polyfunctional cytokines and exhibits a phenotype reminiscent of stem memory cells [136]. The inhibition of Akt has been exploited in the expansion of tumor-infiltrating lymphocytes and the generation of engineered T cells, resulting in more memory-like cells with superior antitumor function [137].

Furthermore, mTOR serves as a crucial downstream signaling module of the PI3K pathway in regulating T cell fate determination, especially in memory CD8 T cell differentiation [138,139]. Mice infected with acute lymphocytic choriomeningitis virus, upon treatment with rapamycin, an mTOR inhibitor, exhibited enhancements in both the quality and quantity of virus-specific CD8 T cells, along with an increased capacity to produce long-lived memory T cells [139,140]. The same strategy was also utilized in the ex vivo expansion of CAR-T cells, where rapamycin-treated CAR-T cells retain a stem cell memory phenotype [141].

The mitogen-activated protein kinase (MAPK) pathway also plays a crucial role activated by TCR stimulation, and it is significant for both T cell development and metabolic reprogramming [20,142]. Blocking MEK enhances the effectiveness of adoptive cell therapy when combined with immune checkpoint blockade, as it decreases TCR-induced T cell death [143]. Additionally, inhibiting MEK boosts mitochondrial respiration fueled by FAO and transforms CD8 T lymphocytes into a state characteristic of stem cell memory [144]. Conversely, MEK inhibition has demonstrated negative effects on T cell immunity, impairing T cell functionality. However, combining agonist immunotherapies, such as α-4-1BB and α-OX-40 antibodies, can counteract the detrimental impact of MEK inhibition on T cell activity and restore T cell function [145].

Manipulation of tumor microenvironment

The presence of an immunosuppressive tumor microenvironment, comprising tumor cells and various myeloid immune populations, not only restricts nutrient availability but also establishes adverse conditions such as acidosis, hypoxia, and the accumulation of immunosuppressive metabolites. These elements collectively induce metabolic alterations in T cells and compromise their effectiveness, particularly within solid tumors. Researchers have concentrated on surmounting these metabolic hurdles and reshaping the microenvironment to foster antitumor responses.

Alteration of hypoxia condition

HIF-1α serves as the master transcriptional regulator controlling cellular responses to hypoxia, leading to increased glucose uptake. The prevalence of Warburg effect and the poor vascularization significantly reduced the oxygen level within the tumor microenvironment. Tumor hypoxia induces an increase in HIF-1α levels in cytotoxic T lymphocytes, leading to decreased differentiation, proliferation, and production of polyfunctional cytokines [51,146]. Hypoxia-experienced cytotoxic T cells demonstrate upregulation of glycolysis-associated genes, subsequently promoting the differentiation of T cells into terminally effector T cells [120,147]. When HIF-1α is genetically eliminated, cytotoxic T cells demonstrate heightened fatty acid catabolism, thereby maintaining their functionality by upregulating peroxisome proliferator-activated receptor α signaling [148]. Moreover, supplying additional oxygen can mitigate hypoxia-induced immunosuppression, enhancing T cell infiltration and the antitumor immune response, whether through endogenously developed mechanisms or via adoptively transferred tumor-specific cytotoxic T lymphocytes [149].

Neutralization of acidic tumor microenvironment

Tumor cells tend to favor aerobic glycolysis as their primary method for generating ATP, resulting in the production of lactate as a byproduct. The significant accumulation of lactate not only creates a low pH environment but also functions as a fuel source for immunosuppressive immune cell populations, such as regulatory T cells [71]. Lactate produced by tumor cells can blunt T cell activation by preventing nuclear factor of activated T cells (NFAT) in T cells, which further attenuates IFNγ secretion [70]. Another study additionally illustrated that the absence of nicotinamide adenine dinucleotide (NAD+) in conditions with high lactate levels hinders NAD + -dependent enzymatic reactions in the glycolysis pathway, particularly affecting glyceraldehyde 3-phosphate dehydrogenase (GAPDH) and 3-phosphoglycerate dehydrogenase (PGDH). This significantly decreases the generation of downstream glycolytic intermediates, such as glucose-derived serine. Adding serine as a supplement can restore T cell proliferation in environments rich in lactate [150]. Alternatively, to address tumor acidosis, utilizing bicarbonate monotherapy to neutralize acidity in conjunction with adoptive cell therapy produces enhanced antitumor immune responses [151]. On the contrary, a compelling study found that minimal levels of lactate enhance the stemness of CD8 T cells by increasing the expression of TCF-1. However, the accumulation of tumor acidity can attenuate the enhancement of CD8 T cell stemness induced by lactate [152].

Depletion of immunosuppressive metabolites

Tryptophan, an essential amino acid, plays a crucial role in regulating peripheral immune tolerance through its catabolism [77]. Kynurenine, a byproduct of tryptophan metabolism, primarily originates from enzymatic processes catalyzed by the key rate-limiting enzyme, indoleamine 2,3-dioxygenase 1 (IDO1). Cancer cells often enhance tryptophan catabolism by increasing the expression of IDO1, leading to a depletion of tryptophan and an elevation of kynurenine levels within the tumor microenvironment [153]. Kynurenine directly impairs the function of effector T cells by inducing cell cycle arrest and apoptosis [154]. Through the administration of PEGylated kynureninase, which converts kynurenine into a nonfunctional form and reduces its presence within the tumor microenvironment, there is an augmentation in the infiltration of polyfunctional CD8 lymphocytes, resulting in improved tumor control [155]. This study additionally showed that removing harmful byproducts could be more effective than depleting essential nutrients, which serve as fuel for both cancer cells and effector T cells.

Extracellular adenosine, a potent immunosuppressive metabolite, is produced by cell surface ectonucleotidases CD39 and CD73, which are overexpressed in various types of cancer [156]. CD39 catalyzes the conversion of extracellular ATP into ADP and AMP, which are then further converted into adenosine by CD73. Stimulation of the A2A adenosine receptor (A(2A) R) by adenosine facilitates tumor immune evasion by diminishing T cell infiltration, secretion of polyfunctional cytokines, and inducing T cell anergy [87,157,158]. Adenosine inhibits both oxidative phosphorylation and glycolysis, predominantly impacting central memory T cells. This occurs through a reduction in the mTORC1 pathway via heightened PKA activation, ultimately leading to impaired TCR signaling. Supplementing caffeine or ZM 241385, A2A receptor antagonists, diminishes A2AR-mediated immunosuppression, enhancing metabolic fitness, and improving tumor recognition, thus bolstering tumor control [87,158].

Inhibition of cancer metabolism

Therapeutic modalities targeting intrinsic cancer cell metabolism have been developed for decades [159]. Some have advanced to clinical trials. However, drugs targeting cancer metabolism would inevitably affect other nontumor cell populations within TME, including T cells, which often failed to take into consideration when initially designing such drugs. Tumor cells and T cells share many common nutrient substrates, such as glucose and glutamine. When targeting these pathways, questions remain about how to deliver tumor-specific metabolic control while sparing metabolic-sensitive T cells in the TME.

One strategy to tackle the on-target, off-tumor problem is exploiting cancer cell metabolic vulnerability, which is often conferred by genetic mutations. For example, the nuclear factor erythroid 2-related factor 2 (NRF2) is a master regulator of cellular response to oxidative stress and is suppressed by Kelch-like ECH-associated protein 1 (KEAP1)-dependent protein degradation in normal condition [160]. Mutations in KEAP/NRF2 axis are found in many types of solid tumors, resulting in NRF2 hyperactivity and increased demand for glutamine for glutathione synthesis [161]. Cancer cells harboring mutations in KEAP/NRF2 genes are particularly prone to inhibition of glutamine metabolism. GLS1 (glutaminase) inhibitors such as IACS-6274 and CB-839 have been evaluated in multiple clinical trials, either as a single agent or in combination with immunotherapy [162–164]. On the other hand, the effects of GLS1 inhibition on T cells are less clear. Though in vitro CD8 T cell activation in the presence of CB-839 has been shown to promote their effector function [23], a recent report demonstrated using a lung cancer model with high expression of GLS that the GLS1 inhibitor CB-839 impaired CD8 T cell activation and failed to synergize with anti-PD1 therapy [165]. Another study using glutamine transport inhibitor, V-9032 showed improved T cell activation and reduced tumor growth in a triple-negative breast cancer model [47]. These contrary results demonstrate the importance of isolating therapeutic effects of metabolic inhibitors on tumor cells and immune cells and also suggest context-dependent effects in different cancers.

Another example is the high dependency of glycolysis in van Hippel-Lindau (pVHL) mutant cancers. Loss of pVHL activates hypoxia-inducible factor (HIF) and upregulates glycolysis [166]. Therefore, inhibiting LDH presents a promising approach to treating tumors harboring mutations in pVHL-HIF axis. In addition, LDH inhibition could reduce lactate concentration in TME, mitigating lactate-mediated functional suppression on T cells [167]. NCI-737 is developed as a potent LDHA inhibitor and impairs tumor growth in preclinical models of Ewing sarcoma [168]. Moreover, NCI-737 synergizes with IL-21 in CD8 T cells to rewire T cell metabolism to OXPHOS and enhance memory stem cell formation. When pretreated with NCI-737, adoptively transferred CD8 T cells display enhanced antitumor activity [99]. However, in these studies, NCI-737 has only been tested either in SCID mice or in an adoptive T cell therapy setting. Its in vivo efficacy in immunocompetent mice or in humans is not examined. Furthermore, in the Ewing sarcoma study, mice displayed hemolysis when treated with high doses of NCI-737, phenocopying genetic loss of LDHA. This on-target toxicity effect narrows its therapeutic window in clinical settings. Strategies, such as targeted delivery or combination therapy with other agents, need to be considered when moving into clinical development.

When manipulating cancer cell metabolism, effects in the immune cell compartment should not be neglected. Testing metabolic inhibitors in immunocompetent mice could enlighten their clinical development. When designing clinical trials, evaluating patients' tumor-infiltrating immune environment should be included to provide additional understanding of the impact of such modalities.

Lymphodepletion

Non-myeloablative (NMA) chemotherapy with fludarabine and cyclophosphamide has been widely employed in patients before undergoing CAR-T and TIL therapy [169]. The advantages of lymphodepletion preconditioning include restructuring the blood environment to improve adoptive cell engraftment and promote their sustained viability, eradicating suppressive immune cell populations to stimulate and amplify adoptive cells, and reshaping the microenvironment to facilitate adoptive cell homing and provoke antitumor responses [96,170].

The success of lymphodepletion in CAR-T therapy can be attributed to the enhanced availability of circulating homeostatic cytokines, such as IL-2, IL-7, and IL-15, to the infused T cells [171]. Besides, low-dose chemotherapy preconditioning prior to CAR-T therapy has been shown to elevate serum IL-15 levels, which is associated with better clinical outcome [172]. On the contrary, lymphodepletion preceding CAR-T therapy leads to a reduction in IDO expression within tumor cells, subsequently lowering the production of immunosuppressive metabolites in the tumor microenvironment [173]. Therefore, preconditioning with lymphodepletion can reshape the tumor microenvironment and rewire the metabolism of infused T cells, resulting in increased T cell proliferation, cytokine secretion, and cytotoxicity.

While low-dose chemotherapy for preconditioning has been employed in certain T cell products, lymphodepletion suppresses the host immune system, which may compromise its intrinsic antitumor capabilities and result in significant adverse effects such as lymphopenia, anemia, and increased susceptibility to infections [174]. Several studies have indicated that stimulating T cells from a patient's peripheral blood with specific peptide epitopes derived from neoantigens or tumor-associated antigens before reinfusing into the patient can result in promising clinical effectiveness for individuals with solid tumors, even in the absence of prior lymphodepletion [175,176]. Additionally, CAR-T cells equipped with IL-12 secretion capability have demonstrated the ability to effectively eliminate tumors in transgenic mouse models without prior conditioning [177,178]. Therefore, optimizing the combination of suitable dosages of chemotherapy with different T cell products can maximize the effectiveness of adoptive cell therapy.

Immune checkpoint blockade

T cells possess various inhibitory receptors, such as programmed cell death protein 1 (PD-1), cytotoxic T-lymphocyte association protein 4 (CTLA-4), lymphocyte activation gene-3 (LAG-3), and mucin domain-containing protein 3 (TIM3), for maintaining self-tolerance [179]. Upon receiving signals from these inhibitory receptors, T cells become exhausted and dysfunctional, characterized by reduced proliferation capacity, cytokine secretion, and unresponsiveness. However, tumor cells have hijacked such pathways for immune evasion by expressing corresponding ligands that dampen T cell priming and activation. Moreover, these inhibitory signals alter metabolic reprogramming during T cell activation [180]. Immune checkpoint blockades (ICBs) have not only perturbed this inhibitory circuit but also reprogrammed T cell metabolism, resulting in a substantial change in the conventional approach to cancer treatment [181]. They stimulate the immune system to unleash its antitumor potential.

Upon receiving PD-1 signaling, T cells downregulate Akt activity and reduce phosphatidylinositol 3-kinase (PI3K) activity induced by CD28 [180]. This further leads to reduced glycolysis and amino acid metabolism, while promoting fatty acid β-oxidation through elevating AMPK activity [182]. The metabolic shift supports their survival through catabolism but diminishes their cytotoxic capabilities. Conversely, tumor cells expressing PD-L1 exhibit increased glycolytic activity, resulting in nutrient competition between tumor cells and T cells, ultimately leading to T cell hyporesponsiveness [54]. ICB can significantly enhance T cell function, elevate the availability of glucose within the tumor microenvironment, and rewire the T cell metabolism to elicit antitumor activity [56,180,183,184].

CD28 and CTLA4 are complete with the same ligands, CD80 and CD86, with CTLA4 exhibiting a higher affinity for them [185]. CTLA-4 signaling acts as a negative regulator, attenuating T cell activation by inhibiting the CD28-mediated co-stimulation [9,186,187]. Upon activation, CTLA4 recruits protein phosphatase 2 (PP2), subsequently leading to the inactivation of PI3K/Akt/mTOR signaling pathways [180]. This not only diminishes glycolysis by suppressing the expression of the GLUT1 but also decreases glutaminolysis by downregulating glutamine transporters (SNAT1, SNAT2) [180,182]. Blocking of CTLA4 remodels T cell metabolism and has achieved clinical success in treating both hematological malignancies and solid tumors [181,188].

LAG-3 is also expressed at a low level in T cells and plays a significant role in immune regulation [189,190]. LAG-3 signaling reduces naïve T cell proliferation by inhibiting IL-7-induced STAT5 activation. T cells deficient in LAG-3 show heightened oxidative phosphorylation and glycolytic metabolism [191]. Additionally, LAG-3 inhibits calcium influx induced by CD3/TCR activation pathway and causes a decrease in cell proliferation and cytokine secretion [192]. TIM-3 is an inhibitory receptor associated with T cell exhaustion [193]. The expression of TIM-3 decreases glycolytic metabolism and diminishes the expression of GLUT1 and the release of lactate [194]. Targeting TIM-3 and PD-1 pathways restores T cell cytokine secretion and antitumor capacity [195].

In summary, directing efforts toward inhibiting signaling pathways in T cells can enhance T cell functionality by boosting their metabolic resilience within the immunosuppressive tumor microenvironment. Numerous clinical trials have integrated metabolic interventions with immune checkpoint blockade to enhance the efficacy of either endogenously developed or adoptively transferred tumor-specific T cells.

Conclusion

The intricate nature of the solid tumor microenvironment offers numerous signals that facilitate immune evasion by tumor cells, including diminishing the metabolic fitness in T cells while nourishing Tregs. Recent studies have focused on engineering T cells to increase T cell function in the TME. Understanding the complexity and dynamics of the immune system and the immunosuppressive tumor microenvironment can provide valuable insights for devising innovative therapeutic approaches and combination strategies.

References

[1] van der Windt GJ, Pearce EL. Metabolic switching and fuel choice during T-cell differentiation and memory development. Immunol Rev 2012;249:27–42. https://doi.org/10.1111/j.1600-065X.2012.01150.x.

[2] Buck MD, O'Sullivan D, Pearce EL. T cell metabolism drives immunity. J Exp Med 2015;212:1345–60. https://doi.org/10.1084/jem.20151159.

[3] Chang CH, et al. Posttranscriptional control of T cell effector function by aerobic glycolysis. Cell 2013;153:1239–51. https://doi.org/10.1016/j.cell.2013.05.016.

[4] Wang R, et al. The transcription factor Myc controls metabolic reprogramming upon T lymphocyte activation. Immunity 2011;35:871–82. https://doi.org/10.1016/j.immuni.2011.09.021.
[5] Pearce EL, et al. Enhancing CD8 T-cell memory by modulating fatty acid metabolism. Nature 2009;460:103–7. https://doi.org/10.1038/nature08097.
[6] Raud B, McGuire PJ, Jones RG, Sparwasser T, Berod L. Fatty acid metabolism in CD8(+) T cell memory: challenging current concepts. Immunol Rev 2018;283:213–31. https://doi.org/10.1111/imr.12655.
[7] van der Windt GJ, et al. Mitochondrial respiratory capacity is a critical regulator of CD8+ T cell memory development. Immunity 2012;36:68–78. https://doi.org/10.1016/j.immuni.2011.12.007.
[8] Rathmell JC, Vander Heiden MG, Harris MH, Frauwirth KA, Thompson CB. In the absence of extrinsic signals, nutrient utilization by lymphocytes is insufficient to maintain either cell size or viability. Mol Cell 2000;6:683–92. https://doi.org/10.1016/s1097-2765(00)00066-6.
[9] Frauwirth KA, et al. The CD28 signaling pathway regulates glucose metabolism. Immunity 2002;16:769–77. https://doi.org/10.1016/s1074-7613(02)00323-0.
[10] Vander Heiden MG, Cantley LC, Thompson CB. Understanding the Warburg effect: the metabolic requirements of cell proliferation. Science 2009;324:1029–33. https://doi.org/10.1126/science.1160809.
[11] Bachem A, et al. Microbiota-derived short-chain fatty acids promote the memory potential of antigen-activated CD8(+) T cells. Immunity 2019;51:285–297.e285. https://doi.org/10.1016/j.immuni.2019.06.002.
[12] Xu S, et al. Uptake of oxidized lipids by the scavenger receptor CD36 promotes lipid peroxidation and dysfunction in CD8(+) T cells in tumors. Immunity 2021;54:1561–1577.e1567. https://doi.org/10.1016/j.immuni.2021.05.003.
[13] Ma X, et al. Cholesterol induces CD8(+) T cell exhaustion in the tumor microenvironment. Cell Metab 2019;30:143–156.e145. https://doi.org/10.1016/j.cmet.2019.04.002.
[14] Lim SA, Su W, Chapman NM, Chi H. Lipid metabolism in T cell signaling and function. Nat Chem Biol 2022;18:470–81. https://doi.org/10.1038/s41589-022-01017-3.
[15] Lacher SM, et al. HMG-CoA reductase promotes protein prenylation and therefore is indispensible for T-cell survival. Cell Death Dis 2017;8, e2824. https://doi.org/10.1038/cddis.2017.221.
[16] O'Sullivan D, et al. Memory CD8(+) T cells use cell-intrinsic lipolysis to support the metabolic programming necessary for development. Immunity 2014;41:75–88. https://doi.org/10.1016/j.immuni.2014.06.005.
[17] Newton R, Priyadharshini B, Turka LA. Immunometabolism of regulatory T cells. Nat Immunol 2016;17:618–25. https://doi.org/10.1038/ni.3466.
[18] Pacella I, et al. Fatty acid metabolism complements glycolysis in the selective regulatory T cell expansion during tumor growth. Proc Natl Acad Sci USA 2018;115:e6546–55. https://doi.org/10.1073/pnas.1720113115.
[19] Chapman NM, et al. mTOR coordinates transcriptional programs and mitochondrial metabolism of activated T(reg) subsets to protect tissue homeostasis. Nat Commun 2018;9:2095. https://doi.org/10.1038/s41467-018-04392-5.
[20] Carr EL, et al. Glutamine uptake and metabolism are coordinately regulated by ERK/MAPK during T lymphocyte activation. J Immunol 2010;185:1037–44. https://doi.org/10.4049/jimmunol.0903586.
[21] Nakaya M, et al. Inflammatory T cell responses rely on amino acid transporter ASCT2 facilitation of glutamine uptake and mTORC1 kinase activation. Immunity 2014;40:692–705. https://doi.org/10.1016/j.immuni.2014.04.007.
[22] Klysz D, et al. Glutamine-dependent α-ketoglutarate production regulates the balance between T helper 1 cell and regulatory T cell generation. Sci Signal 2015;8:ra97. https://doi.org/10.1126/scisignal.aab2610.
[23] Johnson MO, et al. Distinct regulation of Th17 and Th1 cell differentiation by glutaminase-dependent metabolism. Cell 2018;175:1780–95. e1719 https://doi.org/10.1016/j.cell.2018.10.001.
[24] Desdin-Mico G, Soto-Heredero G, Mittelbrunn M. Mitochondrial activity in T cells. Mitochondrion 2018;41: 51–7. https://doi.org/10.1016/j.mito.2017.10.006.
[25] Quintana A, et al. T cell activation requires mitochondrial translocation to the immunological synapse. Proc Natl Acad Sci USA 2007;104:14418–23. https://doi.org/10.1073/pnas.0703126104.
[26] Simula L, et al. Mitochondrial metabolism sustains CD8+ T cell migration for an efficient infiltration into solid tumors. Nat Commun 2024;15:2203. https://doi.org/10.1038/s41467-024-46377-7.
[27] Buck MD, et al. Mitochondrial dynamics controls T cell fate through metabolic programming. Cell 2016;166: 63–76. https://doi.org/10.1016/j.cell.2016.05.035.
[28] Ron-Harel N, et al. Mitochondrial biogenesis and proteome remodeling promote one-carbon metabolism for T cell activation. Cell Metab 2016;24:104–17. https://doi.org/10.1016/j.cmet.2016.06.007.

[29] Song M, et al. IRE1α-XBP1 controls T cell function in ovarian cancer by regulating mitochondrial activity. Nature 2018;562:423–8. https://doi.org/10.1038/s41586-018-0597-x.
[30] Baixauli F, et al. The mitochondrial fission factor dynamin-related protein 1 modulates T-cell receptor signalling at the immune synapse. EMBO J 2011;30:1238–50. https://doi.org/10.1038/emboj.2011.25.
[31] Franchina DG, Dostert C, Brenner D. Reactive oxygen species: involvement in T cell signaling and metabolism. Trends Immunol 2018;39:489–502. https://doi.org/10.1016/j.it.2018.01.005.
[32] Sena LA, et al. Mitochondria are required for antigen-specific T cell activation through reactive oxygen species signaling. Immunity 2013;38:225–36. https://doi.org/10.1016/j.immuni.2012.10.020.
[33] Previte DM, et al. Reactive oxygen species are required for driving efficient and sustained aerobic glycolysis during CD4+ T cell activation. PLoS One 2017;12, e0175549. https://doi.org/10.1371/journal.pone.0175549.
[34] Mueller DL. Mechanisms maintaining peripheral tolerance. Nat Immunol 2010;11:21–7. https://doi.org/10.1038/ni.1817.
[35] Abbas AK, et al. Regulatory T cells: recommendations to simplify the nomenclature. Nat Immunol 2013;14:307–8. https://doi.org/10.1038/ni.2554.
[36] Yadav M, et al. Neuropilin-1 distinguishes natural and inducible regulatory T cells among regulatory T cell subsets in vivo. J Exp Med 2012;209(1713–1722):s1711–9. https://doi.org/10.1084/jem.20120822.
[37] Thornton AM, et al. Expression of Helios, an Ikaros transcription factor family member, differentiates thymic-derived from peripherally induced Foxp3+ T regulatory cells. J Immunol 2010;184:3433–41. https://doi.org/10.4049/jimmunol.0904028.
[38] Wang H, et al. CD36-mediated metabolic adaptation supports regulatory T cell survival and function in tumors. Nat Immunol 2020;21:298–308. https://doi.org/10.1038/s41590-019-0589-5.
[39] Smith PM, Howitt MR, Panikov N, Michaud M, Gallini CA, Bohlooly-Y M, Glickman JN, Garrett WS. The microbial metabolites, short-chain fatty acids, regulate colonic Treg cell homeostasis. Science 2013;341:569–73. https://doi.org/10.1126/science.1241165.
[40] Berod L, et al. De novo fatty acid synthesis controls the fate between regulatory T and T helper 17 cells. Nat Med 2014;20:1327–33. https://doi.org/10.1038/nm.3704.
[41] Miska J, et al. HIF-1α is a metabolic switch between glycolytic-driven migration and oxidative phosphorylation-driven immunosuppression of Tregs in glioblastoma. Cell Rep 2019;27:226–237.e224. https://doi.org/10.1016/j.celrep.2019.03.029.
[42] Angelin A, et al. Foxp3 reprograms T cell metabolism to function in low-glucose, high-lactate environments. Cell Metab 2017;25:1282–93. e1287 https://doi.org/10.1016/j.cmet.2016.12.018.
[43] Kurniawan H, et al. Glutathione restricts serine metabolism to preserve regulatory T cell function. Cell Metab 2020;31:920–36. e927 https://doi.org/10.1016/j.cmet.2020.03.004.
[44] Yu X, et al. SENP3 maintains the stability and function of regulatory T cells via BACH2 deSUMOylation. Nat Commun 2018;9:3157. https://doi.org/10.1038/s41467-018-05676-6.
[45] Roychoudhuri R, et al. BACH2 represses effector programs to stabilize T(reg)-mediated immune homeostasis. Nature 2013;498:506–10. https://doi.org/10.1038/nature12199.
[46] Arner EN, Rathmell JC. Metabolic programming and immune suppression in the tumor microenvironment. Cancer Cell 2023;41:421–33. https://doi.org/10.1016/j.ccell.2023.01.009.
[47] Edwards DN, et al. Selective glutamine metabolism inhibition in tumor cells improves antitumor T lymphocyte activity in triple-negative breast cancer. J Clin Invest 2021;131. https://doi.org/10.1172/jci140100.
[48] Goffaux G, Hammami I, Jolicoeur M. A dynamic metabolic flux analysis of myeloid-derived suppressor cells confirms immunosuppression-related metabolic plasticity. Sci Rep 2017;7:9850. https://doi.org/10.1038/s41598-017-10464-1.
[49] Jian SL, et al. Glycolysis regulates the expansion of myeloid-derived suppressor cells in tumor-bearing hosts through prevention of ROS-mediated apoptosis. Cell Death Dis 2017;8, e2779. https://doi.org/10.1038/cddis.2017.192.
[50] Reinfeld BI, et al. Cell-programmed nutrient partitioning in the tumour microenvironment. Nature 2021;593:282–8. https://doi.org/10.1038/s41586-021-03442-1.
[51] Nakagawa Y, et al. Effects of extracellular pH and hypoxia on the function and development of antigen-specific cytotoxic T lymphocytes. Immunol Lett 2015;167:72–86. https://doi.org/10.1016/j.imlet.2015.07.003.
[52] Fu X, et al. Spatial patterns of tumour growth impact clonal diversification in a computational model and the TRACERx renal study. Nat Ecol Evol 2022;6:88–102. https://doi.org/10.1038/s41559-021-01586-x.

[53] Gatenby RA, Gillies RJ. Why do cancers have high aerobic glycolysis? Nat Rev Cancer 2004;4:891–9. https://doi.org/10.1038/nrc1478.

[54] Chang CH, et al. Metabolic competition in the tumor microenvironment is a driver of Cancer progression. Cell 2015;162:1229–41. https://doi.org/10.1016/j.cell.2015.08.016.

[55] Ho PC, et al. Phosphoenolpyruvate is a metabolic checkpoint of anti-tumor T cell responses. Cell 2015;162: 1217–28. https://doi.org/10.1016/j.cell.2015.08.012.

[56] Cham CM, Driessens G, O'Keefe JP, Gajewski TF. Glucose deprivation inhibits multiple key gene expression events and effector functions in CD8+ T cells. Eur J Immunol 2008;38:2438–50. https://doi.org/10.1002/eji.200838289.

[57] Cham CM, Gajewski TF. Glucose availability regulates IFN-gamma production and p70S6 kinase activation in CD8+ effector T cells. J Immunol 2005;174:4670–7. https://doi.org/10.4049/jimmunol.174.8.4670.

[58] Zhao E, et al. Cancer mediates effector T cell dysfunction by targeting micro RNAs and EZH2 via glycolysis restriction. Nat Immunol 2016;17:95–103. https://doi.org/10.1038/ni.3313.

[59] Cormerais Y, et al. The glutamine transporter ASCT2 (SLC1A5) promotes tumor growth independently of the amino acid transporter LAT1 (SLC7A5). J Biol Chem 2018;293:2877–87. https://doi.org/10.1074/jbc.RA117.001342.

[60] Wang Q, et al. Targeting ASCT2-mediated glutamine uptake blocks prostate cancer growth and tumour development. J Pathol 2015;236:278–89. https://doi.org/10.1002/path.4518.

[61] Geiger R, et al. L-arginine modulates T cell metabolism and enhances survival and anti-tumor activity. Cell 2016;167:829–42. e813 https://doi.org/10.1016/j.cell.2016.09.031.

[62] Fletcher M, et al. L-arginine depletion blunts antitumor T-cell responses by inducing myeloid-derived suppressor cells. Cancer Res 2015;75:275–83. https://doi.org/10.1158/0008-5472.Can-14-1491.

[63] Bingisser RM, Tilbrook PA, Holt PG, Kees UR. Macrophage-derived nitric oxide regulates T cell activation via reversible disruption of the Jak3/STAT5 signaling pathway. J Immunol 1998;160:5729–34.

[64] Mazzoni A, et al. Myeloid suppressor lines inhibit T cell responses by an NO-dependent mechanism. J Immunol 2002;168:689–95. https://doi.org/10.4049/jimmunol.168.2.689.

[65] Macphail SE, et al. Nitric oxide regulation of human peripheral blood mononuclear cells: critical time dependence and selectivity for cytokine versus chemokine expression. J Immunol 2003;171:4809–15. https://doi.org/10.4049/jimmunol.171.9.4809.

[66] Claps G, et al. The multiple roles of LDH in cancer. Nat Rev Clin Oncol 2022;19:749–62. https://doi.org/10.1038/s41571-022-00686-2.

[67] Garcia CK, Goldstein JL, Pathak RK, Anderson RG, Brown MS. Molecular characterization of a membrane transporter for lactate, pyruvate, and other monocarboxylates: implications for the Cori cycle. Cell 1994;76:865–73. https://doi.org/10.1016/0092-8674(94)90361-1.

[68] Le Floch R, et al. CD147 subunit of lactate/H+ symporters MCT1 and hypoxia-inducible MCT4 is critical for energetics and growth of glycolytic tumors. Proc Natl Acad Sci USA 2011;108:16663–8. https://doi.org/10.1073/pnas.1106123108.

[69] Fischer K, et al. Inhibitory effect of tumor cell-derived lactic acid on human T cells. Blood 2007;109: 3812–9. https://doi.org/10.1182/blood-2006-07-035972.

[70] Brand A, et al. LDHA-associated lactic acid production blunts tumor Immunosurveillance by T and NK cells. Cell Metab 2016;24:657–71. https://doi.org/10.1016/j.cmet.2016.08.011.

[71] Watson MJ, et al. Metabolic support of tumour-infiltrating regulatory T cells by lactic acid. Nature 2021;591: 645–51. https://doi.org/10.1038/s41586-020-03045-2.

[72] Kumagai S, et al. Lactic acid promotes PD-1 expression in regulatory T cells in highly glycolytic tumor microenvironments. Cancer Cell 2022;40:201–18. e209 https://doi.org/10.1016/j.ccell.2022.01.001.

[73] Zhao JL, et al. Notch-mediated lactate metabolism regulates MDSC development through the Hes1/MCT2/c-Jun axis. Cell Rep 2022;38, 110451. https://doi.org/10.1016/j.celrep.2022.110451.

[74] Yang X, et al. Lactate-modulated immunosuppression of myeloid-derived suppressor cells contributes to the radioresistance of pancreatic cancer. Cancer Immunol Res 2020;8:1440–51. https://doi.org/10.1158/2326-6066.Cir-20-0111.

[75] Petrelli F, et al. Prognostic role of lactate dehydrogenase in solid tumors: a systematic review and meta-analysis of 76 studies. Acta Oncol 2015;54:961–70. https://doi.org/10.3109/0284186X.2015.1043026.

[76] Walenta S, et al. High lactate levels predict likelihood of metastases, tumor recurrence, and restricted patient survival in human cervical cancers. Cancer Res 2000;60:916–21.

[77] Fallarino F, et al. T cell apoptosis by tryptophan catabolism. Cell Death Differ 2002;9:1069–77. https://doi.org/10.1038/sj.cdd.4401073.

[78] Siska PJ, et al. Kynurenine induces T cell fat catabolism and has limited suppressive effects in vivo. EBioMedicine 2021;74, 103734. https://doi.org/10.1016/j.ebiom.2021.103734.

[79] Opitz CA, et al. An endogenous tumour-promoting ligand of the human aryl hydrocarbon receptor. Nature 2011;478:197–203. https://doi.org/10.1038/nature10491.

[80] Frumento G, et al. Tryptophan-derived catabolites are responsible for inhibition of T and natural killer cell proliferation induced by indoleamine 2,3-dioxygenase. J Exp Med 2002;196:459–68. https://doi.org/10.1084/jem.20020121.

[81] DiNatale BC, et al. Kynurenic acid is a potent endogenous aryl hydrocarbon receptor ligand that synergistically induces interleukin-6 in the presence of inflammatory signaling. Toxicol Sci 2010;115:89–97. https://doi.org/10.1093/toxsci/kfq024.

[82] Bessede A, et al. Aryl hydrocarbon receptor control of a disease tolerance defence pathway. Nature 2014;511:184–90. https://doi.org/10.1038/nature13323.

[83] Liu Y, et al. Tumor-repopulating cells induce PD-1 expression in CD8(+) T cells by transferring kynurenine and AhR activation. Cancer Cell 2018;33:480–94. e487 https://doi.org/10.1016/j.ccell.2018.02.005.

[84] Mezrich JD, et al. An interaction between kynurenine and the aryl hydrocarbon receptor can generate regulatory T cells. J Immunol 2010;185:3190–8. https://doi.org/10.4049/jimmunol.0903670.

[85] Campesato LF, et al. Blockade of the AHR restricts a Treg-macrophage suppressive axis induced by L-kynurenine. Nat Commun 2020;11:4011. https://doi.org/10.1038/s41467-020-17750-z.

[86] Maj T, et al. Oxidative stress controls regulatory T cell apoptosis and suppressor activity and PD-L1-blockade resistance in tumor. Nat Immunol 2017;18:1332–41. https://doi.org/10.1038/ni.3868.

[87] Ohta A, et al. A2A adenosine receptor protects tumors from antitumor T cells. Proc Natl Acad Sci USA 2006;103:13132–7. https://doi.org/10.1073/pnas.0605251103.

[88] Ohta A, et al. Hypoxia-induced and A2A adenosine receptor-independent T-cell suppression is short lived and easily reversible. Int Immunol 2014;26:83–91. https://doi.org/10.1093/intimm/dxt045.

[89] Fong L, et al. Adenosine 2A receptor blockade as an immunotherapy for treatment-refractory renal cell Cancer. Cancer Discov 2020;10:40–53. https://doi.org/10.1158/2159-8290.Cd-19-0980.

[90] Lokshin A, et al. Adenosine-mediated inhibition of the cytotoxic activity and cytokine production by activated natural killer cells. Cancer Res 2006;66:7758–65. https://doi.org/10.1158/0008-5472.Can-06-0478.

[91] Deaglio S, et al. Adenosine generation catalyzed by CD39 and CD73 expressed on regulatory T cells mediates immune suppression. J Exp Med 2007;204:1257–65. https://doi.org/10.1084/jem.20062512.

[92] Sulkowski PL, et al. 2-Hydroxyglutarate produced by neomorphic IDH mutations suppresses homologous recombination and induces PARP inhibitor sensitivity. Sci Transl Med 2017;9. https://doi.org/10.1126/scitranslmed.aal2463.

[93] Reiter-Brennan C, Semmler L, Klein A. The effects of 2-hydroxyglutarate on the tumorigenesis of gliomas. Contemp Oncol (Pozn) 2018;22:215–22. https://doi.org/10.5114/wo.2018.82642.

[94] Bunse L, et al. Suppression of antitumor T cell immunity by the oncometabolite (R)-2-hydroxyglutarate. Nat Med 2018;24:1192–203. https://doi.org/10.1038/s41591-018-0095-6.

[95] Notarangelo G, et al. Oncometabolite d-2HG alters T cell metabolism to impair CD8(+) T cell function. Science 2022;377:1519–29. https://doi.org/10.1126/science.abj5104.

[96] Dudley ME, et al. Cancer regression and autoimmunity in patients after clonal repopulation with antitumor lymphocytes. Science 2002;298:850–4. https://doi.org/10.1126/science.1076514.

[97] Gaffen SL, Liu KD. Overview of interleukin-2 function, production and clinical applications. Cytokine 2004;28:109–23. https://doi.org/10.1016/j.cyto.2004.06.010.

[98] Kalia V, et al. Prolonged interleukin-2Ralpha expression on virus-specific CD8+ T cells favors terminal-effector differentiation in vivo. Immunity 2010;32:91–103. https://doi.org/10.1016/j.immuni.2009.11.010.

[99] Hermans D, et al. Lactate dehydrogenase inhibition synergizes with IL-21 to promote CD8(+) T cell stemness and antitumor immunity. Proc Natl Acad Sci USA 2020;117:6047–55. https://doi.org/10.1073/pnas.1920413117.

[100] Ahmadzadeh M, et al. Tumor antigen-specific CD8 T cells infiltrating the tumor express high levels of PD-1 and are functionally impaired. Blood 2009;114:1537–44. https://doi.org/10.1182/blood-2008-12-195792.

[101] Hinrichs CS, et al. Adoptively transferred effector cells derived from naive rather than central memory CD8+ T cells mediate superior antitumor immunity. Proc Natl Acad Sci USA 2009;106:17469–74. https://doi.org/10.1073/pnas.0907448106.

[102] Tan JT, et al. IL-7 is critical for homeostatic proliferation and survival of naive T cells. Proc Natl Acad Sci USA 2001;98:8732–7. https://doi.org/10.1073/pnas.161126098.

[103] Schluns KS, Kieper WC, Jameson SC, Lefrançois L. Interleukin-7 mediates the homeostasis of naïve and memory CD8 T cells in vivo. Nat Immunol 2000;1:426–32. https://doi.org/10.1038/80868.

[104] Cui G, et al. IL-7-induced glycerol transport and TAG synthesis promotes memory CD8+ T cell longevity. Cell 2015;161:750–61. https://doi.org/10.1016/j.cell.2015.03.021.

[105] Cieri N, et al. IL-7 and IL-15 instruct the generation of human memory stem T cells from naive precursors. Blood 2013;121:573–84. https://doi.org/10.1182/blood-2012-05-431718.

[106] Giuffrida L, et al. IL-15 preconditioning augments CAR T cell responses to checkpoint blockade for improved treatment of solid tumors. Mol Ther 2020;28:2379–93. https://doi.org/10.1016/j.ymthe.2020.07.018.

[107] Xu Y, et al. Closely related T-memory stem cells correlate with in vivo expansion of CAR.CD19-T cells and are preserved by IL-7 and IL-15. Blood 2014;123:3750–9. https://doi.org/10.1182/blood-2014-01-552174.

[108] Hinrichs CS, et al. IL-2 and IL-21 confer opposing differentiation programs to CD8+ T cells for adoptive immunotherapy. Blood 2008;111:5326–33. https://doi.org/10.1182/blood-2007-09-113050.

[109] Loschinski R, et al. IL-21 modulates memory and exhaustion phenotype of T-cells in a fatty acid oxidation-dependent manner. Oncotarget 2018;9:13125–38. https://doi.org/10.18632/oncotarget.24442.

[110] Alvarez-Fernández C, Escribà-Garcia L, Vidal S, Sierra J, Briones J. A short CD3/CD28 costimulation combined with IL-21 enhance the generation of human memory stem T cells for adoptive immunotherapy. J Transl Med 2016;14:214. https://doi.org/10.1186/s12967-016-0973-y.

[111] Sabatino M, et al. Generation of clinical-grade CD19-specific CAR-modified CD8+ memory stem cells for the treatment of human B-cell malignancies. Blood 2016;128:519–28. https://doi.org/10.1182/blood-2015-11-683847.

[112] Wang J, et al. Histone deacetylase inhibitors and IL21 cooperate to reprogram human effector CD8(+) T cells to memory T cells. Cancer Immunol Res 2020;8:794–805. https://doi.org/10.1158/2326-6066.Cir-19-0619.

[113] Li Y, et al. Targeting IL-21 to tumor-reactive T cells enhances memory T cell responses and anti-PD-1 antibody therapy. Nat Commun 2021;12:951. https://doi.org/10.1038/s41467-021-21241-0.

[114] Bian Y, et al. Cancer SLC43A2 alters T cell methionine metabolism and histone methylation. Nature 2020;585:277–82. https://doi.org/10.1038/s41586-020-2682-1.

[115] Rodriguez PC, et al. L-arginine consumption by macrophages modulates the expression of CD3 zeta chain in T lymphocytes. J Immunol 2003;171:1232–9. https://doi.org/10.4049/jimmunol.171.3.1232.

[116] Leone RD, et al. Glutamine blockade induces divergent metabolic programs to overcome tumor immune evasion. Science 2019;366:1013–21. https://doi.org/10.1126/science.aav2588.

[117] Wang T, et al. Inosine is an alternative carbon source for CD8(+)-T-cell function under glucose restriction. Nat Metab 2020;2:635–47. https://doi.org/10.1038/s42255-020-0219-4.

[118] Klysz DD, et al. Inosine induces stemness features in CAR-T cells and enhances potency. Cancer Cell 2024;42:266–282.e268. https://doi.org/10.1016/j.ccell.2024.01.002.

[119] Qiu J, et al. Acetate promotes T cell effector function during glucose restriction. Cell Rep 2019;27:2063–2074.e2065. https://doi.org/10.1016/j.celrep.2019.04.022.

[120] Sukumar M, et al. Inhibiting glycolytic metabolism enhances CD8+ T cell memory and antitumor function. J Clin Invest 2013;123:4479–88. https://doi.org/10.1172/jci69589.

[121] Klein Geltink RI, et al. Metabolic conditioning of CD8(+) effector T cells for adoptive cell therapy. Nat Metab 2020;2:703–16. https://doi.org/10.1038/s42255-020-0256-z.

[122] Davoodzadeh Gholami M, et al. Exhaustion of T lymphocytes in the tumor microenvironment: significance and effective mechanisms. Cell Immunol 2017;322:1–14. https://doi.org/10.1016/j.cellimm.2017.10.002.

[123] Scharping NE, et al. The tumor microenvironment represses T cell mitochondrial biogenesis to drive Intratumoral T cell metabolic insufficiency and dysfunction. Immunity 2016;45:374–88. https://doi.org/10.1016/j.immuni.2016.07.009.

[124] Chowdhury PS, Chamoto K, Kumar A, Honjo T. PPAR-induced fatty acid oxidation in T cells increases the number of tumor-reactive CD8(+) T cells and facilitates anti-PD-1 therapy. Cancer Immunol Res 2018;6:1375–87. https://doi.org/10.1158/2326-6066.Cir-18-0095.

[125] Yu YR, et al. Disturbed mitochondrial dynamics in CD8(+) TILs reinforce T cell exhaustion. Nat Immunol 2020;21:1540–51. https://doi.org/10.1038/s41590-020-0793-3.

[126] Vardhana SA, et al. Impaired mitochondrial oxidative phosphorylation limits the self-renewal of T cells exposed to persistent antigen. Nat Immunol 2020;21:1022–33. https://doi.org/10.1038/s41590-020-0725-2.
[127] Finisguerra V, et al. Metformin improves cancer immunotherapy by directly rescuing tumor-infiltrating CD8 T lymphocytes from hypoxia-induced immunosuppression. J Immunother Cancer 2023;11. https://doi.org/10.1136/jitc-2022-005719.
[128] Petersen CT, et al. Improving T-cell expansion and function for adoptive T-cell therapy using ex vivo treatment with PI3Kδ inhibitors and VIP antagonists. Blood Adv 2018;2:210–23. https://doi.org/10.1182/bloodadvances.2017011254.
[129] Dwyer CJ, et al. Ex vivo blockade of PI3K gamma or delta signaling enhances the antitumor potency of adoptively transferred CD8(+) T cells. Eur J Immunol 2020;50:1386–99. https://doi.org/10.1002/eji.201948455.
[130] Bowers JS, et al. PI3Kδ inhibition enhances the antitumor fitness of adoptively transferred CD8(+) T cells. Front Immunol 2017;8:1221. https://doi.org/10.3389/fimmu.2017.01221.
[131] Abu Eid R, et al. Enhanced therapeutic efficacy and memory of tumor-specific CD8 T cells by ex vivo PI3K-δ inhibition. Cancer Res 2017;77:4135–45. https://doi.org/10.1158/0008-5472.Can-16-1925.
[132] Zheng W, et al. PI3K orchestration of the in vivo persistence of chimeric antigen receptor-modified T cells. Leukemia 2018;32:1157–67. https://doi.org/10.1038/s41375-017-0008-6.
[133] Funk CR, et al. PI3Kδ/γ inhibition promotes human CART cell epigenetic and metabolic reprogramming to enhance antitumor cytotoxicity. Blood 2022;139:523–37. https://doi.org/10.1182/blood.2021011597.
[134] Crompton JG, et al. Akt inhibition enhances expansion of potent tumor-specific lymphocytes with memory cell characteristics. Cancer Res 2015;75:296–305. https://doi.org/10.1158/0008-5472.Can-14-2277.
[135] Urak R, et al. Ex vivo Akt inhibition promotes the generation of potent CD19CAR T cells for adoptive immunotherapy. J Immunother Cancer 2017;5:26. https://doi.org/10.1186/s40425-017-0227-4.
[136] Mousset CM, et al. Ex vivo AKT-inhibition facilitates generation of polyfunctional stem cell memory-like CD8(+) T cells for adoptive immunotherapy. Onco Targets Ther 2018;7, e1488565. https://doi.org/10.1080/2162402x.2018.1488565.
[137] van der Waart AB, et al. Inhibition of Akt signaling promotes the generation of superior tumor-reactive T cells for adoptive immunotherapy. Blood 2014;124:3490–500. https://doi.org/10.1182/blood-2014-05-578583.
[138] Chi H. Regulation and function of mTOR signalling in T cell fate decisions. Nat Rev Immunol 2012;12:325–38. https://doi.org/10.1038/nri3198.
[139] Araki K, et al. mTOR regulates memory CD8 T-cell differentiation. Nature 2009;460:108–12. https://doi.org/10.1038/nature08155.
[140] He S, et al. Characterization of the metabolic phenotype of rapamycin-treated CD8+ T cells with augmented ability to generate long-lasting memory cells. PLoS One 2011;6, e20107. https://doi.org/10.1371/journal.pone.0020107.
[141] Alizadeh D, et al. IL15 enhances CAR-T cell antitumor activity by reducing mTORC1 activity and preserving their stem cell memory phenotype. Cancer Immunol Res 2019;7:759–72. https://doi.org/10.1158/2326-6066.Cir-18-0466.
[142] Adachi K, Davis MM. T-cell receptor ligation induces distinct signaling pathways in naive vs. antigen-experienced T cells. Proc Natl Acad Sci USA 2011;108:1549–54. https://doi.org/10.1073/pnas.1017340108.
[143] Ebert PJR, et al. MAP kinase inhibition promotes T cell and anti-tumor activity in combination with PD-L1 checkpoint blockade. Immunity 2016;44:609–21. https://doi.org/10.1016/j.immuni.2016.01.024.
[144] Verma V, et al. MEK inhibition reprograms CD8(+) T lymphocytes into memory stem cells with potent antitumor effects. Nat Immunol 2021;22:53–66. https://doi.org/10.1038/s41590-020-00818-9.
[145] Dushyanthen S, et al. Agonist immunotherapy restores T cell function following MEK inhibition improving efficacy in breast cancer. Nat Commun 2017;8:606. https://doi.org/10.1038/s41467-017-00728-9.
[146] Fang H-Y, et al. Hypoxia-inducible factors 1 and 2 are important transcriptional effectors in primary macrophages experiencing hypoxia. Blood 2009;114:844–59. https://doi.org/10.1182/blood-2008-12-195941.
[147] Gropper Y, et al. Culturing CTLs under hypoxic conditions enhances their cytolysis and improves their antitumor function. Cell Rep 2017;20:2547–55. https://doi.org/10.1016/j.celrep.2017.08.071.
[148] Zhang Y, et al. Enhancing CD8(+) T cell fatty acid catabolism within a metabolically challenging tumor microenvironment increases the efficacy of melanoma immunotherapy. Cancer Cell 2017;32:377–391.e379. https://doi.org/10.1016/j.ccell.2017.08.004.
[149] Hatfield SM, et al. Immunological mechanisms of the antitumor effects of supplemental oxygenation. Sci Transl Med 2015;7:277ra230. https://doi.org/10.1126/scitranslmed.aaa1260.

[150] Quinn 3rd WJ, et al. Lactate limits T cell proliferation via the NAD(H) redox state. Cell Rep 2020;33, 108500. https://doi.org/10.1016/j.celrep.2020.108500.

[151] Pilon-Thomas S, et al. Neutralization of tumor acidity improves antitumor responses to immunotherapy. Cancer Res 2016;76:1381–90. https://doi.org/10.1158/0008-5472.CAN-15-1743.

[152] Feng Q, et al. Lactate increases stemness of CD8+T cells to augment anti-tumor immunity. Nat Commun 2022;13:4981. https://doi.org/10.1038/s41467-022-32521-8.

[153] Munn DH, et al. Inhibition of T cell proliferation by macrophage tryptophan catabolism. J Exp Med 1999;189:1363–72. https://doi.org/10.1084/jem.189.9.1363.

[154] Grohmann U, Fallarino F, Puccetti P. Tolerance, DCs and tryptophan: much ado about IDO. Trends Immunol 2003;24:242–8. https://doi.org/10.1016/s1471-4906(03)00072-3.

[155] Triplett TA, et al. Reversal of indoleamine 2,3-dioxygenase-mediated cancer immune suppression by systemic kynurenine depletion with a therapeutic enzyme. Nat Biotechnol 2018;36:758–64. https://doi.org/10.1038/nbt.4180.

[156] Vijayan D, Young A, Teng MWL, Smyth MJ. Targeting immunosuppressive adenosine in cancer. Nat Rev Cancer 2017;17:709–24. https://doi.org/10.1038/nrc.2017.86.

[157] Zarek PE, et al. A2A receptor signaling promotes peripheral tolerance by inducing T-cell anergy and the generation of adaptive regulatory T cells. Blood 2008;111:251–9. https://doi.org/10.1182/blood-2007-03-081646.

[158] Mastelic-Gavillet B, et al. Adenosine mediates functional and metabolic suppression of peripheral and tumor-infiltrating CD8(+) T cells. J Immunother Cancer 2019;7:257. https://doi.org/10.1186/s40425-019-0719-5.

[159] Stine ZE, Schug ZT, Salvino JM, Dang CV. Targeting cancer metabolism in the era of precision oncology. Nat Rev Drug Discov 2022;21:141–62. https://doi.org/10.1038/s41573-021-00339-6.

[160] Rojo de la Vega M, Chapman E, Zhang DD. NRF2 and the hallmarks of cancer. Cancer Cell 2018;34:21–43. https://doi.org/10.1016/j.ccell.2018.03.022.

[161] Romero R, et al. Keap1 loss promotes Kras-driven lung cancer and results in dependence on glutaminolysis. Nat Med 2017;23:1362–8. https://doi.org/10.1038/nm.4407.

[162] Yap TA, et al. First-in-human biomarker-driven phase I trial of the potent and selective glutaminase-1 (GLS1) inhibitor IACS-6274 (IPN60090) in patients (pts) with molecularly selected advanced solid tumors. J Clin Oncol 2021. https://doi.org/10.1200/JCO.2021.39.15_suppl.3001.

[163] Gross MI, et al. Antitumor activity of the glutaminase inhibitor CB-839 in triple-negative breast cancer. Mol Cancer Ther 2014;13:890–901. https://doi.org/10.1158/1535-7163.Mct-13-0870.

[164] Soth MJ, et al. Discovery of IPN60090, a clinical stage selective Glutaminase-1 (GLS-1) inhibitor with excellent pharmacokinetic and physicochemical properties. J Med Chem 2020;63:12957–77. https://doi.org/10.1021/acs.jmedchem.0c01398.

[165] Best SA, et al. Glutaminase inhibition impairs CD8 T cell activation in STK11−/Lkb1-deficient lung cancer. Cell Metab 2022;34:874–887.e876. https://doi.org/10.1016/j.cmet.2022.04.003.

[166] Srinivasan R, Ricketts CJ, Sourbier C, Linehan WM. New strategies in renal cell carcinoma: targeting the genetic and metabolic basis of disease. Clin Cancer Res 2015;21:10–7. https://doi.org/10.1158/1078-0432.Ccr-13-2993.

[167] Cascone T, et al. Increased tumor glycolysis characterizes immune resistance to adoptive T cell therapy. Cell Metab 2018;27:977–987.e974. https://doi.org/10.1016/j.cmet.2018.02.024.

[168] Yeung C, et al. Targeting glycolysis through inhibition of lactate dehydrogenase impairs tumor growth in preclinical models of Ewing sarcoma. Cancer Res 2019;79:5060–73. https://doi.org/10.1158/0008-5472.can-19-0217.

[169] Wallen H, et al. Fludarabine modulates immune response and extends in vivo survival of adoptively transferred CD8 T cells in patients with metastatic melanoma. PLoS One 2009;4, e4749. https://doi.org/10.1371/journal.pone.0004749.

[170] Muranski P, et al. Increased intensity lymphodepletion and adoptive immunotherapy- -how far can we go? Nat Clin Pract Oncol 2006;3:668–81. https://doi.org/10.1038/ncponc0666.

[171] Gattinoni L, et al. Removal of homeostatic cytokine sinks by lymphodepletion enhances the efficacy of adoptively transferred tumor-specific CD8+ T cells. J Exp Med 2005;202:907–12. https://doi.org/10.1084/jem.20050732.

[172] Kochenderfer JN, et al. Lymphoma remissions caused by anti-CD19 chimeric antigen receptor T cells are associated with high serum interleukin-15 levels. J Clin Oncol 2017;35:1803–13. https://doi.org/10.1200/jco.2016.71.3024.

[173] Ninomiya S, et al. Tumor indoleamine 2,3-dioxygenase (IDO) inhibits CD19-CAR T cells and is downregulated by lymphodepleting drugs. Blood 2015;125:3905–16. https://doi.org/10.1182/blood-2015-01-621474.

[174] Bechman N, Maher J. Lymphodepletion strategies to potentiate adoptive T-cell immunotherapy—what are we doing; where are we going? Expert Opin Biol Ther 2021;21:627–37. https://doi.org/10.1080/14712598.2021.1857361.

[175] Li D, et al. A pilot study of lymphodepletion intensity for peripheral blood mononuclear cell-derived neoantigen-specific CD8+T cell therapy in patients with advanced solid tumors. Nat Commun 2023;14:3447. https://doi.org/10.1038/s41467-023-39225-7.

[176] Yee C, Lizee G, Schueneman AJ. Endogenous T-cell therapy: clinical experience. Cancer J 2015;21:492–500. https://doi.org/10.1097/ppo.0000000000000158.

[177] Kueberuwa G, Kalaitsidou M, Cheadle E, Hawkins RE, Gilham DE. CD19 CAR T cells expressing IL-12 eradicate lymphoma in fully Lymphoreplete mice through induction of host immunity. Mol Ther Oncolytics 2018;8:41–51. https://doi.org/10.1016/j.omto.2017.12.003.

[178] Pegram HJ, et al. Tumor-targeted T cells modified to secrete IL-12 eradicate systemic tumors without need for prior conditioning. Blood 2012;119:4133–41. https://doi.org/10.1182/blood-2011-12-400044.

[179] Sharpe AH, Pauken KE. The diverse functions of the PD1 inhibitory pathway. Nat Rev Immunol 2018;18:153–67. https://doi.org/10.1038/nri.2017.108.

[180] Parry RV, et al. CTLA-4 and PD-1 receptors inhibit T-cell activation by distinct mechanisms. Mol Cell Biol 2005;25:9543–53. https://doi.org/10.1128/mcb.25.21.9543-9553.2005.

[181] Sharma P, Allison JP. The future of immune checkpoint therapy. Science 2015;348:56–61. https://doi.org/10.1126/science.aaa8172.

[182] Patsoukis N, et al. PD-1 alters T-cell metabolic reprogramming by inhibiting glycolysis and promoting lipolysis and fatty acid oxidation. Nat Commun 2015;6:6692. https://doi.org/10.1038/ncomms7692.

[183] Spranger S, et al. Mechanism of tumor rejection with doublets of CTLA-4, PD-1/PD-L1, or IDO blockade involves restored IL-2 production and proliferation of CD8(+) T cells directly within the tumor microenvironment. J Immunother Cancer 2014;2:3. https://doi.org/10.1186/2051-1426-2-3.

[184] West EE, et al. PD-L1 blockade synergizes with IL-2 therapy in reinvigorating exhausted T cells. J Clin Invest 2013;123:2604–15. https://doi.org/10.1172/jci67008.

[185] Pardoll DM. The blockade of immune checkpoints in cancer immunotherapy. Nat Rev Cancer 2012;12:252–64. https://doi.org/10.1038/nrc3239.

[186] Walunas TL, et al. CTLA-4 can function as a negative regulator of T cell activation. Immunity 1994;1:405–13. https://doi.org/10.1016/1074-7613(94)90071-x.

[187] Kennedy PT, et al. Soluble CTLA-4 attenuates T cell activation and modulates anti-tumor immunity. Mol Ther 2024;32:457–68. https://doi.org/10.1016/j.ymthe.2023.11.028.

[188] Beckermann KE, Dudzinski SO, Rathmell JC. Dysfunctional T cell metabolism in the tumor microenvironment. Cytokine Growth Factor Rev 2017;35:7–14. https://doi.org/10.1016/j.cytogfr.2017.04.003.

[189] Woo SR, et al. Immune inhibitory molecules LAG-3 and PD-1 synergistically regulate T-cell function to promote tumoral immune escape. Cancer Res 2012;72:917–27. https://doi.org/10.1158/0008-5472.Can-11-1620.

[190] Grosso JF, et al. LAG-3 regulates CD8+ T cell accumulation and effector function in murine self- and tumor-tolerance systems. J Clin Invest 2007;117:3383–92. https://doi.org/10.1172/jci31184.

[191] Previte DM, et al. Lymphocyte activation Gene-3 maintains mitochondrial and metabolic quiescence in naive CD4(+) T cells. Cell Rep 2019;27:129–141.e124. https://doi.org/10.1016/j.celrep.2019.03.004.

[192] Hannier S, Tournier M, Bismuth G, Triebel F. CD3/TCR complex-associated lymphocyte activation gene-3 molecules inhibit CD3/TCR signaling. J Immunol 1998;161:4058–65.

[193] Wolf Y, Anderson AC, Kuchroo VK. TIM3 comes of age as an inhibitory receptor. Nat Rev Immunol 2020;20:173–85. https://doi.org/10.1038/s41577-019-0224-6.

[194] Lee MJ, et al. Association of TIM-3 expression with glucose metabolism in Jurkat T cells. BMC Immunol 2020;21:48. https://doi.org/10.1186/s12865-020-00377-6.

[195] Sakuishi K, et al. Targeting Tim-3 and PD-1 pathways to reverse T cell exhaustion and restore anti-tumor immunity. J Exp Med 2010;207:2187–94. https://doi.org/10.1084/jem.20100643.

CHAPTER

2

Links between the DNA damage response and immune response in cancer

Agnes Malysa[a,b], *Jianxun Jim Song*[c], *Gerold Bepler*[b], *and Xiaohong Mary Zhang*[b]

[a]Cancer Biology Graduate Program, Karmanos Cancer Institute, Wayne State University School of Medicine, Detroit, MI, United States [b]Department of Oncology, Karmanos Cancer Institute, Wayne State University School of Medicine, Detroit, MI, United States [c]Department of Microbial Pathogenesis & Immunology, School of Medicine, University of Texas A&M, Bryan, TX, United States

A SHORT SUMMARY OF THE CHAPTER'S CONTENTS

This chapter first describes the discovery of immune checkpoint molecules and the impact of this discovery on cancer treatment. Subsequently, it describes various types of DNA damage and DNA damage response (DDR) cascade, including five DNA repair mechanisms, cisplatin-affected DNA repair pathways, intersection of DDR and tumor immunology, and the cGAS-STING pathway and cisplatin. We focus on how cisplatin affects DNA repair and elicits innate immune responses through the cGAS-STING pathways.

Abstract

Immune checkpoint blockade-based therapy has opened the door to a new era of cancer treatment. Numerous immune checkpoint monoclonal antibodies have been approved by the US FDA and used to treat nearly 20 types of cancer. Despite this therapy bringing hope to many cancer patients, the response rates are still very low. Although the levels of PD-L1 and tumor mutational burden (TMB) have been utilized to stratify patients in the clinic, more biomarkers are urgently needed to select responders and nonresponders. DNA damage-based radiation and

https://doi.org/10.1016/B978-0-443-13827-0.00014-7

chemotherapy have shown benefits for patients when they are used in combination with immune checkpoint inhibitors. Deficiencies of certain genes or proteins in the DNA damage response/DNA repair pathways were found to be associated with responsiveness to immune checkpoint inhibitors. Therefore, many inhibitors against certain DNA damage response/DNA repair proteins are being used in combination with immune checkpoint inhibitors in hundreds of clinical trials. In this chapter, we will review DNA damage response sensors, transducers, and effectors, as well as five different types of DNA repair mechanisms. We will also focus on how cisplatin affects DNA repair and elicits innate immune responses through the cGAS-STING pathways. A deeper understanding of the relationship between the DNA damage response and responsiveness to the immune checkpoint blockade will aid in designing better combination treatment strategies for cancer patients.

Abbreviations

53BP1	p53-binding protein 1
ATM	ataxia-telangiectasia mutated
ATR	ATM- and Rad3-related
B7	B-cell surface antigen 7
B7-H1	B7 homolog 1
BER	base excision repair
BRAC1	BReast CAncer gene 1
CD28	cluster of differentiation 28
CDC25A	cell division cycle 25A
Cdk1	cyclin-dependent kinase 1
cGAS-STING	cyclic GMP-AMP synthase-stimulator of interferon genes
Chk1	checkpoint kinase 1
Chk2	checkpoint kinase 2
CTLA-4	cytotoxic T lymphocyte-associated protein 4
DDR	DNA damage response
dMMR	mismatch repair deficient
DNA	deoxyribonucleic acid
DSB	double strand breaks
ERCC4	excision repair 4, endonuclease catalytic subunit
FDA	Food & Drug Administration
HR	homologous recombination
ICB	immune checkpoint blockade
ICLs	interstrand cross-links
IFNs	interferons
IR	ionizing radiation
IRF3	interferon regulatory factor 3
LAG3	lymphocyte-activation gene 3
MDC1	mediator of DNA damage checkpoint protein 1
MMR	mismatch repair
MSI-H	microsatellite instability-high
NBN	nibrin
NER	nucleotide excision repair
NHEJ	nonhomologous end joining
NSCLC	nonsmall cell lung cancer
PARP1	poly [ADP-ribose] polymerase 1
PD-1	programmed cell death 1
PD-L1	programmed cell death legend 1
POLE	DNA polymerase epsilon catalytic subunit
PRR	pattern recognition receptor
RFC2-5	replication factor C subunit 2–5

ROS reactive oxygen species
SSB single strand breaks
TMB tumor mutational burden
US United States
UV ultraviolet

Conflict of interest

No potential conflicts of interest were disclosed.

Introduction

In recent years, immune checkpoint blockade (ICB)-based immunotherapy has revolutionized cancer treatment [1–3]. Although exploiting the immune system to treat cancer has long been a dream for cancer immunologists, their previous attempts, such as vaccine-based and cytokine-based therapies, only gained limited clinical significance [4,5]. With accumulating knowledge on T cell regulation gained by basic immunologists and timely collaboration among basic immunologists, physicians, and pharmaceutical companies, ICB-based immunotherapy finally emerged and surpassed all other types of immunotherapies to become a standard of care for the treatment of many cancers, including melanoma and nonsmall cell lung cancer (NSCLC). This type of therapy, which unleashes the power of cytotoxic T cells to kill tumors, is unique in two regards. First, ICB-based immunotherapy targets the immune cells and the tumor microenvironment, while traditional chemoradiotherapy and targeted therapy target cancer cells. Second, immunotherapy could induce a durable clinical response, while most traditional therapies will inevitably cause resistance as new oncogenic mutations or pathways will often appear after the therapy.

Back in the early 1990s, American immunologist Dr. James P. Allison, who had a longstanding interest in T cell regulation, discovered CD28 as a costimulator for CD8+ T cells [6]. At that time, many thought another T cell surface receptor, called cytotoxic T lymphocyte-associated protein 4 (CTLA-4), was also a costimulatory molecule as it is homologous to CD28 and binds to B7 as CD28 does [7,8]. But the work from Dr. Jeffery Bluestone and Dr. James P. Allison conclusively proved that CTLA-4 is in fact a negative regulator for T cells and serves as a brake for T cell functions [9–11]. With this information, Dr. Allison immediately thought that inhibition of CTLA-4 could be used to treat cancer. In 1996, the follow-up work from Dr. Allison's group showed that anti-CTLA-4 could reject tumors and reduce tumor growth in mice as a monotherapy [12] (Fig. 1). With this exciting result, the anti-CTLA-4 monoclonal antibody, named ipilimumab, was promptly raised against human CTLA-4 protein. In 2011, the US Food & Drug Administration (FDA) approved it for treating patients with melanoma [13]. Many melanoma patients, including metastatic patients, who received anti-CTLA-4 lived more than 10 years after the treatment [14].

However, only a subset of melanoma patients responded to ipilimumab, indicating that multiple inhibitory molecules for T cells exist. In a cohort of BRAF wild-type melanoma patients, when ipilimumab was administered in combination with another antibody (nivolumab) against a checkpoint inhibitor called programmed cell death 1 (PD-1), the response rate was as high as 61%, while the response rate for ipilimumab monotherapy was only 11% [15]. The PD-1

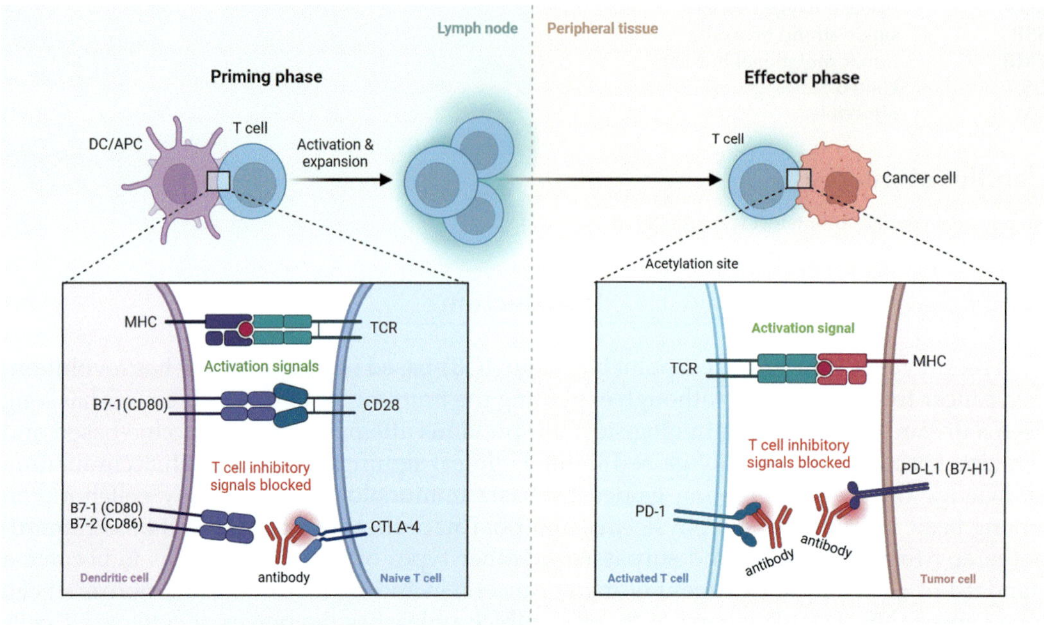

FIG. 1 The mechanism by which the anti-CTLA-4 and anti-PD-1/L1 antibodies block T cell inhibitory signals. The specific recognition of cognate antigenic peptides presented by major histocompatibility complex (MHC) molecules triggers T cell receptor (TCR) signaling, but other costimulatory and coinhibitory molecules are also needed to determine the fate of the T cell. CD28 was discovered as a costimulatory molecule and binds to members of the B7 family. The coinhibitory molecule CTLA-4 binds to B7-1 (also known as CD80) and B7-2 (also known as CD86), which are expressed on professional antigen-presenting cells (APCs). Dendritic cells are a type of APCs, which are the primary drivers of T cell activation and differentiation in lymphoid organs. Anti-CTLA-4 antibodies block the interaction between CTLA-4 and B7, suppressing the inhibitory effect of CTLA-4, leading to T cell activation. Anti-PD-1 and anti-PD-L1 antibodies block the interaction between PD-1 and PD-L1, and as a result, activate T cell functions. *The graph was generated by BioRender.com.*

molecule was discovered from dying T cells by Japanese molecular immunologist Dr. Tasuku Honjo in 1992 [16]. In 1999, a research group led by Dr. Lieping Chen, then at Mayo Clinic, discovered a molecule called B7-H1 as the third member in the B7 family [17]. This new B7 family member turned out to be the ligand of PD-1. A year later, a collaborative effort from the Genetics Institute in Cambridge, MA, USA, a research group led by Dr. Gordon Freeman at the Dana-Faber Cancer Institute, and Dr. Tasuku Honjo's lab led to independent discovery of the PD-1 ligand, named PD-1 ligand 1 (PD-L1) [18] (Fig. 1). Subsequently, a second PD-1 ligand, PD-L2, was identified [19].

To date, a total of 10 monoclonal antibodies against immune checkpoint molecules, including two anti-CTLA-4 antibodies, four anti-PD-1 antibodies, three anti-PD-L1 antibodies, and one anti-LAG3 antibody, have been approved by the US FDA and are widely used in treating nearly 20 types of cancer (Table 1). LAG3 is another immune checkpoint molecule [20]. Notably, Dr. James P. Allison and Dr. Tasuku Honjo shared the Nobel Prize in Physiology or Medicine in 2018 for their contributions to cancer immunotherapy through their studies on CTLA-4 and PD-1.

TABLE 1 FDA-approved checkpoint inhibitors.

	Antibody name	Trade name	Year approved
Anti-CTLA-4	ipilimumab	Yervoy	2011
Anti-CTLA-4	tremelimumab	IMJUDO	2022
Anti-PD-1	pembrolizumab	Keytruda	2014
Anti-PD-1	nivolumab	Opdivo	2014
Anti-PD-1	cemiplimab	Libtayo	2018
Anti-PD-1	dostarlimab	Jemperli	2021
Anti-PD-L1	avelumab	Bavencio	2015
Anti-PD-L1	atezolizumab	Tecentriq	2016
Anti-PD-L1	durvalumab	Imfinzi	2016
Anti-LAG-3	relatimab	Opdualag	2022

Despite the unprecedented success of ICB-based immunotherapy, the patients who respond to the therapy are still limited due to the innate and acquired resistance [21]. To increase the response rates, DNA damage-based chemotherapy and radiation were used in combination with immunotherapy in numerous clinical trials, such as PACIFIC and Keynote 189 trials in lung cancer patients [22–24]. Emerging evidence has shown that DNA repair deficiency could elicit an immune response to enhance the efficacy of immunotherapy. Notably, DNA mismatch repair (MMR) deficiency predicted clinical benefit of immune checkpoint blockade with pembrolizumab in a Phase II clinical trial conducted in 2015 in colorectal cancer patients [25]. Subsequently, the FDA approved pembrolizumab on May 23, 2017, for the treatment of adult and pediatric patients with unresectable or metastatic, microsatellite instability-high (MSI-H), or mismatch repair deficient (dMMR) solid tumors. In addition to MMR genes, mutations in seven other types of DNA repair genes (ATM, ATR, POLE, ERCC4, NBN, RAD50, and PARP1) are significantly associated with improved overall survival in ICB-treated patients and have significant interaction with treatment [26]. Therefore, it becomes apparent that the DNA damage response/DNA repair is associated with ICB-based immunotherapy response. Because cisplatin is a major drug being used in treating many cancers [27], in this chapter, we will review literature on the DNA damage response/DNA repair with an emphasis on cisplatin-induced DNA damage and repair and how it links to the cGAS-STING pathway.

Various types of DNA damage

In general, DNA is constantly under attack by intrinsic and extrinsic factors that can compromise its structure and function. If this damage cannot be repaired, it can lead to errors that result in aberrant cell growth. DNA damage can occur through endogenous factors such as replication errors and oxidative damage by reactive oxygen species (ROS) or exogenous factors like chemicals/toxins, radiation, or environmental stress [28]. In the presence of such

DNA-damaging agents, it is vital for cells to respond to these assaults in a timely fashion to minimize further disruption to the DNA structure. The ability of the cells to assess the damage and determine how to fix it is known as the DNA damage response (DDR) [29]. If DNA damage is not fixed in a timely fashion, genomic instability will occur and lead to tumorigenesis.

The DDR connects multiple signaling pathways that direct proteins that respond to and repair DNA damage [30]. To understand the DDR, it is necessary to understand how cells grow. Cells utilize an orchestrated process, called the cell cycle, which promotes replication of DNA and division of cells and allows cell growth [31]. To protect cells from damage, the cell cycle has checkpoints throughout. Within these checkpoints are proteins that are responsible for sensing different types of DNA damage and relaying signals to the correct repair pathways [30,32].

The DDR begins with the presence of DNA damage by either endogenous or exogenous means. Focus here will be on how exogenous DNA damage leads to disruptions in the DNA structure and triggers the DDR. Exogenous DNA damage can be organized by the following modifications to the DNA helix structure: alkylation, oxidation, DNA cross-links, and of interest to this chapter, DNA strand breaks, which include single and double strand breaks [33] (Fig. 2).

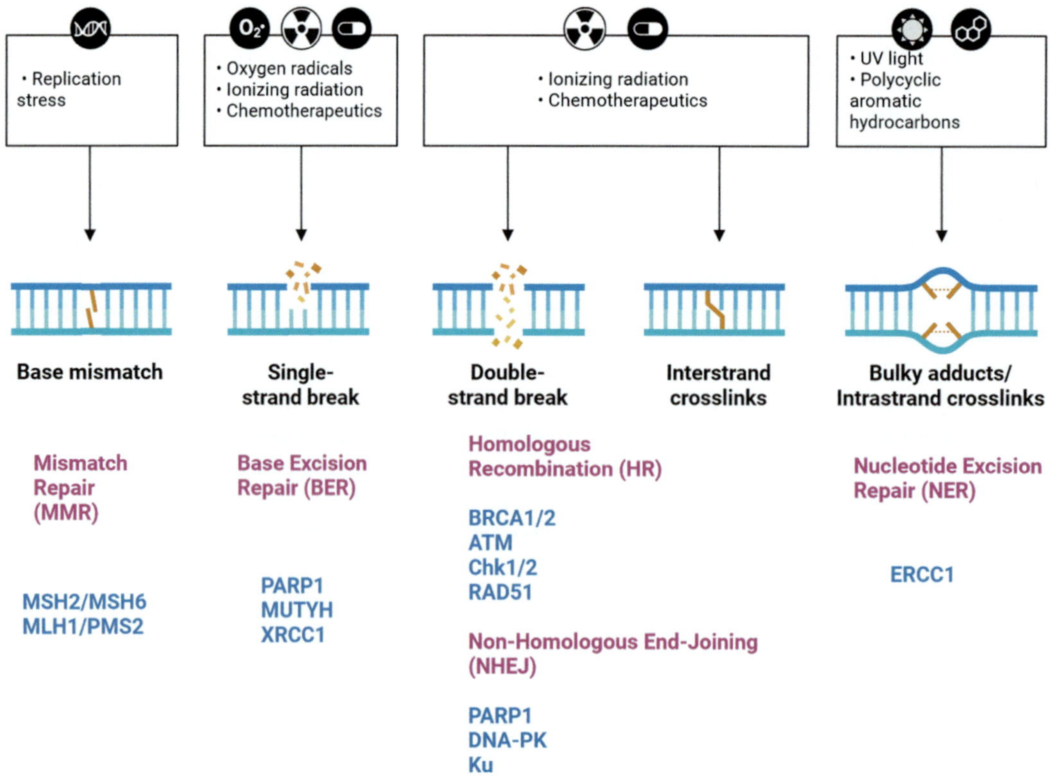

FIG. 2 Various types of DNA damage, DNA repair pathways, and key proteins involved in the repair processes are shown. *The graph was generated by BioRender.com.*

Alkylation DNA damage occurs when guanosine is attacked by free radicals, which causes the mispairing of DNA bases [34]. The alkylating agents could add alkyl groups to the major and minor grooves of DNA bases, where methylation is the most common outcome of alkylation [34,35]. Oxidative damage is caused by the attack of ROS that can occur due to ultraviolet conditions [36]. ROS could add hydroxyl groups to DNA bases, mainly guanosine. Like alkylation, this modification to DNA bases causes mispairing, which in turn disrupts later processes like DNA replication [37].

Another type of DNA damage is DNA cross-linking, which occurs when a covalent link forms between two DNA bases [38]. If cross-linking occurs on the same strand of DNA, it is referred to as an intrastrand cross-link, whereas if the cross-link occurs between opposite strands, it is referred to as an interstrand cross-link [38]. When DNA cross-links are present, they interfere with DNA replication and cellular metabolism, which, if not repaired, can lead to cell death [39]. Despite the negative effects of cross-linked DNA, cross-linking agents, notably cisplatin, have been developed for treating cancer. These will be discussed later in this chapter. The last type of DNA damage that will be discussed here is DNA strand breaks, which come in two forms: single strand breaks (SSBs), which occur when one of the DNA strands is broken, and double strand breaks (DSBs) when both strands of the DNA helix are broken [40]. SSBs are considered the most common type of strand break and can occur in many ways, as by-products from repair pathways or oxidative stress [40,41]. Unrepaired SSBs will cause disruption in DNA replication through replication fork collapse and stalling of RNA polymerase [42]. DSBs, which are not as frequent, arise from breaks in two complementary DNA strands; however, these types of lesions are very toxic and could lead to cell death [43].

How the DNA damage response cascade unfolds

Normal functioning cells grow and divide by proceeding through the cell cycle. The cell cycle has designated checkpoints that become activated when something aberrant is detected. When this happens, the DDR process will begin. If SSBs or DSBs are sensed at a particular checkpoint, DNA damage sensors will activate a DNA damage transducer to relay the DDR signal to arrest the cell cycle at these checkpoints so that DNA repair can commence. Once this is done, the cell cycle can resume. The DDR can be broken down into three components: DNA damage sensors, transducers/effectors, and repair pathways [30]. The following sections will go into some depth into these three components to lay the foundation for understanding the DNA damage response.

DNA damage sensors

When considering SSBs, literature has begun to understand how DNA sensors function with the DDR. When Rad9, Rad1, and Hus1 sense SSBs, they will form a clamp that is loaded onto the DNA by Rad17/RFC2–5. This process will activate ATR, which is a DNA damage transducer that will be discussed later [44,45]. There is also another complex, Rad26/Rad3, that works simultaneously yet independently from Rad9, Rad1, and Hus1 to promote checkpoint activation [46].

DSBs, on the other hand, are sensed by the Mre11-Rad50-NBS1 (MRN) complex, which promotes ATM autophosphorylation, where ATM is a master DNA damage transducer that will be mentioned later [32]. Another very important DNA damage sensor is PARP1, which is considered the first responder for DNA damage. While it is mostly known for its work in DSB repair, PARP1 is also known to be responsible for sensing SSBs [47]. Specifically, cells lacking PARP have defective DSB repair or will accumulate SSBs during DNA replication [48].

Cell cycle checkpoints and DNA damage transducers and effectors

There are three checkpoints in the cell cycle: G1 (which occurs at the G1/S border), intra-S phase (before DNA replication occurs), and G2 (which occurs at the G2/M border before cell division occurs) [49]. There are DNA damage transducers within these checkpoints (ATM, ATR, p53, and p21) that will transduce the damage signal and stimulate the checkpoint cell cycle arrest to proceed with the DDR.

Ataxia-telangiectasia mutated (ATM) and Ataxia and rad-related (ATR)

ATM and ATR are master kinases that regulate the DNA damage response network by relaying the DNA damage signal to potentiate the DDR [50]. ATM and ATR govern the G1, intra-S, and G2 checkpoints, where ATM responds to DSBs and chromatin disturbance and ATR responds to stalled replication forks [51]. To relay the DNA damage signal, ATM and ATR phosphorylate downstream mediator proteins (BRCA1, 53BP1, and MDC1), which transmit the checkpoint activation signal to downstream effector proteins [50,52]. Some notable downstream effector proteins are Chk2 (phosphorylated by ATM) and Chk1 (phosphorylated by ATR) that are responsible for amplifying the DNA damage signals from ATM and ATR [53]. ATM phosphorylates Chk2, and once activated, Chk2 relays the DNA damage signal, where it will act on multiple substrates, like p53 and others that are involved in gene transcription, apoptosis, and cell cycle progression. ATR activates Chk1, which causes the degradation of CDC25A, a phosphatase that promotes origin firing in late S phase and leads to cell cycle arrest [54]. ATM and ATR have parallel pathways where they can signal each other's effector proteins, specifically that ATM activates Chk1 during ionizing radiation (IR), and similarly, ATR can activate Chk2 under ultraviolet and IR conditions [55,56].

p53

p53 is known as the "guardian of the genome" because it transduces the DNA damage signal, which induces cell cycle arrest to promote DNA repair. If the DNA damage cannot be repaired, p53 will induce apoptosis [57]. In normal cells, p53 levels are low due to its binding to MDM2, a ubiquitin E3 ligase of p53. At the time of DNA damage, MDM2 is inhibited and p53 levels accumulate so that it can respond to the DNA damage [58,59]. In addition to regulating the G1 checkpoint through p21, p53 controls the G2 checkpoint by decreasing levels of cyclin B1 to cause cell cycle arrest and is activated by ATM and ATR [60,61].

p21

p21 is a cyclin-dependent kinase inhibitor that binds to and inhibits CDK1, CDK2. and CDK4/6, where it regulates the G1 to S transition [62]. When there is DNA damage, p21 interacts with PCNA, which is a critical regulator of DNA replication and inhibits DNA synthesis [62,63]. It has been observed that p21 can affect different types of DNA repair pathways by either promoting or inhibiting them [64]. p21 is also responsible for regulating the G2/M checkpoint through its activation by p53 upon DNA damage [65].

DNA repair pathways

Now that the DDR has been relayed, the next step is to initiate DNA repair pathways and restore DNA integrity to prevent negative repercussions such as tumorigenesis [66]. Based on the DNA lesion, DNA repair comprises five major pathways, including mismatch repair (MMR), homologous recombination (HR), nonhomologues end joining (NHEJ), base excision repair (BER), and nucleotide excision repair (NER) (Fig. 2). In 2015, the Nobel Prize in Chemistry was awarded to Dr. Tomas Lindahl, Dr. Paul Modrich, and Dr. Aziz Sancar for their mechanistic studies of BER, MMR, and NER, respectively.

Base excision repair (BER)

This repair pathway is necessary for repairing abnormal bases and individual base lesions that are caused by oxidation, deamination, and alkylation [67]. BER is initiated by different DNA glycosylases (depending on the type of damage the individual nucleotide encompasses), where BER will remove the damaged base and DNA polymerase will fill the DNA gap made by BER [68].

Nucleotide excision repair (NER)

Bulky DNA adducts caused by UV light and other mutagenic agents are repaired by NER [69]. The difference between BER and NER is that NER removes sections of DNA, whereas BER removes individual modified bases [70]. The process of NER removes a short section of DNA that contains the lesions, and much like BER, DNA polymerase will use the undamaged parts of the DNA to rebuild the missing DNA section [71].

Mismatch repair (MMR)

This repair pathway is responsible for repairing mismatched base substitutions and removing inserted and deleted DNA bases that were missed by DNA polymerase [72]. This repair pathway is referred to as the guardian of genomic integrity because its repair process is free of errors, which ensures a high quality of DNA repair. This process is critical because studies have shown that MMR deficiencies lead to spontaneous tumors [73]. Although DNA damage does not initiate MMR, MMR can influence which repair pathways cells use because MMR-deficient cells confer resistance to UV and alkylation damage [74].

Homologous recombination (HR)

This repair pathway repairs ICLs and DSBs, as well as broken or stalled forks [75]. In HR repair, the DSB is resected, which produces DNA ends with 3′ single-stranded DNA tails. The 3′ ends invade homologous DNA, forming a DNA crossover, which acts as a primer to initiate new DNA synthesis [76]. During the repair process, homologous base pairs are exchanged and moved further away from the initial crossover site until the crossover end is cleaved out, marking that the damage has been repaired [77].

Nonhomologous end joining (NHEJ)

Similar to HR, this pathway repairs DSBs, but it repairs lesions by ligating the broken ends without a homologous template [78]. Specifically, this repair pathway recognizes, resects, polymerizes, and ligates a wide range of DNA end configurations, which often contain mutations [78]. NHEJ is active during the cell cycle and is used more frequently than HR because there is no requirement for a repair template, which makes the process of NHEJ an order of magnitude faster than HR [79].

The DNA damage response and repair pathways are critical for understanding how normal functioning cells repair themselves in the presence of DNA damage. However, studying the DDR has also given insight into how to exploit these repair mechanisms to develop strategies to kill tumors. One way to do this is to design drugs, like what has been done with platinum-based chemotherapy. These drugs are known to cause both inter- and intrastrand cross-links and other DNA modifications that distort the DNA structure [80]. The following section will focus on how cisplatin affects the DDR response.

Cisplatin-affected DNA repair pathways

DNA repair pathways are essential for maintaining genomic integrity; however, when there are defects in these pathways, DNA damage can accumulate. If the damage is tolerated, the cell will continue to divide and pass on DNA that is rife with errors despite having repair pathways to correct these mistakes [81]. The replication and division of such unstable genomes is a characteristic of cancer cells, where positive and negative attributes get passed on and help cancer cells evade repair mechanisms or make them deficient in repair processes. Repair-deficient cancer cells make for ideal cisplatin targets because these cancer cells will not survive the damage induced by cisplatin. However, due to the redundancy in the DNA repair pathways, cancer cells have the option to use other means to repair themselves so they can continue to propagate, which leads to the development of future generations of cancer cells that are resistant to drugs like cisplatin [82,83]. Despite cisplatin (and its derivatives) being highly efficient at targeting cancer cells, cisplatin resistance remains a major challenge in cisplatin-based chemotherapy [83]. Once cisplatin causes DNA lesions, many of the DNA repair pathways, including NER, MMR, and HR/NHEJ, work together to repair cisplatin-induced DNA damage [84].

NER

NER can detect and repair cisplatin-induced DNA adducts using both global genome repair and transcription-coupled repair [85]. And both pathways use the endonuclease ERCC1/XPF to cleave out DNA lesions [86]. The commonplace use of cisplatin is forcing cancer cells to adapt once they no longer respond to NER because they carry a higher burden of DNA damage. The relationship between NER and cisplatin is such that deficiencies in ERCC1/XPF enhance cisplatin efficacy in cancer cells, which suggests the importance of NER in cisplatin resistance [86]. This relationship extends even further because there is a correlation between cisplatin resistance and UV resistance, which circles back to the ability of ERCC1/XPF to remove these types of adducts [87,88].

MMR

MMR recognizes lesions induced by cisplatin, but it is unable to repair these lesions because it can only replace DNA bases opposite to the cisplatin adduct [83]. Since MMR cannot properly repair the lesion, MMR will perform futile repair, where it will restart the repair process and continue to try to remove the adduct. This type of repair leads to DSBs, which will trigger apoptosis [89]. It is suggested that this MMR futile cycle is why MMR-deficient cells confer cisplatin resistance [90]. It has also been observed that MMR proteins can protect cisplatin adducts from NER repair, so it suggested that because of this, MMR may contribute to cisplatin resistance in tumors [91].

HR/NHEJ

While both of these pathways repair DSBs, they also play a role in the repair of interstrand cross-links (ICLs) caused by cisplatin [83]. In the case of activated HR, histone H2AX is phosphorylated at DNA lesions (phosphorylated H2AX is called γH2AX), which activates the recruitment of HR proteins and promotes repair [92]. In NHEJ, Ku proteins detect DNA lesions and actively ligate the broken DNA strands without the use of a template, like in HR repair [93]. In a cisplatin-resistant condition, the promotion of γH2AX inhibits Ku-mediated apoptosis, and together, these events contribute to drug resistance [94]. However, it was also observed that a lack of γH2AX is associated with cisplatin-resistant tumors, which suggests that cisplatin resistance is not dependent on the amount of cisplatin-induced adducts [95]. Additionally, NHEJ repair tends to occur in G0/G1 cells, whereas HR repair occurs during S/G2, when cells are actively undergoing DNA replication. This suggests that the nuances of these pathways may dictate cisplatin resistance, where slow-growing cells may become targets [96,97].

In sum, there is no simple answer for how cisplatin resistance affects DNA repair pathways because there are many gaps of knowledge for understanding how cancer cells exploit different aspects of the DNA repair pathways to tolerate and accumulate cisplatin-induced damage.

Intersection of DDR and tumor immunology

The immune response intersects with the DDR through innate immunity when it is triggered by the presence of SSBs and DSBs [98]. This specific immune signaling pathway, cGAS-STING, is turned on when DNA accumulates in the cytoplasm, where cGAS (cyclic GMP-AMP synthase), a pattern recognition receptor (PRR), acts as a DNA sensor [99]. Upon detection, cGAS dimerizes and forms cGAMP, which subsequently binds to STING (stimulator for interferon genes). This interaction causes the activation of type I interferons (IFNs), IRF3, and NFκB, which then promotes the expression of inflammatory genes, like IFN-β [100]. Fork stalling and other replicative stresses can cause an increase in cytoplasmic DNA, and literature has observed that protein regulators from the cell cycle can activate the cGAS-STING pathway [101]. It has also been observed that cGAS-STING misexpression can cause disruptions to the cell cycle, where knockdown of cGAS-STING causes decreases in p21 expression and an increase in micronuclei, which are DNA structures that are associated with genetic instability [102]. In addition, it is believed that cGAS can decrease proliferation by interacting with nuclear DNA in the S phase [103].

Micronuclei formation does occur in the DDR due to radiation; however, from this perspective, cGAS-STING activation has been shown to stimulate antitumor characteristics [104]. Literature has suggested that cGAS-STING bridges the immune response with DDR by enhancing the effect of IR and causing the activation of the adaptive immune response through $CD8^{+}$ cytotoxic T cell activation [104]. Additionally, it is thought that the immune system is suppressed by radiation by a negative feedback loop, where it has been observed that PD-L1 is upregulated in the presence of radiation [105].

Alternatively, pharmacological inhibitors for ATM and ATR reduce DNA repair ability and lead to cytoplasmic DNA accumulation, which in turn stimulates STING signaling and induces an interferon response [101,106]. Interestingly, Chk1 inhibitors are not effective at promoting STING-stimulated interferon response [107], whereas Chk2 inhibition causes the activation of the cGAS-STING pathway [106].

From the perspective of DNA repair pathways, cGAS does inhibit homologous recombination (HR) by binding with PARP1 to prevent DNA repair, which will promote tumorigenesis [108]. cGAS can also inhibit nonhomologous end joining (NHEJ) repair by interacting with CDK1 and preventing protein factors, like RNF8, from protecting exposed DNA ends [109]. Surprisingly, Ku, a predominate protein in NHEJ, also acts as a DNA sensor (similar to cGAS) and can induce a type III interferon response [110]. The mismatch repair pathway (MMR) is also affected by cGAS-STING. It has been observed that MMR-deficient cells accumulate SSBs, which activate cGAS-STING due to the loss of Exo1, which is a vital endonuclease in MMR [111]. There is no evidence as of date that links the GAS-STING pathway with base excision repair (BER) or nucleotide excision repair (NER).

The cGAS-STING pathway and cisplatin

There is a seemingly beneficial relationship between cisplatin and cGAS-STING signaling. Cisplatin induces cGAS-STING signaling, and cGAS-STING promotes antitumor immunity (Fig. 3); however, cGAS-STING also contributes to protumor immunity [112]. Literature has observed the importance of the cGAS-STING signal in the cisplatin-induced immune response, where cisplatin-induced cGAS-STING signals can inhibit cellular proliferation and increase CD8+ T cell infiltration, therefore promoting antitumor immunity [113]. Additionally, it has also been observed that cisplatin-activated cGAS-STING increased PD-L1 expression, which underscores that using cisplatin in combination with checkpoint inhibitors is beneficial [114]. As previously mentioned, inhibition of certain DDR proteins can activate the cGAS-STING pathway, which means the genetic background of DNA repair-deficient tumors could dictate whether the immune response will enhance or diminish antitumor immunity and how they respond to cisplatin. This means that DNA repair-deficient tumors that favor cGAS-STING signaling could be targetable by cisplatin. Literature lauds cGAS-STING agonists as an ideal drug sensitizer that can decrease the risk of drug resistance [114]. However, one should be mindful to impart that kind of promise of a signaling pathway because literature has demonstrated that chronic cGAS-STING signaling can lead to drug

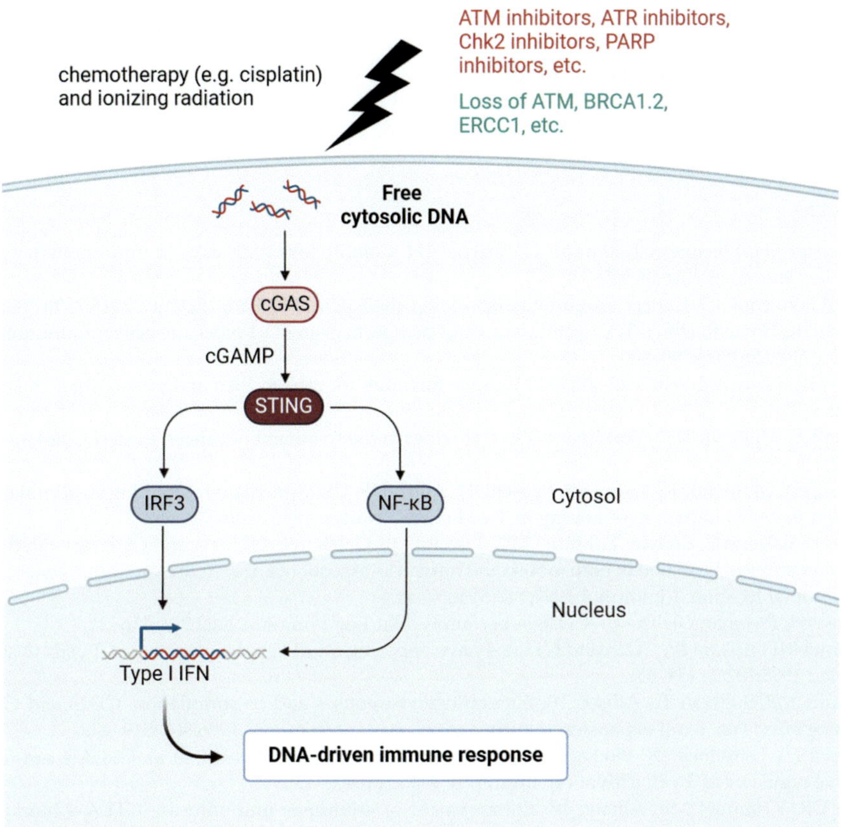

FIG. 3 DNA damage, DDR inhibitors, and genetic deficiency of DDR genes could activate the cGAS-STING pathway, thereby promoting Type I IFN signaling and antitumor immune response. The DNA sensor cGAS could bind to the free cytosolic DNA produced by DNA damage. Then cGAS could activate STING. The latter would stimulate IRF3 and NFκB to promote the Type I IFN signaling and elicit immune response. *The graph was generated by BioRender.com.*

resistance [112,115]. Despite all this, a complete understanding about how cGAS-STING, DDR, and cisplatin converge on one another and how this convergence could be utilized for targeting tumor cells.

Perspectives

In the preimmunotherapy era, our focus was mainly on the tumors per se. The success of immunotherapy has shifted our attention to immune cells and tumor immune microenvironment. The promising strategy of combining chemoradiotherapy and immunotherapy to treat patients prompts us to study the link between DNA damage response and responsiveness of immunotherapy. Future work on DNA damage's influence on cytokine regulation, immune cell infiltration, neoantigen, and tumor immunogenicity as well as checkpoint regulation will shed light on designing better strategies to treat cancer patients.

Acknowledgments

We thank Kylie Roth and Catherine Bai for proofreading the manuscript. This work was in part supported by R21CA256423 from the National Cancer Institute to XZ and GB and W81XWH-22-1-0369 from the Department of Defense to XZ.

References

[1] Baumeister SH, Freeman GJ, Dranoff G, Sharpe AH. Coinhibitory pathways in immunotherapy for cancer. Annu Rev Immunol 2016;34:539–73.
[2] Ribas A, Wolchok JD. Cancer immunotherapy using checkpoint blockade. Science 2018;359(6382):1350–5.
[3] Chamoto K, Hatae R, Honjo T. Current issues and perspectives in PD-1 blockade cancer immunotherapy. Int J Clin Oncol 2020;25(5):790–800.
[4] Morse MA, Gwin 3rd WR, Mitchell DA. Vaccine therapies for cancer: then and now. Target Oncol 2021;16(2):121–52.
[5] Conlon KC, Miljkovic MD, Waldmann TA. Cytokines in the treatment of cancer. J Interf Cytokine Res 2019;39(1):6–21.
[6] Harding FA, McArthur JG, Gross JA, Raulet DH, Allison JP. CD28-mediated signalling co-stimulates murine T cells and prevents induction of anergy in T-cell clones. Nature 1992;356(6370):607–9.
[7] Harper K, Balzano C, Rouvier E, Mattei MG, Luciani MF, Golstein P. CTLA-4 and CD28 activated lymphocyte molecules are closely related in both mouse and human as to sequence, message expression, gene structure, and chromosomal location. J Immunol 1991;147(3):1037–44.
[8] Sharpe AH, Freeman GJ. The B7-CD28 superfamily. Nat Rev Immunol 2002;2(2):116–26.
[9] Krummel MF, Allison JP. CD28 and CTLA-4 have opposing effects on the response of T cells to stimulation. J Exp Med 1995;182(2):459–65.
[10] Krummel MF, Sullivan TJ, Allison JP. Superantigen responses and co-stimulation: CD28 and CTLA-4 have opposing effects on T cell expansion in vitro and in vivo. Int Immunol 1996;8(4):519–23.
[11] Walunas TL, Lenschow DJ, Bakker CY, Linsley PS, Freeman GJ, Green JM, et al. CTLA-4 can function as a negative regulator of T cell activation. Immunity 1994;1(5):405–13.
[12] Leach DR, Krummel MF, Allison JP. Enhancement of antitumor immunity by CTLA-4 blockade. Science 1996;271(5256):1734–6.
[13] Lipson EJ, Drake CG. Ipilimumab: an anti-CTLA-4 antibody for metastatic melanoma. Clin Cancer Res 2011;17(22):6958–62.
[14] Eroglu Z, Kim DW, Wang X, Camacho LH, Chmielowski B, Seja E, et al. Long term survival with cytotoxic T lymphocyte-associated antigen 4 blockade using tremelimumab. Eur J Cancer 2015;51(17):2689–97.
[15] Postow MA, Chesney J, Pavlick AC, Robert C, Grossmann K, McDermott D, et al. Nivolumab and ipilimumab versus ipilimumab in untreated melanoma. N Engl J Med 2015;372(21):2006–17.
[16] Ishida Y, Agata Y, Shibahara K, Honjo T. Induced expression of PD-1, a novel member of the immunoglobulin gene superfamily, upon programmed cell death. EMBO J 1992;11(11):3887–95.
[17] Dong H, Zhu G, Tamada K, Chen L. B7-H1, a third member of the B7 family, co-stimulates T-cell proliferation and interleukin-10 secretion. Nat Med 1999;5(12):1365–9.
[18] Freeman GJ, Long AJ, Iwai Y, Bourque K, Chernova T, Nishimura H, et al. Engagement of the PD-1 immunoinhibitory receptor by a novel B7 family member leads to negative regulation of lymphocyte activation. J Exp Med 2000;192(7):1027–34.
[19] Latchman Y, Wood CR, Chernova T, Chaudhary D, Borde M, Chernova I, et al. PD-L2 is a second ligand for PD-1 and inhibits T cell activation. Nat Immunol 2001;2(3):261–8.
[20] Andrews LP, Marciscano AE, Drake CG, Vignali DA. LAG3 (CD223) as a cancer immunotherapy target. Immunol Rev 2017;276(1):80–96.
[21] Jenkins RW, Barbie DA, Flaherty KT. Mechanisms of resistance to immune checkpoint inhibitors. Br J Cancer 2018;118(1):9–16.
[22] Gandhi L, Rodriguez-Abreu D, Gadgeel S, Esteban E, Felip E, De Angelis F, et al. Pembrolizumab plus chemotherapy in metastatic non-small-cell lung cancer. N Engl J Med 2018;378(22):2078–92.

[23] Antonia SJ, Villegas A, Daniel D, Vicente D, Murakami S, Hui R, et al. Overall survival with Durvalumab after chemoradiotherapy in stage III NSCLC. N Engl J Med 2018;379(24):2342–50.
[24] Antonia SJ. Durvalumab after chemoradiotherapy in stage III non-small-cell lung cancer. Reply. N Engl J Med 2019;380(10):990.
[25] Le DT, Uram JN, Wang H, Bartlett BR, Kemberling H, Eyring AD, et al. PD-1 blockade in tumors with mismatch-repair deficiency. N Engl J Med 2015;372(26):2509–20.
[26] Guo JA, Alshalalfa M, Kim DY, Hoffman HI, Shiau C, Su J, et al. DNA repair and immune checkpoint blockade response. Cancer Genet 2022;264-265:1–4.
[27] Ali R, Aouida M, Alhaj Sulaiman A, Madhusudan S, Ramotar D. Can cisplatin therapy be improved? Pathways that can be targeted. Int J Mol Sci 2022;23(13).
[28] Spampinato CP. Protecting DNA from errors and damage: an overview of DNA repair mechanisms in plants compared to mammals. Cell Mol Life Sci 2017;74(9):1693–709.
[29] Lord CJ, Ashworth A. The DNA damage response and cancer therapy. Nature 2012;481(7381):287–94.
[30] Zhou B, Elledge SJ. The DNA damage response: putting checkpoints in perspective. Nature 2000;408:433–9.
[31] Uzbekov R, Prigent C. A journey through time on the discovery of cell cycle regulation. Cells 2022;11(4).
[32] Huang RX, Zhou PK. DNA damage response signaling pathways and targets for radiotherapy sensitization in cancer. Signal Transduct Target Ther 2020;5(1):60.
[33] Kay J, Thadhani E, Samson L, Engelward B. Inflammation-induced DNA damage, mutations and cancer. DNA Repair (Amst) 2019;83, 102673.
[34] Soll JM, Sobol RW, Mosammaparast N. Regulation of DNA alkylation damage repair: lessons and therapeutic opportunities. Trends Biochem Sci 2017;42(3):206–18.
[35] Xu L, Wang W, Wu J, Shin JH, Wang P, Unarta IC, et al. Mechanism of DNA alkylation-induced transcriptional stalling, lesion bypass, and mutagenesis. Proc Natl Acad Sci U S A 2017;114(34):E7082–91.
[36] Schuch AP, Moreno NC, Schuch NJ, Menck CFM, Garcia CCM. Sunlight damage to cellular DNA: focus on oxidatively generated lesions. Free Radic Biol Med 2017;107:110–24.
[37] Srinivas US, Tan BWQ, Vellayappan BA, Jeyasekharan AD. ROS and the DNA damage response in cancer. Redox Biol 2019;25, 101084.
[38] Huang Y, Li L. DNA crosslinking damage and cancer - a tale of friend and foe. Transl Cancer Res 2013;2(3): 144–54.
[39] Osawa T, Davies D, Hartley JA. Mechanism of cell death resulting from DNA interstrand cross-linking in mammalian cells. Cell Death Dis 2011;2(8), e187.
[40] Ma A, Dai X. The relationship between DNA single-stranded damage response and double-stranded damage response. Cell Cycle 2018;17(1):73–9.
[41] Hossain MA, Lin Y, Yan S. Single-strand break end resection in genome integrity: mechanism and regulation by APE2. Int J Mol Sci 2018;19(8).
[42] Caldecott KW. DNA single-strand break repair and human genetic disease. Trends Cell Biol 2022;32(9):733–45.
[43] White RR, Vijg J. Do DNA double-Strand breaks drive aging? Mol Cell 2016;63(5):729–38.
[44] Venclovas C, Thelen MP. Structure based predictions of Rad1, Rad9, Hus1 and Rad17 participation in sliding clamp and clamp loading complexes. Nucleic Acids Res 2000;28(13):2481–93.
[45] Hwang BJ, Jin J, Gunther R, Madabushi A, Shi G, Wilson GM, et al. Association of the Rad9-Rad1-Hus1 checkpoint clamp with MYH DNA glycosylase and DNA. DNA Repair (Amst) 2015;31:80–90.
[46] Andersen KR. Insights into Rad3 kinase recruitment from the crystal structure of the DNA damage checkpoint protein Rad26. J Biol Chem 2017;292(20):8149–57.
[47] Wang Y, Luo W, Wang Y. PARP-1 and its associated nucleases in DNA damage response. DNA Repair (Amst) 2019;81, 102651.
[48] Helleday T. The underlying mechanism for the PARP and BRCA synthetic lethality: clearing up the misunderstandings. Mol Oncol 2011;5(4):387–93.
[49] Chao HX, Poovey CE, Privette AA, Grant GD, Chao HY, Cook JG, et al. Orchestration of DNA damage checkpoint dynamics across the human cell cycle. Cell Syst 2017;5(5):445–59. e5.
[50] Marechal A, Zou L. DNA damage sensing by the ATM and ATR kinases. Cold Spring Harb Perspect Biol 2013;5(9).
[51] Cuadrado M, Martinez-Pastor B, Murga M, Toledo LI, Gutierrez-Martinez P, Lopez E, et al. ATM regulates ATR chromatin loading in response to DNA double-strand breaks. J Exp Med 2006;203(2):297–303.

[52] Blackford AN, Jackson SP. ATM, ATR, and DNA-PK: the trinity at the heart of the DNA damage response. Mol Cell 2017;66(6):801–17.
[53] Ronco C, Martin AR, Demange L, Benhida R. ATM, ATR, CHK1, CHK2 and WEE1 inhibitors in cancer and cancer stem cells. Medchemcomm 2017;8(2):295–319.
[54] Zhang Y, Hunter T. Roles of Chk1 in cell biology and cancer therapy. Int J Cancer 2014;134(5):1013–23.
[55] Lee KY, Dutta A. Chk1 promotes non-homologous end joining in G1 through direct phosphorylation of ASF1A. Cell Rep 2021;34(4), 108680.
[56] Pabla N, Huang S, Mi QS, Daniel R, Dong Z. ATR-Chk2 signaling in p53 activation and DNA damage response during cisplatin-induced apoptosis. J Biol Chem 2008;283(10):6572–83.
[57] Williams AB, Schumacher B. p53 in the DNA-damage-repair process. Cold Spring Harb Perspect Med 2016;6(5).
[58] Zannini L, Delia D, Buscemi G. CHK2 kinase in the DNA damage response and beyond. J Mol Cell Biol 2014;6 (6):442–57.
[59] Kung CP, Weber JD. It's getting complicated-a fresh look at p53-MDM2-ARF triangle in tumorigenesis and cancer therapy. Front Cell Dev Biol 2022;10, 818744.
[60] Innocente SA, Abrahamson JLA, Cogswell JP, Lee JM. p53 regulates a G2 checkpoint through cyclin B1. Proc Natl Acad Sci USA 1999;96(5):2147–52.
[61] Reinhardt HC, Aslanian AS, Lees JA, Yaffe MB. p53-deficient cells rely on ATM- and ATR-mediated checkpoint signaling through the p38MAPK/MK2 pathway for survival after DNA damage. Cancer Cell 2007;11(2):175–89.
[62] Cazzalini O, Scovassi AI, Savio M, Stivala LA, Prosperi E. Multiple roles of the cell cycle inhibitor p21(CDKN1A) in the DNA damage response. Mutat Res 2010;704(1–3):12–20.
[63] Mansilla SF, de la Vega MB, Calzetta NL, Siri SO, Gottifredi V. CDK-independent and PCNA-dependent functions of p21 in DNA replication. Genes (Basel) 2020;11(6).
[64] Abbas T, Dutta A. p21 in cancer: intricate networks and multiple activities. Nat Rev Cancer 2009;9(6):400–14.
[65] Karimian A, Ahmadi Y, Yousefi B. Multiple functions of p21 in cell cycle, apoptosis and transcriptional regulation after DNA damage. DNA Repair (Amst) 2016;42:63–71.
[66] Li LY, Guan YD, Chen XS, Yang JM, Cheng Y. DNA repair pathways in cancer therapy and resistance. Front Pharmacol 2020;11, 629266.
[67] Maynard S, Schurman SH, Harboe C, de Souza-Pinto NC, Bohr VA. Base excision repair of oxidative DNA damage and association with cancer and aging. Carcinogenesis 2009;30(1):2–10.
[68] Krokan HE, Bjoras M. Base excision repair. Cold Spring Harb Perspect Biol 2013;5(4), a012583.
[69] Scharer OD. Nucleotide excision repair in eukaryotes. Cold Spring Harb Perspect Biol 2013;5(10), a012609.
[70] Casal-Mourino A, Ruano-Ravina A, Torres-Duran M, Parente-Lamelas I, Provencio-Pulla M, Castro-Anon O, et al. Polymorphisms in the BER and NER pathways and their influence on survival and toxicity in never-smokers with lung cancer. Sci Rep 2020;10(1):21147.
[71] Marteijn JA, Lans H, Vermeulen W, Hoeijmakers JH. Understanding nucleotide excision repair and its roles in cancer and ageing. Nat Rev Mol Cell Biol 2014;15(7):465–81.
[72] Hsieh P, Yamane K. DNA mismatch repair: molecular mechanism, cancer, and ageing. Mech Ageing Dev 2008;129(7–8):391–407.
[73] Pena-Diaz J, Jiricny J. Mammalian mismatch repair: error-free or error-prone? Trends Biochem Sci 2012;37 (5):206–14.
[74] Fritzell JA, Narayanan L, Baker SM, Bronner E, Andrew SE, Prolla TA, et al. Role of DNA mismatch repair in the cytotoxicity of ionizing radiation. Cancer Res 1997;57:5143–7.
[75] Li X, Heyer WD. Homologous recombination in DNA repair and DNA damage tolerance. Cell Res 2008;18 (1):99–113.
[76] Daley JM, Gaines WA, Kwon Y, Sung P. Regulation of DNA pairing in homologous recombination. Cold Spring Harb Perspect Biol 2014;6(11), a017954.
[77] Karymov M, Daniel D, Sankey OF, Lyubchenko YL. Holliday junction dynamics and branch migration: single molecule analysis. Proc Natl Acad Sci USA 2005;102(23):8186–91.
[78] Chang HHY, Pannunzio NR, Adachi N, Lieber MR. Non-homologous DNA end joining and alternative pathways to double-strand break repair. Nat Rev Mol Cell Biol 2017;18(8):495–506.
[79] Mao Z, Bozzella M, Seluanov A, Gorbunova V. Comparison of nonhomologous end joining and homologous recombination in human cells. DNA Repair (Amst) 2008;7(10):1765–71.
[80] Rycenga HB, Long DT. The evolving role of DNA inter-strand crosslinks in chemotherapy. Curr Opin Pharmacol 2018;41:20–6.

[81] Chang DJ, Cimprich KA. DNA damage tolerance: when it's OK to make mistakes. Nat Chem Biol 2009;5 (2):82–90.
[82] Shen DW, Pouliot LM, Hall MD, Gottesman MM. Cisplatin resistance: a cellular self-defense mechanism resulting from multiple epigenetic and genetic changes. Pharmacol Rev 2012;64(3):706–21.
[83] Rocha CRR, Silva MM, Quinet A, Cabral-Neto JB, Menck CFM. DNA repair pathways and cisplatin resistance: an intimate relationship. Clinics (Sao Paulo) 2018;73(suppl 1):e478s.
[84] Basu A, Krishnamurthy S. Cellular responses to cisplatin-induced DNA damage. J Nucleic Acids 2010;2010.
[85] Duan M, Ulibarri J, Liu KJ, Mao P. Role of nucleotide excision repair in cisplatin resistance. Int J Mol Sci 2020;21 (23).
[86] Arora S, Kothandapani A, Tillison K, Kalman-Maltese V, Patrick SM. Downregulation of XPF-ERCC1 enhances cisplatin efficacy in cancer cells. DNA Repair (Amst) 2010;9(7):745–53.
[87] Chao CCK, Huang S, Huang H, Lin-Chao S. Cross-resistance to UV radiation of a cisplatin resistance human cell line: overexpression of cellular factors that recgonize UV modified DNA. Mol Cell Biol 1991;11(4):2075–80.
[88] Chu G, Chang E. Cisplatin-resistant cells express increased levels of a factor that recognizes damaged DNA. Proc Natl Acad Sci USA 1990;87:3324–7.
[89] Li GM. Mechanisms and functions of DNA mismatch repair. Cell Res 2008;18(1):85–98.
[90] Sawant A, Kothandapani A, Zhitkovich A, Sobol RW, Patrick SM. Role of mismatch repair proteins in the processing of cisplatin interstrand cross-links. DNA Repair (Amst) 2015;35:126–36.
[91] Moggs JG, Szmkowski DE, Yamada M, Karran P, Wood RD. Differential human nucleotide excision repair of paired and mispaired cisplatin DNA adducts. Nucleic Acids Res 1997;25(3):480–90.
[92] Collins PL, Purman C, Porter SI, Nganga V, Saini A, Hayer KE, et al. DNA double-strand breaks induce H2Ax phosphorylation domains in a contact-dependent manner. Nat Commun 2020;11(1):3158.
[93] Fattah F, Lee EH, Weisensel N, Wang Y, Lichter N, Hendrickson EA. Ku regulates the non-homologous end joining pathway choice of DNA double-strand break repair in human somatic cells. PLoS Genet 2010;6(2), e1000855.
[94] Wang A, Ning Z, Lu C, Gao W, Liang J, Yan Q, et al. USP22 induces cisplatin resistance in lung adenocarcinoma by regulating gammaH2AX-mediated DNA damage repair and Ku70/Bax-mediated apoptosis. Front Pharmacol 2017;8:274.
[95] Oliver TG, Mercer KL, Sayles LC, Burke JR, Mendus D, Lovejoy KS, et al. Chronic cisplatin treatment promotes enhanced damage repair and tumor progression in a mouse model of lung cancer. Genes Dev 2010;24(8):837–52.
[96] Qi Y, Warmenhoven JW, Henthorn NT, Ingram SP, Xu XG, Kirkby KJ, et al. Mechanistic modelling of slow and fast NHEJ DNA repair pathways following radiation for G0/G1 normal tissue cells. Cancers (Basel) 2021;13(9).
[97] Duan G, Tang Q, Yan H, Xie L, Wang Y, Zheng XE, et al. A strategy to delay the development of cisplatin resistance by maintaining a certain amount of cisplatin-sensitive cells. Sci Rep 2017;7(1):432.
[98] Zhang H, Chen Y, Jiang Y, Mao Z. DNA double-strand break repair and nucleic acid-related immunity. Acta Biochim Biophys Sin Shanghai 2022;54(6):828–35.
[99] Reislander T, Groelly FJ, Tarsounas M. DNA damage and cancer immunotherapy: a STING in the tale. Mol Cell 2020;80(1):21–8.
[100] Decout A, Katz JD, Venkatraman S, Ablasser A. The cGAS-STING pathway as a therapeutic target in inflammatory diseases. Nat Rev Immunol 2021;21(9):548–69.
[101] Long ZJ, Wang JD, Xu JQ, Lei XX, Liu Q. cGAS/STING cross-talks with cell cycle and potentiates cancer immunotherapy. Mol Ther 2022;30(3):1006–17.
[102] Basit A, Cho MG, Kim EY, Kwon D, Kang SJ, Lee JH. The cGAS/STING/TBK1/IRF3 innate immunity pathway maintains chromosomal stability through regulation of p21 levels. Exp Mol Med 2020;52(4):643–57.
[103] Bai J, Liu F. Nuclear cGAS: sequestration and beyond. Protein Cell 2022;13(2):90–101.
[104] Storozynsky Q, Hitt MM. The impact of radiation-induced DNA damage on cGAS-STING-mediated immune responses to cancer. Int J Mol Sci 2020;21(22).
[105] Wang NH, Lei Z, Yang HN, Tang Z, Yang MQ, Wang Y, et al. Radiation-induced PD-L1 expression in tumor and its microenvironment facilitates cancer-immune escape: a narrative review. Ann Transl Med 2022; 10(24):1406.
[106] Wang L, Yang L, Wang C, Zhao W, Ju Z, Zhang W, et al. Inhibition of the ATM/Chk2 axis promotes cGAS/ STING signaling in ARID1A-deficient tumors. J Clin Invest 2020;130(11):5951–66.
[107] Brooks T, Wayne J, Massey AJ. Checkpoint kinase 1 (Chk1) inhibition fails to activate the stimulator of interferon genes (STING) innate immune signalling in a human coculture cancer system. Mol Biomed 2021;2(1):19.
[108] Liu H, Zhang H, Wu X, Ma D, Wu J, Wang L, et al. Nuclear cGAS suppresses DNA repair and promotes tumorigenesis. Nature 2018;563(7729):131–6.

[109] Liu H, Wang F, Cao Y, Dang Y, Ge B. The multifaceted functions of cGAS. J Mol Cell Biol 2022;14(5).
[110] Zhang X, Brann TW, Zhou M, Yang J, Oguariri RM, Lidie KB, et al. Cutting edge: Ku70 is a novel cytosolic DNA sensor that induces type III rather than type I IFN. J Immunol 2011;186(8):4541–5.
[111] Guan J, Lu C, Jin Q, Lu H, Chen X, Tian L, et al. MLH1 deficiency-triggered DNA hyperexcision by exonuclease 1 activates the cGAS-STING pathway. Cancer Cell 2021;39(1):109–21 e5.
[112] Kwon J, Bakhoum SF. The cytosolic DNA-sensing cGAS-STING pathway in cancer. Cancer Discov 2020; 10(1):26–39.
[113] Fu G, Wu Y, Zhao G, Chen X, Xu Z, Sun J, et al. Activation of cGAS-STING signal to inhibit the proliferation of bladder cancer: the immune effect of cisplatin. Cells 2022;11(19).
[114] Li A, Yi M, Qin S, Song Y, Chu Q, Wu K. Activating cGAS-STING pathway for the optimal effect of cancer immunotherapy. J Hematol Oncol 2019;12(1):35.
[115] Zheng J, Mo J, Zhu T, Zhuo W, Yi Y, Hu S, et al. Comprehensive elaboration of the cGAS-STING signaling axis in cancer development and immunotherapy. Mol Cancer 2020;19(1):133.

CHAPTER

3

Improving adoptive T cell therapy by positive conversion of negative signaling

Jin-Qing Liu, Bolin Zhao, Cho-Hao Lin, and Xue-Feng Bai

Department of Pathology, College of Medicine and Comprehensive Cancer Center, Ohio State University, Columbus, OH, United States

Abstract

Adoptive T cell transfer therapy (ACT) is an emerging cancer treatment that involves expanding tumor-infiltrating T lymphocytes (TILs) or harvesting and genetically modifying a patient's T lymphocytes to target cancer cells. This personalized immunotherapy approach has shown remarkable effectiveness in treating various types of cancer, including solid tumors and leukemia. With ongoing advancements in technology and research, ACT offers great promise in improving cancer treatment outcomes and surpassing current treatment limitations. However, tumor microenvironment (TME) factors, such as inhibitory receptor-ligand interactions, can directly dampen T cell effector functions and are some major obstacles for the success of ACT. Hence, developing novel approaches to improve ACT therapy will greatly benefit cancer patients. The positive conversion of negative signaling for T cell therapy of cancer refers to the concept of manipulating the inhibitory signaling in T cells into positive one so that when the T cells see the same ligand, the antitumor activity is promoted rather than inhibited. In this essay, we provide an overview of a list of such inhibitory receptors (CTLA4, PD-1, CD200R, TIGIT, Tim-3, and Fas) in tumors, their positive conversion in T cells, and their status of preclinical and clinical development.

Abbreviations

ACT	adoptive T cell transfer therapy
BCMA	B-cell maturation antigen
CARs	chimeric antigen receptors
CLL	chronic lymphocytic leukemia
CTL	$CD8^+$ cytotoxic T lymphocytes.
CTLA-4	cytotoxic T-lymphocyte antigen-4
MM	multiple myeloma
PD-1	programmed death-1

https://doi.org/10.1016/B978-0-443-13827-0.00010-X

TCRs	T cell receptors
TIGIT	T cell immunoreceptor with Ig and ITIM domains
TILs	tumor-infiltrating lymphocytes
TIM-3	T cell immunoglobulin mucin domain molecule 3
TME	tumor microenvironment

Conflict of interest

No potential conflicts of interest were disclosed.

Introduction

Adoptive T cell transfer therapy (ACT), involving the ex vivo selection and expansion of tumor-infiltrating lymphocytes (TILs), or engineering T cells with chimeric antigen receptors (CARs) or T cell receptors (TCRs), followed by adoptive transfer into cancer patients, has shown significant clinical promise for the treatment of established cancers. For instance, recent clinical trials using ACT of TIL T cells have resulted in objective responses in patients [1–5]. Engineering T cells with CAR or TCR has permitted the treatment of a broad range of malignancies [6–10]. However, most ACT trials for solid tumors have not met such great expectations, with durable, complete responses remaining infrequent. Thus, improvements to this approach are needed. Current understanding of the determinants of successful $CD8^+$ T cell ACT includes but is not limited to (1) persistence of transferred T cells [11], (2) differentiation status of transferred T cells [12], (3) telomere length [13], and (4) lymphodepleting conditions [14]. Developing better methods to generate long-lasting T cells that can more effectively eradicate large established tumors in vivo will significantly improve the therapeutic efficacy of ACT therapy in human cancer. In addition to T cell-intrinsic factors, tumor microenvironment factors also influence T cell functions. Cancer cells often use various mechanisms [15] to suppress T cells, including the upregulation of ligands for inhibitory checkpoints such as programmed death-1 (PD-1), cytotoxic T-lymphocyte antigen-4 (CTLA-4) and CD200, and secreting immune suppressive cytokines such as TGF-β. Thus, endorsing the abilities of T cells to overcome these inhibitory signals is also important for ACT therapy. The positive conversion of negative signaling for ACT therapy of cancer refers to the concept of manipulating the inhibitory signaling in T cells so that when the modified T cells see inhibitory molecules in the tumor microenvironment (TME), the T cells become more activated and exhibit more potent antitumor activity [16]. In this paper, we will discuss several such strategies and their status of preclinical and clinical development.

Positive conversion of PD-1 signaling for ACT therapy

PD-1 is a protein that is found on the surface of certain immune cells, particularly T cells. It plays a role in regulating the immune response and preventing the body from attacking its own tissues. However, in some cases, PD-1 can also allow cancer cells to evade the immune system, since they often can express programmed death ligand 1 (PD-L1) that binds to PD-1 and prevents T cells from attacking them. By engaging PD-1 on activated T cells, $PD\text{-}L1^+$ tumors directly render tumor-specific T cells, including adoptively transferred T cells, functionally exhausted. Therefore, researchers have been developing drugs that block PD-1 or its binding partner, PD-L1, to boost the immune response against cancer. These drugs, known as checkpoint inhibitors, have shown remarkable success in treating certain types of cancer, particularly melanoma and lung cancer [17,18]. They can help to activate T cells and allow them to recognize and attack cancer cells more effectively. Overall, PD-1/PD-L1 checkpoint inhibitors represent a promising avenue for cancer immunotherapy that has already shown significant clinical benefit. However, they can also cause side effects if the immune system becomes overly active. As a strategy to overcome tumor PD-L1 effects on adoptively transferred T cells, Jensen and colleagues converted PD-1 to a T cell costimulatory receptor by exchanging its transmembrane and cytoplasmic tail with that of CD28. Rather than becoming exhausted upon engagement of PDL1$^+$ tumors, they showed that CD8$^+$ cytotoxic T lymphocytes (CTL) genetically modified to express this PD1:CD28 chimera exhibited enhanced functions including increased levels of ERK phosphorylation, proliferation, and enhanced expression of cytokines and effector molecule granzyme B [19]. Kobold et al. [20] went on to test the efficacy of this strategy in a syngeneic mouse tumor model. They found that treatment of mice with established tumors with PD1:CD28 fusion receptor–transduced OT-1T cells mediated complete tumor regression. Similarly, Cohen et al. [21] used a similar strategy and tested the function of PD1:CD28 chimera-modified human T cells in two xenograft models of human melanoma tumors and showed that the engineered human T cells demonstrated superior antitumor function. In mouse tumor models of ACT, PD1:28 engineering was also shown to enable low-avidity T cells to proliferate stronger and prevent PD-L1 upregulation and Th2 polarization in the tumor microenvironment [22]. Moon et al. [23] applied this strategy to CAR-T cells and used them to treat mice bearing large established solid tumors, and they observed significant regression in tumor volume due to enhanced CAR TIL infiltration and decreased susceptibility to tumor-induced hypofunction. Additionally, in a gastric tumor model, the PD1:CD28 engineering of c-MET recognizing CAR-Ts was shown to enhance their killing ability, reduce the release of IL-6, and enhance their long-term antitumor activity [24]. The PD1:CD28-engineered CAR-T cells were used to treat glioblastoma patients in a phase I clinical trial and were deemed safe [25]. More recently, CD19-PD-1:CD28-CAR-T cells were used to treat patients with relapsed/refractory diffuse large B-cell lymphoma [26]. After the second treatment, 3 out of 6 patients achieved complete remissions, and one patient had stable disease. The duration of the responses ranged from 8 to 25 months. These data suggest that CD19-PD-1:CD28-CAR-T cells can elicit potent and durable anticancer responses and can be used in the post-CD19-CAR-T failure setting.

Positive conversion of CD200R signaling for ACT therapy

CD200R, initially identified as a myeloid-restricted, inhibitory immune receptor [27], was later also discovered to be expressed within the lymphoid lineage, especially in human tumor-infiltrating effector T cells [28–31]. CD200R binds to its ligand CD200, which is often broadly expressed in human cancer cells. For instance, CD200 has been shown to be overexpressed in human neuroblastoma [31], melanoma [32], ovarian cancer [33], and majorities of hematopoietic malignancies [34,35], and has been thought to promote growth. For instance, CD200 expression in leukemic cells was shown to be associated with decreased survival of patients [35], and high CD200 levels were linked to decreased numbers of $CD8^+$ T cells and increased regulatory T cells [36]. In solid tumors, CD200 expression in pediatric neuroblastoma [31], CNS tumor [37], and pancreatic tumor [38] was shown to induce immune suppression. CD200 in cancer stem cells was shown to promote tumor initiation [39] and metastasis [40]. In $CD200^{-/-}$ mice, reduced carcinogen-induced tumor development was observed [41]. $CD200R^{-/-}$ mice showed decreased growth and metastasis of $CD200^+$ EMT6 tumors [42]. In recent years, CD200R has emerged as a promising target for T cell therapy for cancer. Preclinical studies [43,44] have suggested that blockade of CD200R signaling either through genetic disruption or monoclonal antibodies can enhance the antitumor activity of T cells by reversing immunosuppression within the tumor microenvironment. Currently, blockade of CD200-CD200R axis is proposed to be an immunotherapy for $CD200^+$ human cancers. This was based on initial studies performed in a SCID-hu model, where established tumors can be rejected by adoptively transferred peripheral blood mononuclear cells upon CD200 blockade [43,44]. Recently, Mahadevan et al. [45] published the results of their phase I clinical trial where they administered a novel recombinant humanized monoclonal antibody Samalizumab (ALXN6000) that blocks CD200 in chronic lymphocytic leukemia (CLL) and multiple myeloma (MM) patients. Samalizumab treatment reduced tumor burden in some CLL patients but did not prove effective in MM patients. Due to an unexpected observation of depletion of activated $CD4^+$ T cells, maximal tolerized dose was not achieved in the clinical trial [45]. Since CD200 has a broad expression in both cancer cells and normal tissues, CD200 may not be an easily saturated target, and adverse events will likely arise. Thus, targeting CD200 may not be feasible. In contrast, CD200R has restricted expression in TAMCs and some subsets of T cells in TME, targeting CD200R appears to be a more feasible approach. In this regard, two humanized mAbs that block CD200R (23ME-00610 and HBM1047) have recently been developed (SITC 2022) with the former in phase1/2a clinical trial (NCT05199272). In addition to CD200R blockade, positive conversion of CD200R signaling in T cells may be another potential immunotherapy. In this regard, Greenberg et al. [16] have generated CD200R: CD28-engineered mouse T cells and showed that cells transduced with the optimized CD200R-CD28 fusion construct exhibited enhanced proliferation and effector function in response to $CD200^+$ leukemic cells in vitro. In adoptive therapy of disseminated leukemia, CD200R-CD28-transduced leukemia-specific $CD8^+$ T cells eradicated otherwise lethal disease more efficiently than wild-type cells and bypassed the requirement for interleukin-2 administration to sustain in vivo activity. Moreover, the transduction of human primary T cells with the equivalent human CD200R-CD28 increased proliferation and cytokine production in response to $CD200^+$ leukemia cells, supporting clinical translation. More recently, Tang et al. [46] tested the CD200R-CD28 switch in CAR-T cells recognizing B-cell maturation antigen (BCMA) and found that the CD200R-CD28 switch potently enhanced the polyfunctionality of CAR-T cells and improved cytotoxicity, proliferative capacity, CAR-T cell metabolism, and performance in a chronic antigen exposure assay. While this strategy has an apparent impact on the ACT therapy of leukemia, where TCR- and CAR-T-based ACT were well established [47], it is of particular interest to the immunotherapy of patients with solid tumors where CD200 expression is overexpressed [48], and

CD200–CD200R interaction plays a role in inhibition of T cells. Additionally, it would be interesting to know if this novel switch can be utilized in the design of next generation of CAR-T therapy targeting CD200-overexpressed solid tumors.

Positive conversion of other inhibitory pathways for ACT therapy

After T cell activation, inhibitory receptors, also called checkpoint molecules, including cytotoxic T-lymphocyte-associated protein 4 (CTLA4), T cell immunoglobulin mucin domain molecule 3 (TIM-3), and T cell immunoreceptor with Ig and ITIM domains (TIGIT), are upregulated to maintain T cell homeostasis by preventing T cell-mediated cytotoxicity and prompting T cell exhaustion. Moreover, the inflammatory tumor microenvironment usually upregulates ligands for these checkpoints on tumor cells to dampen T cell activation signaling, which might lead to anergy or exhaustion of T cells in the tumor microenvironment [49]. **CTLA4** (cytotoxic T-lymphocyte-associated protein 4) has emerged as a crucial target in cancer immunotherapy. This protein is known to act as an immune checkpoint regulator, and its role in dampening the immune system has been a major challenge for cancer immunotherapy. The development of immune checkpoint inhibitors that target CTLA4 has been shown to restore the activity of T cells against tumor cells, leading to improved antitumor responses. Clinical studies have demonstrated the effectiveness of CTLA4 blockade in treating various types of cancer, including melanoma, bladder cancer, and lung cancer. However, due to the potential for autoimmune side effects [50,51], careful management and monitoring are required when using CTLA4 inhibitors. Choi et al. [52] developed a CTLA4-CD28 chimera that consists of extracellular and transmembrane domains of CTLA4, linked with cytoplasmic CD28 domain. Overexpression of CTLA4-CD28 fusion protein in T cells delivered stimulatory signals rather than inhibitory signals of CTLA4 and significantly enhanced T cell reactivity, particularly in CD4$^+$ T cells. CTLA4-CD28 modification of CD4 T cells also significantly enhanced the antitumor effect of unmodified CD8 T cells, and the gene modification of both CD8 and CD4 T cells further maximized the antitumor effect of T cells in mouse tumor models. **TIGIT** is also a coinhibitory receptor expressed by T and NK cells, which, upon binding to its ligand (e.g., CD155), diminish cytokine production and effector function. Hoogi et al. [53] first designed a TIGIT-CD28 switch receptor that increased activation and cytokine production by human T cells and CAR-modified T cells. Moreover, the TIGIT-CD28 switch also modified antitumor activity in vivo in xenograft tumor models. **Tim-3** is an inhibitory receptor expressed on various immune cells, including T cells [54]. The binding of Tim-3 to its ligand, such as Galectin-9 [55], leads to the suppression of Th1-type immune responses. In recent years, Tim-3 has emerged as a promising target for cancer immunotherapy [56]. Gao et al. [57] demonstrated that the Tim3-CD28 switch improved proliferative capacity, reduced exhaustion of CAR-T cells, and enhanced their antitumor activity in a B lymphoma model. Importantly, the Tim3-CD28 switch substantially prolonged the persistence of CAR-T cells via the IL-21/Stat3 axis. Feuchtinger et al. [58] also generated Tim3-CD28-modified anti-CD19 CAR-T cells and found that the Tim3-CD28 switch can increase proliferation, activation, and cytotoxic capacity of anti-CD19 CAR-T cells.

In addition to coinhibitory receptors/checkpoint molecules, recent studies also suggest that positive conversion of death signal may also enhance ACT of cancer. For example,

Fas ligand is commonly overexpressed in TMEs and induces apoptosis in tumor-infiltrating, Fas receptor-positive lymphocytes [59,60]. Greenberg et al. [61,62] developed a Fas-4-1BB switch that replaces the Fas intracellular tail with costimulatory 4-1BB. Fas-4-1BB-engineered murine T cells exhibited increased survival, proliferation, antitumor function, and altered metabolism in vitro. In vivo, Fas-4-1BB ACT eradicated leukemia and significantly improved survival in the aggressive KPC pancreatic cancer model and ovarian model. Fas-4-1BB expression also enhanced primary human T cell function in vitro. Thus, the Fas-4-1BB switch is a novel strategy to improve multiple T cell functions and enhance ACT against solid tumors and hematologic malignancies.

Summary statement and future perspective

ACT is an emerging cancer treatment that involves expanding tumor-infiltrating T lymphocytes (TILs) or harvesting and genetically modifying a patient's T lymphocytes to target cancer cells. This personalized immunotherapy approach has shown remarkable effectiveness in treating various types of cancer, including solid tumors and leukemia. However, TME factors, such as inhibitory receptor-ligand interactions, can directly dampen T cell effector functions and appear to be some major obstacles to the success of ACT. The positive conversion of negative signaling (exemplified by the CD200R-CD28 switch) for T cell therapy of cancer shows great premise (Fig. 1). Among all the major switches (CTLA4, PD-1, CD200R, TIGIT, Tim-3, FAS) investigated so far, the PD1:CD28 switch-engineered T cells, including CAR-T, cells are best studied, and PD1:CD28-modified CAR-T cells have been tested in humans with promising results. Given the wide expression of CD200 in many tumor types, it would be interesting to test the efficacy of CD200R-CD28-modified T cells in cancers that overexpress CD200. Given CTLA4 switch competes for binding to CD80 and CD86, and tumors do not normally express these molecules, it is predicted that the CTLA4-CD28 switch may not be feasible for ACT therapy. Thus, further evaluation of these switches in trials of human cancer patients is essential for their wide use in ACT therapy. Finally, the engineering approach of turning a negative signal in TME into a positive one also applies to cytokine receptors. In TME, the presence of immune suppressive cytokines such as TGF-β inhibits T cell

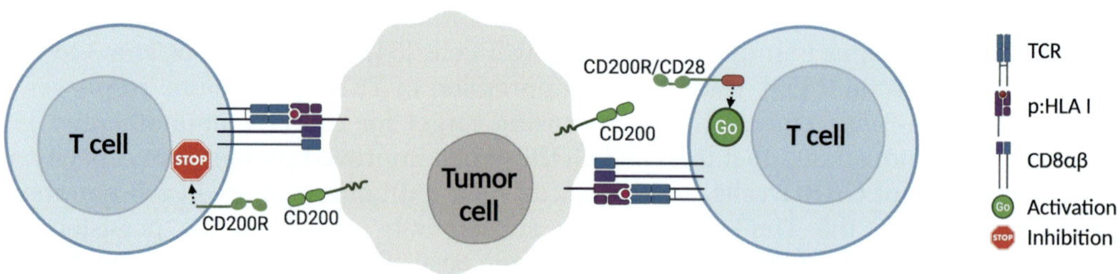

FIG. 1 Disruption or positive conversion of CD200R signaling in T cells. CD200-CD200R interaction between tumor cell and T cell inhibits T cell effector functions. Disruption of CD200R signaling by expression of a truncated CD200R (removal of intracellular domain, left) or expression of a CD200R-CD28 switch (right) will improve T cell effector functions in TME.

functions, and T cells engineered to express a dominant-negative TGF-β receptor showed improved functions in TME [63]. Thus, designing novel cytokine switches will allow T cells to overcome immune suppression to persist and survive with improved effector functions in TME.

References

[1] June CH. Principles of adoptive T cell cancer therapy. J Clin Invest 2007;117(5):1204–12.

[2] Klebanoff CA, Rosenberg SA, Restifo NP. Prospects for gene-engineered T cell immunotherapy for solid cancers. Nat Med 2016;22(1):26–36.

[3] Chandran SS, Somerville RPT, Yang JC, Sherry RM, Klebanoff CA, Goff SL, Wunderlich JR, Danforth DN, Zlott D, Paria BC, et al. Treatment of metastatic uveal melanoma with adoptive transfer of tumour-infiltrating lymphocytes: a single-Centre, two-stage, single-arm, phase 2 study. Lancet Oncol 2017;18(6):792–802.

[4] Tran E, Robbins PF, Lu YC, Prickett TD, Gartner JJ, Jia L, Pasetto A, Zheng Z, Ray S, Groh EM, et al. T-cell transfer therapy targeting mutant KRAS in Cancer. N Engl J Med 2016;375(23):2255–62.

[5] Tran E, Turcotte S, Gros A, Robbins PF, Lu YC, Dudley ME, Wunderlich JR, Somerville RP, Hogan K, Hinrichs CS, et al. Cancer immunotherapy based on mutation-specific CD4+ T cells in a patient with epithelial cancer. Science 2014;344(6184):641–5.

[6] Porter DL, Levine BL, Kalos M, Bagg A, June CH. Chimeric antigen receptor-modified T cells in chronic lymphoid leukemia. N Engl J Med 2011;365(8):725–33.

[7] Robbins PF, Morgan RA, Feldman SA, Yang JC, Sherry RM, Dudley ME, Wunderlich JR, Nahvi AV, Helman LJ, Mackall CL, et al. Tumor regression in patients with metastatic synovial cell sarcoma and melanoma using genetically engineered lymphocytes reactive with NY-ESO-1. J Clin Oncol 2011;29(7):917–24.

[8] Porter DL, Hwang WT, Frey NV, Lacey SF, Shaw PA, Loren AW, Bagg A, Marcucci KT, Shen A, Gonzalez V, et al. Chimeric antigen receptor T cells persist and induce sustained remissions in relapsed refractory chronic lymphocytic leukemia. Sci Transl Med 2015;7(303), 303ra139.

[9] Heczey A, Louis CU, Savoldo B, Dakhova O, Durett A, Grilley B, Liu H, Wu MF, Mei Z, Gee A, et al. CAR T cells administered in combination with Lymphodepletion and PD-1 inhibition to patients with neuroblastoma. Mol Ther 2017;25(9):2214–24.

[10] Stadtmauer EA, Fraietta JA, Davis MM, Cohen AD, Weber KL, Lancaster E, Mangan PA, Kulikovskaya I, Gupta M, Chen F, et al. CRISPR-engineered T cells in patients with refractory cancer. Science 2020;367(6481).

[11] Rosenberg SA, Yang JC, Sherry RM, Kammula US, Hughes MS, Phan GQ, Citrin DE, Restifo NP, Robbins PF, Wunderlich JR, et al. Durable complete responses in heavily pretreated patients with metastatic melanoma using T-cell transfer immunotherapy. Clin Cancer Res 2011;17(13):4550–7.

[12] Hinrichs CS, Borman ZA, Cassard L, Gattinoni L, Spolski R, Yu Z, Sanchez-Perez L, Muranski P, Kern SJ, Logun C, et al. Adoptively transferred effector cells derived from naive rather than central memory CD8+ T cells mediate superior antitumor immunity. Proc Natl Acad Sci U S A 2009;106(41):17469–74.

[13] Hinrichs CS, Borman ZA, Gattinoni L, Yu Z, Burns WR, Huang J, Klebanoff CA, Johnson LA, Kerkar SP, Yang S, et al. Human effector CD8+ T cells derived from naive rather than memory subsets possess superior traits for adoptive immunotherapy. Blood 2011;117(3):808–14.

[14] Gattinoni L, Finkelstein SE, Klebanoff CA, Antony PA, Palmer DC, Spiess PJ, Hwang LN, Yu Z, Wrzesinski C, Heimann DM, et al. Removal of homeostatic cytokine sinks by lymphodepletion enhances the efficacy of adoptively transferred tumor-specific CD8+ T cells. J Exp Med 2005;202(7):907–12.

[15] Anderson KG, Stromnes IM, Greenberg PD. Obstacles posed by the tumor microenvironment to T cell activity: a case for synergistic therapies. Cancer Cell 2017;31(3):311–25.

[16] Oda SK, Daman AW, Garcia NM, Wagener F, Schmitt TM, Tan X, Chapuis AG, Greenberg PD. A CD200R-CD28 fusion protein appropriates an inhibitory signal to enhance T-cell function and therapy of murine leukemia. Blood 2017;130(22):2410–9.

[17] Carlino MS, Larkin J, Long GV. Immune checkpoint inhibitors in melanoma. Lancet 2021;398(10304):1002–14.

[18] Mazieres J, Drilon A, Lusque A, Mhanna L, Cortot AB, Mezquita L, Thai AA, Mascaux C, Couraud S, Veillon R, et al. Immune checkpoint inhibitors for patients with advanced lung cancer and oncogenic driver alterations: results from the IMMUNOTARGET registry. Ann Oncol 2019;30(8):1321–8.

[19] Prosser ME, Brown CE, Shami AF, Forman SJ, Jensen MC. Tumor PD-L1 co-stimulates primary human CD8(+) cytotoxic T cells modified to express a PD1:CD28 chimeric receptor. Mol Immunol 2012;51(3–4):263–72.

[20] Kobold S, Grassmann S, Chaloupka M, Lampert C, Wenk S, Kraus F, Rapp M, Duwell P, Zeng Y, Schmollinger JC, et al. Impact of a new fusion receptor on PD-1-mediated immunosuppression in adoptive T cell therapy. J Natl Cancer Inst 2015;107(8).

[21] Ankri C, Shamalov K, Horovitz-Fried M, Mauer S, Cohen CJ. Human T cells engineered to express a programmed death 1/28 costimulatory retargeting molecule display enhanced antitumor activity. J Immunol 2013;191(8):4121–9.

[22] Schlenker R, Olguin-Contreras LF, Leisegang M, Schnappinger J, Disovic A, Ruhland S, Nelson PJ, Leonhardt H, Harz H, Wilde S, et al. Chimeric PD-1:28 receptor upgrades low-avidity T cells and restores effector function of tumor-infiltrating lymphocytes for adoptive cell therapy. Cancer Res 2017;77(13):3577–90.

[23] Liu X, Ranganathan R, Jiang S, Fang C, Sun J, Kim S, Newick K, Lo A, June CH, Zhao Y, et al. A chimeric switch-receptor targeting PD1 augments the efficacy of second-generation CAR T cells in advanced solid tumors. Cancer Res 2016;76(6):1578–90.

[24] Chen C, Gu YM, Zhang F, Zhang ZC, Zhang YT, He YD, Wang L, Zhou N, Tang FT, Liu HJ, et al. Construction of PD1/CD28 chimeric-switch receptor enhances anti-tumor ability of c-Met CAR-T in gastric cancer. Onco Targets Ther 2021;10(1):1901434.

[25] Guo JX, Wu CX, Wang PF, Li ZJ, Han S, Jin W, Lin ZX. Bioactivity and safety of chimeric switch receptor T cells in glioblastoma patients. Front Biosci 2019;24(6):1158–66.

[26] Liang Y, Liu H, Lu Z, Lei W, Zhang C, Li P, Liang A, Young KH, Qian W. CD19 CAR-T expressing PD-1/CD28 chimeric switch receptor as a salvage therapy for DLBCL patients treated with different CD19-directed CAR T-cell therapies. J Hematol Oncol 2021;14(1):26.

[27] Barclay AN, Wright GJ, Brooke G, Brown MH. CD200 and membrane protein interactions in the control of myeloid cells. Trends Immunol 2002;23(6):285–90.

[28] Wright GJ, Cherwinski H, Foster-Cuevas M, Brooke G, Puklavec MJ, Bigler M, Song Y, Jenmalm M, Gorman D, McClanahan T, et al. Characterization of the CD200 receptor family in mice and humans and their interactions with CD200. J Immunol 2003;171(6):3034–46.

[29] Rijkers ES, de Ruiter T, Baridi A, Veninga H, Hoek RM, Meyaard L. The inhibitory CD200R is differentially expressed on human and mouse T and B lymphocytes. Mol Immunol 2008;45(4):1126–35.

[30] Rygiel TP, Meyaard L. CD200R signaling in tumor tolerance and inflammation: a tricky balance. Curr Opin Immunol 2012;24(2):233–8.

[31] Xin C, Zhu J, Gu S, Yin M, Ma J, Pan C, Tang J, Zhang P, Liu Y, Bai XF, et al. CD200 is overexpressed in neuroblastoma and regulates tumor immune microenvironment. Cancer Immunol Immunother 2020;69(11): 2333–43.

[32] Petermann KB, Rozenberg GI, Zedek D, Groben P, McKinnon K, Buehler C, Kim WY, Shields JM, Penland S, Bear JE, et al. CD200 is induced by ERK and is a potential therapeutic target in melanoma. J Clin Invest 2007;117(12):3922–9.

[33] Moreaux J, Veyrune JL, Reme T, De Vos J, Klein B. CD200: a putative therapeutic target in cancer. Biochem Biophys Res Commun 2008;366(1):117–22.

[34] Tonks A, Hills R, White P, Rosie B, Mills KI, Burnett AK, Darley RL. CD200 as a prognostic factor in acute myeloid leukaemia. Leukemia 2007;21(3):566–8.

[35] Moreaux J, Hose D, Reme T, Jourdan E, Hundemer M, Legouffe E, Moine P, Bourin P, Moos M, Corre J, et al. CD200 is a new prognostic factor in multiple myeloma. Blood 2006;108(13):4194–7.

[36] Pallasch CP, Ulbrich S, Brinker R, Hallek M, Uger RA, Wendtner CM. Disruption of T cell suppression in chronic lymphocytic leukemia by CD200 blockade. Leuk Res 2009;33(3):460–4.

[37] Moertel CL, Xia J, LaRue R, Waldron NN, Andersen BM, Prins RM, Okada H, Donson AM, Foreman NK, Hunt MA, et al. CD200 in CNS tumor-induced immunosuppression: the role for CD200 pathway blockade in targeted immunotherapy. J Immunother Cancer 2014;2(1):46.

[38] Choueiry F, Torok M, Shakya R, Agrawal K, Deems A, Benner B, Hinton A, Shaffer J, Blaser BW, Noonan AM, et al. CD200 promotes immunosuppression in the pancreatic tumor microenvironment. J Immunother Cancer 2020;8(1).

[39] Colmont CS, Benketah A, Reed SH, Hawk NV, Telford WG, Ohyama M, Udey MC, Yee CL, Vogel JC, Patel GK. CD200-expressing human basal cell carcinoma cells initiate tumor growth. Proc Natl Acad Sci U S A 2013; 110(4):1434–9.

[40] Stumpfova M, Ratner D, Desciak EB, Eliezri YD, Owens DM. The immunosuppressive surface ligand CD200 augments the metastatic capacity of squamous cell carcinoma. Cancer Res 2010;70(7):2962–72.

[41] Rygiel TP, Karnam G, Goverse G, van der Marel AP, Greuter MJ, van Schaarenburg RA, Visser WF, Brenkman AB, Molenaar R, Hoek RM, et al. CD200-CD200R signaling suppresses anti-tumor responses independently of CD200 expression on the tumor. Oncogene 2012;31(24):2979–88.

[42] Podnos A, Clark DA, Erin N, Yu K, Gorczynski RM. Further evidence for a role of tumor CD200 expression in breast cancer metastasis: decreased metastasis in CD200R1KO mice or using CD200-silenced EMT6. Breast Cancer Res Treat 2012;136(1):117–27.

[43] Kretz-Rommel A, Qin F, Dakappagari N, Ravey EP, McWhirter J, Oltean D, Frederickson S, Maruyama T, Wild MA, Nolan MJ, et al. CD200 expression on tumor cells suppresses antitumor immunity: new approaches to cancer immunotherapy. J Immunol 2007;178(9):5595–605.

[44] Kretz-Rommel A, Qin F, Dakappagari N, Cofiell R, Faas SJ, Bowdish KS. Blockade of CD200 in the presence or absence of antibody effector function: implications for anti-CD200 therapy. J Immunol 2008;180(2):699–705.

[45] Mahadevan D, Lanasa MC, Farber C, Pandey M, Whelden M, Faas SJ, Ulery T, Kukreja A, Li L, Bedrosian CL, et al. Phase I study of samalizumab in chronic lymphocytic leukemia and multiple myeloma: blockade of the immune checkpoint CD200. J Immunother Cancer 2019;7(1):227.

[46] Tang Y, Liu W, Kadu S, Johnson O, Hasanali ZS, Kelly AR, Shestov AA, Greenblat E, Holmes M, Wang LP, et al. Exploiting the CD200-CD200R immune checkpoint Axis in multiple myeloma to enhance CAR-T therapy. Blood 2023.

[47] Chapuis AG, Egan DN, Bar M, Schmitt TM, McAfee MS, Paulson KG, Voillet V, Gottardo R, Ragnarsson GB, Bleakley M, et al. T cell receptor gene therapy targeting WT1 prevents acute myeloid leukemia relapse post-transplant. Nat Med 2019;25(7):1064–72.

[48] Liu JQ, Hu A, Zhu J, Yu J, Talebian F, Bai XF. CD200-CD200R pathway in the regulation of tumor immune microenvironment and immunotherapy. Adv Exp Med Biol 2020;1223:155–65.

[49] Tang R, Rangachari M, Kuchroo VK. Tim-3: a co-receptor with diverse roles in T cell exhaustion and tolerance. Semin Immunol 2019;42, 101302.

[50] Villadolid J, Amin A. Immune checkpoint inhibitors in clinical practice: update on management of immune-related toxicities. Transl Lung Cancer Res 2015;4(5):560–75.

[51] Wang DY, Salem JE, Cohen JV, Chandra S, Menzer C, Ye F, Zhao S, Das S, Beckermann KE, Ha L, et al. Fatal toxic effects associated with immune checkpoint inhibitors: a systematic review and Meta-analysis. JAMA Oncol 2018;4(12):1721–8.

[52] Shin JH, Park HB, Oh YM, Lim DP, Lee JE, Seo HH, Lee SJ, Eom HS, Kim IH, Lee SH, et al. Positive conversion of negative signaling of CTLA4 potentiates antitumor efficacy of adoptive T-cell therapy in murine tumor models. Blood 2012;119(24):5678–87.

[53] Hoogi S, Eisenberg V, Mayer S, Shamul A, Barliya T, Cohen CJ. A TIGIT-based chimeric co-stimulatory switch receptor improves T-cell anti-tumor function. J Immunother Cancer 2019;7(1):243.

[54] Monney L, Sabatos CA, Gaglia JL, Ryu A, Waldner H, Chernova T, Manning S, Greenfield EA, Coyle AJ, Sobel RA, et al. Th1-specific cell surface protein Tim-3 regulates macrophage activation and severity of an autoimmune disease. Nature 2002;415(6871):536–41.

[55] Zhu C, Anderson AC, Schubart A, Xiong H, Imitola J, Khoury SJ, Zheng XX, Strom TB, Kuchroo VK. The Tim-3 ligand galectin-9 negatively regulates T helper type 1 immunity. Nat Immunol 2005;6(12):1245–52.

[56] Anderson AC. Tim-3: an emerging target in the cancer immunotherapy landscape. Cancer Immunol Res 2014;2 (5):393–8.

[57] Zhao S, Wang C, Lu P, Lou Y, Liu H, Wang T, Yang S, Bao Z, Han L, Liang X, et al. Switch receptor T3/28 improves long-term persistence and antitumor efficacy of CAR-T cells. J Immunother Cancer 2021;9(12).

[58] Blaeschke F, Ortner E, Stenger D, Mahdawi J, Apfelbeck A, Habjan N, Weisser T, Kaeuferle T, Willier S, Kobold S, et al. Design and evaluation of TIM-3-CD28 checkpoint fusion proteins to improve anti-CD19 CAR T-cell function. Front Immunol 2022;13, 845499.

[59] Motz GT, Santoro SP, Wang LP, Garrabrant T, Lastra RR, Hagemann IS, Lal P, Feldman MD, Benencia F, Coukos G. Tumor endothelium FasL establishes a selective immune barrier promoting tolerance in tumors. Nat Med 2014;20(6):607–15.

[60] Peter ME, Hadji A, Murmann AE, Brockway S, Putzbach W, Pattanayak A, Ceppi P. The role of CD95 and CD95 ligand in cancer. Cell Death Differ 2015;22(5):885–6.

[61] Oda SK, Anderson KG, Ravikumar P, Bonson P, Garcia NM, Jenkins CM, Zhuang S, Daman AW, Chiu EY, Bates BM, et al. A Fas-4-1BB fusion protein converts a death to a pro-survival signal and enhances T cell therapy. J Exp Med 2020;217(12).

[62] Anderson KG, Oda SK, Bates BM, Burnett MG, Rodgers Suarez M, Ruskin SL, Greenberg PD. Engineering adoptive T cell therapy to co-opt Fas ligand-mediated death signaling in ovarian cancer enhances therapeutic efficacy. J Immunother Cancer 2022;10(3).

[63] Foster AE, Dotti G, Lu A, Khalil M, Brenner MK, Heslop HE, Rooney CM, Bollard CM. Antitumor activity of EBV-specific T lymphocytes transduced with a dominant negative TGF-beta receptor. J Immunother 2008;31 (5):500–5.

CHAPTER

4

TGF-β in tumor microenvironment, metabolism, and immunotherapy

Haotian Bai[a,b] *and Yisong Y. Wan*[b]

[a]Department of Molecular Microbiology and Immunology, Johns Hopkins Bloomberg School of Public Health, Baltimore, MD, United States [b]Department of Microbiology and Immunology, Lineberger Comprehensive Cancer Center, Chapel Hill, NC, United States

Abstract

Transforming growth factor-beta (TGF-β) plays important roles in regulating various cellular processes including survival, metabolism, proliferation, differentiation, and function. Dysregulation of TGF-β signaling contributes to a variety of human diseases, especially to the development of tumors and cancer. TGF-β acts as a tumor suppressor at the early stage of tumorigenesis to inhibit proliferation and promote the apoptosis of malignant cells. However, at the advanced stage of tumor development, TGF-β promotes tumor growth by enhancing cellular transformation, immune evasion, and tumor metastasis. In this regard, TGF-β signaling has been recognized as an important therapeutic target to treat advanced tumors. Various therapeutic strategies have been and are being developed and explored at preclinical and clinical stages to treat cancers. Here, we aim to discuss the basic biology of TGF-β and how it participates in tumorigenesis and cancer development as well as the current landscape of TGF-β-based strategies to treat cancers.

Abbreviations

ACVR2 activin receptor type 2
ALK activin receptor-like kinases
AMH anti-Mullerian hormone
AMHR2 AMH type-2 receptor
AP-1 activator protein 1
ASOs antisense oligonucleotides
ATF1 activating transcription factor 1
ATF3 activating transcription factor 3
BA bintrafusp alfa
BAD BCL-2 antagonist of cell death
BCL-2 B-cell lymphoma 2
BMPR2 BMP receptor type 2
BMPs bone morphogenetic proteins

https://doi.org/10.1016/B978-0-443-13827-0.00017-2

CAFs	cancer-associated fibroblasts
CAV-1	caveolin-1
CDK	cyclin-dependent kinase
CTC	circulating tumor cell
CTLA-4	cytotoxic T-lymphocyte-associated protein 4
CTLs	cytotoxic T lymphocytes
CXCR3	chemokine receptor CXC motif 3
DAPK	death-associated protein kinase
DNMT3B	DNA methyltransferase 3 beta
DP1	dimerization partner 1
ECM	extracellular matrix
EMT	epithelial-mesenchymal transition
EndMT	endothelial-mesenchymal transition
ERK	extracellular-signal regulated kinase
FAPα	fibroblast activation protein-α
FASN	fatty acid synthase
Fox	forkhead family member
Foxp3	forkhead box P3
GA3P	glyceraldehyde 3-phosphate
GARP	glycoprotein-A repetitions predominant protein
GDFs	growth and differentiation factors
GLS1	glutaminase 1
GS	glycine-serine rich
GSK-3	glycogen synthase kinase-3
Gsk3-β	glycogen synthase kinase-3 beta
HK2	hexokinase 2
HLA	human leukocyte antigens
Id	inhibitor of differentiation proteins
IFN-γ	interferon gamma
IL	interleukin
IL-1	interleukin 1
ITGA2	integrin alpha 2
ITGB6	integrin beta 6
LAD	lung adenocarcinoma
LAP	latency-associated peptide
LBP	latent TGF-β-binding protein
LLC	large latent complex
LRRC33	leucine-rich repeat-containing protein 33
mAbs	monoclonal neutralizing antibodies
MDSCs	myeloid-derived suppressive cells
MEK	MAP kinase/ERK kinase
MET	mesenchymal-epithelial transition
MHC	major histocompatibility complex
MMP	matrix metalloproteinase
mTOR	mammalian target of rapamycin
mTORC	mTOR complex
mUC	metastatic urothelial cancer
NCR	natural cytotoxicity receptors
NF-κB	nuclear factor kappa B
NK	natural killer cells
NKG2D	natural killer group 2, member D
NSCLC	non-small cell lung cancer
NSDHL	NADPH steroid dehydrogenase-like protein

OXPHOS	oxidative phosphorylation
PD-1	programmed death-1
PDA	pancreatic ductal adenocarcinoma
PD-L1	programmed death-ligand 1
PHF14	plant homeodomain finger protein 14
PI3K	phosphoinositide 3-kinase
PKM2	pyruvate kinase M2
pVHL	von Hippel-Lindau tumor suppressor
ROR1	receptor tyrosine kinase-like orphan receptor 1
R-SMAD	receptor-regulated SMAD
SBE	SMAD-binding element
SLC	small latent complex
SLC7A5	solute carrier family 7 member 5
SMURF1	SMAD ubiquitination-related factor 1
SPH2s	Src homology region 2 domain-containing phosphatases
ST3GAL5	ST3 beta-galactoside alpha-2,3-sialyltransferase 5
STAT	signal transducer and activator of transcription
TAB1	TGF-β-activated kinase 1 binding protein 1
TAK1	TGF-β-activated kinase 1
TAMs	tumor-associated macrophages
TANs	tumor-associated neutrophils
TCR	T cell receptor
TFCP2	transcription factor CP2
TGFBR	TGF-β receptor
TGF-β	transforming growth factor β
Th	T helper cells
TIEG1	TGF-β-inducible early response gene 1
TIM-3	T cell immunoglobulin mucin-3
TNF-α	tumor necrosis factor-alpha
Tregs	regulatory T cells
TRFs	tumor necrosis factor receptor
TSK	TGF-β signaling kinases
TβRI	TGF-β receptor type I
TβRII	TGF-β receptor type II
USP	ubiquitin-specific peptidase
VEGF	vascular endothelial growth factor

Conflict of interest

No potential conflicts of interest were disclosed.

Introduction

Transforming growth factor β (TGF-β) plays pleiotropic roles in cancer progression and regression through regulating proliferation and differentiation of many cell types, including fibroblasts, endothelial, epithelial, mesenchymal, hematopoietic cells, and immune cells [1–5]. TGF-β is a member of the TGF-β superfamily that plays important roles as central regulators of embryogenesis and tissue morphogenesis [6,7]. TGF-β actively participates in a variety of physiological processes, including cell growth, proliferation, differentiation, and apoptosis [2,8]. The dysregulation or malfunction of TGF-β causes various immune disorders, such as autoimmune diseases, fibrotic diseases, and many types of cancers [1,3,9].

TGF-β has been found to regulate cancer development in a stage-dependent manner. In premalignant cells, TGF-β acts as a tumor suppressor by inducing apoptosis and inhibiting cell proliferation. In advanced tumors, however, TGF-β promotes tumor progression and metastasis in the tumor microenvironment (TME) by inducing epithelial-mesenchymal transition (EMT), tumor metabolic microenvironment reprogramming, immune evasion, and angiogenesis [1,2,10]. It was found that tumor cells can escape the inhibitory effect of TGF-β due to the accumulation of mutations in TGF-β signaling pathway components or due to the decoupling from TGF-β signaling to deactivate TGF-β's antitumor function and conversely to promote cancer development [1,10].

TGF-β-mediated immune regulation also plays an important role in controlling the fate of premalignant cells. In the innate immune system, TGF-β is important in regulating the functions of natural killer (NK) cells, tumor-associated macrophages (TAMs), and myeloid-derived suppressive cells (MDSCs) [11–13]. In the adaptive immune system, TGF-β inhibits the proliferation and differentiation of antitumor T cells such as Th1 cells and cytotoxic T lymphocytes (CTLs), while promoting the inhibitory cells such as regulatory T cells (Tregs) to suppress immunity and facilitate the immune evasion of tumor cells [1].

In recent years, cancer treatments via targeting TGF-β have exhibited promising outcomes; an increasing number of studies and clinical trials are being conducted. Here, we aim to comprehensively discuss the regulatory mechanisms of the TGF-β signaling pathway in the tumor microenvironment (TME) for cancer development as well as current anticancer strategies of targeting TGF-β signaling pathways.

TGF-β superfamily

TGF-β superfamily is expressed in vertebrates [14]. In humans, the TGF-β superfamily consists of 33 members, within several subfamilies, including TGF-βs, bone morphogenetic proteins (BMPs), growth and differentiation factors (GDFs), activins, NODAL, and anti-Mullerian hormone (AMH) [15]. Most of the TGF-β superfamily members are dimeric molecules that are stabilized by disulfide bonds [16,17]. Three TGF-β isomers, TGF-β1, TGF-β2, and TGF-β3, are synthesized as a ~ 50-kDa proprotein, consisting of a secretion signal peptide, a ~ 250-residue N-terminal prodomain termed latency-associated peptide (LAP), and a ~ 110-residue mature active cytokine peptide called growth factor (GF) in the C-terminal [15,18,19]. The signal peptide directs the proprotein to the endoplasmic reticulum to be further processed. Within the lumen of the endoplasmic reticulum, the signal peptide is removed from proprotein, which promotes a homodimer formation via three interchain disulfide bonds [18]. Before being further processed, the dimeric LAP of the pro-TGF-β homodimer will covalently bind with latent TGF-β-binding protein (LTBP) via a disulfide bond in the endoplasmic reticulum [20]. Then, the complex is transferred to the trans-Golgi network, and the dimeric LAP will be cleaved by a Furin protease [19] (Fig. 1). Following cleavage by Furin, the active dimeric TGF-β and LAP homodimer still tightly interact to form a small latent complex (SLC) via strong noncovalent interactions [19]. It can further interact with the LTBP to form a large latent complex (LLC) before external secretion. After being secreted, LLC keeps a latent state called pro-TGF-β or L-TGF-β and facilitates the dimeric LAP to keep the latency by encircling all the contact sites of the active TGF-β in a ring-shaped structure to avoid binding with the TGF-β receptors [21]. LTBPs are important components of the extracellular matrix (ECM), which plays a role as a bridge between the latent TGF-β complex and other extracellular matrix components, such as fibrillin and fibronectin (Fig. 1). In addition, LTBPs actively participate in TGF-β activation through integrins or other factors, such as protease [20]. Notably, the latent TGF-β complexes are secreted in both SLC and LLC forms, and not all the SLCs bind with LTBPs. Some SLCs will bind with other molecules like glycoprotein-A repetitions predominant protein (GARP)/leucine-rich repeat-containing protein 32 (LRRC32) and LRRC33 in different cell types [22] (Fig. 1). Compared to the LTBP, GARP has a transmembrane domain and will interact with LAP on the cell surface rather than in an extracellular matrix [23,24]. The latent, GARP-bound TGF-β complexes were found in endothelial cells, Tregs, and platelets. Knocking down of GARP causes inhibition of TGF-β in Tregs [25]. LRRC33 is associated with the activation of TGF-β in myeloid lineage cells such as macrophages, dendritic cells, and microglia [26].

TGF-β activation is under strict regulation. In the context where L-TGF-β is stored in the extracellular matrix through LTBP, the active TGF-β ligands can be released from L-TGF-β from the LLC by cleavage via serine proteases, including kallikreins, plasmin [27], cathepsin D [18], and matrix metalloproteinases (MMPs) such as MMP-2 [28], MMP-9 [29], and MMP-28 [30]. However, the most significant mechanism to activate L-TGF-β is based on integrin activity, particularly the αVβ6 and αVβ8 integrins [31]. The LAPs of L-TGF-β1 and L-TGF-β3 have an Arg-Gly-Asp (RGD) motif, which has high affinity toward αVβ6 and αVβ8 integrins [21]. Dysfunction of RGD motif in the LAP will lead to similar phenotypes as *Tgfb1*$^{-/-}$ mice, including vasculogenesis defects, multiorgan inflammation, and a lack of Langerhans cells, indicating the significance of the RGD motif-binding integrins in TGF-β activation [32]. Upon binding to the RGD motif, the αVβ6 integrins transfer the tension generated by actomyosin-mediated cell contraction to the L-TGF-βs, resulting in the unfolding of the ring-shaped LAP to release the active TGF-β peptide [21,31]. The LTBP is tightly associated with the extracellular matrix to function as a stiff scaffold for this process [21]. In other contexts where the L-TGF-β is stored on the plasma membrane through GARP or LRRC33, both integrins αvβ6 and αvβ8 can regulate the release of active TGF-βs as well [33]. In addition, there are proteases capable of cleaving the surface GARP or LRRC33 to activate TGF-β [34]. Since TGF-β activation is a required step for TGF-β signaling, it has great implications in cancer. For instance, the expression of integrins αvβ6 can predict poor prognosis for colorectal cancer (CRC) [35], and the enforced expression of GARP is a feature of breast, colon, and lung cancers [36].

TGF-β signaling pathway

The TGF-β signaling pathway is an epitome of the membrane-to-nucleus signaling process [1], which consists of ligand-receptor binding, transcription factor activation and subcellular translocation, and transcriptional regulation (Fig. 1). The activation and regulation of TGF-β signaling have been systematically discussed [17]. Here we focus on the interaction between TGF-β signaling and cancer development.

Upon being released from the latent complex, the active TGF-βs bind with membrane-associated TGF-β receptors. All the TGF-β receptors are featured with a short cysteine-rich extracellular domain for ligand binding, a transmembrane domain, and a cytoplasmic kinase

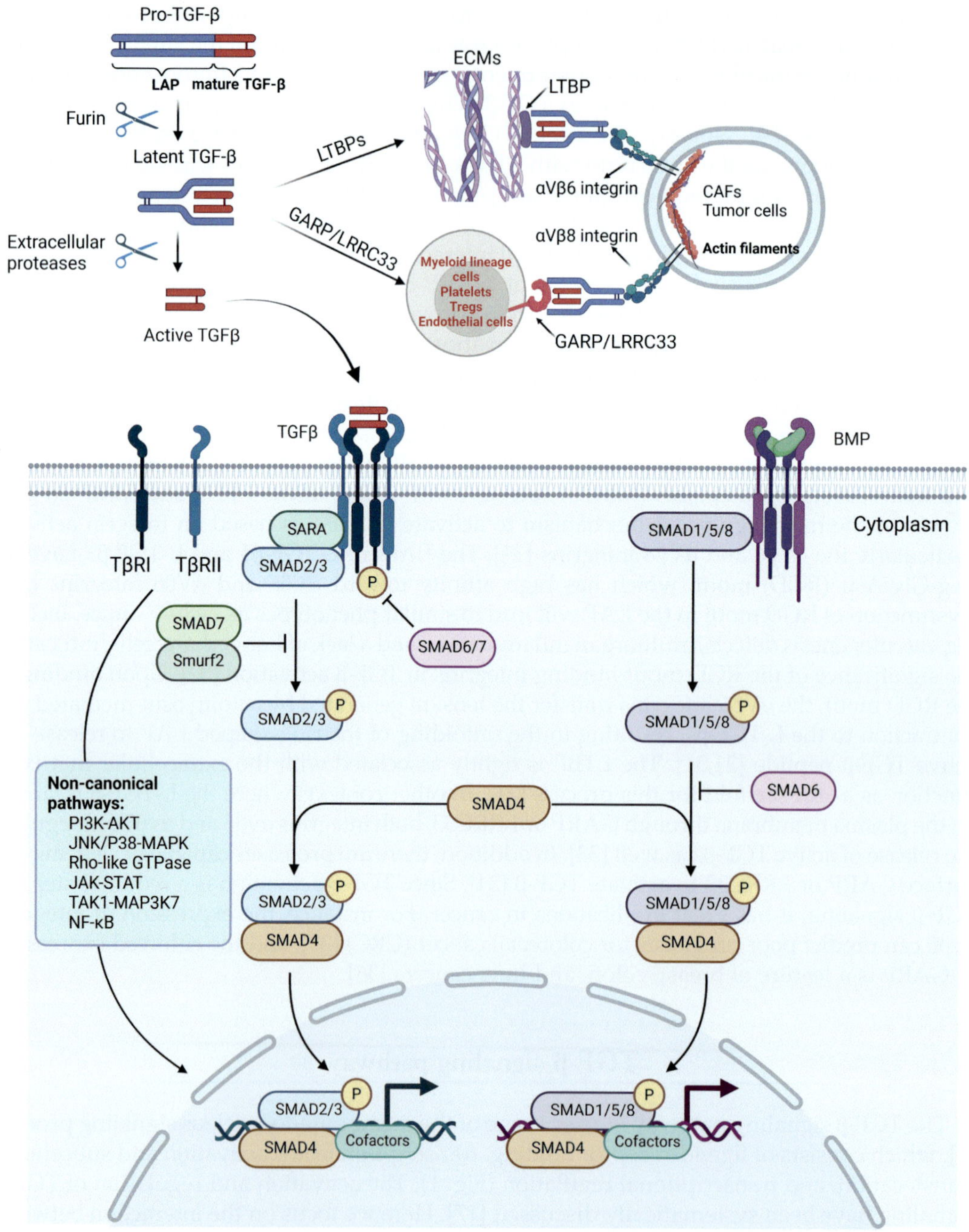

FIG. 1 TGF-β and TGF-β signaling pathways. A series of processes are required for TGF-β to mature and be activated. The canonical TGF-β signaling can be activated through TGF-β binding to TβRI and TβRII, followed by SMAD2 and SMAD3 phosphorylation and dissociation from SARA. SMAD2/3 heterooligomerizes and associates with SMAD4. SMAD complexes translocate into the nucleus to mediate target gene expression. The noncanonical TGF-β signaling is inducible by various signaling pathways. BMP signaling shares a high level of similarity to TGF-β signaling.

domain [16,37]. The cytoplasmic kinase domain is a dual-specificity kinase that has both serine/threonine kinase activity and tyrosine kinase activity [38]. Based on sequence and function similarities, the receptors can be categorized into type I receptors and type II receptors (Fig. 1). The mammalian genome encodes seven type I receptors and five type II receptors. Type I receptors, also termed activin receptor-like kinases (ALKs), range from ALK1 to ALK7. Type II receptors are TGFβ receptor type 2 (TGFBR2), AMH type-2 receptor (AMHR2), activin receptor type 2 (ACVR2), ACVR2B, and BMP receptor type 2 (BMPR2) [17]. Among these receptors, the TGFBR1/TβRI (ALK5) and TGFBR2/TβRII are the specific receptors of TGF-β1, TGF-β2, and TGF-β3. Once active homo-dimeric TGF-βs bind with the receptors, two TGFBR1 and two TGFBR2 form a hetero-tetrameric complex [39]. The constitutively active kinase domain of TGFBR2 will then phosphorylate a glycine–serine-rich (GS) domain within the juxta-membrane region immediately N-terminal to the kinase domain in TGFBR1 [40]. The phosphorylation of the GS domain will activate the kinase of TGFBR1, leading to the propagation of downstream signaling pathways mediated by SMAD-dependent canonical and SMAD-independent noncanonical pathways.

SMAD-dependent canonical pathways

The canonical TGF-β signaling pathway functions via SMAD protein phosphorylation and complex formation. The SMAD family proteins (SMAD1–8) have three conserved domains: an N-terminal Mad homology 1 (MH1) domain, a proline-rich linker domain, and a C-terminal MH2 domain [41]. The MH1 domain has a hairpin structure and can mediate binding with DNA. The MH2 domain is related to multiple processes such as TGF-β receptor-mediated phosphorylation, homo- and hetero-oligomerization of SMAD, and interaction with partner DNA-binding cofactors [8,39,40].

The eight SMAD proteins can be categorized into three types based on their functional differences. SMAD1, 2, 3, 5, 8 are known as receptor-regulated SMAD (R-SMAD), which are targets of active TGF-β superfamily receptors. The SMAD4 is not a receptor substrate but will cofunction with R-SMAD, consequently called co-mediator SMAD (co-SMAD). And the SMAD6 and SMAD7 play an inhibitory function in the signaling pathway, defined as inhibitory SMAD (I-SMAD).

To initiate the SMAD pathway, the active type I receptor kinase will phosphorylate two serine (Ser) residues of a C-terminal Ser-X-Ser motif (SXS motif) in the MH2 domain of R-SMADs [42]. The SMAD2, 3 are phosphorylated by type I receptors for the TGF-β subfamily, while SMAD1, 5, and 8 are mainly phosphorylated by type I receptors for BMP superfamily [40] (Fig. 1). Upon phosphorylation, the active SMAD2/3 will dissociate with type I receptor and form a heterotrimeric complex with the co-mediator SMAD4 through the interaction between their MH2 domains [43]. The SMAD heterotrimer then translocates to the nucleus and regulates corresponding gene transcription. For the SMAD 2/3/4 complex, the MH1 domains of SMAD3 and SMAD4 will mediate the binding with a targeted DNA sequence called SMAD-binding element (SBE) that is a palindromic sequence 5′-GTCT-AGAC-3′ [44,45]. Two β-hairpins of MH1 domains in SMAD3 and SMAD4 will insert into the major groove of SBE, while SMAD2 does not play a role in binding with the SBE due to two additional segments before the β-hairpin motif (10-residual sequence called CAG and a 30-residual sequence called TID) [46]. However, the SMAD complex itself cannot bind with

DNA tightly; other transcription factors, such as forkhead (Fox) family member FoxH1, can cooperate with the SMADs to mediate the TGF-β signaling [47]. Also, the coactivators p300 and CBP (cyclic AMP response element-binding protein) can strengthen the binding between SMAD2,3 and DNA by acetylating the Lys residues in the MH1 domain of SMAD proteins [48]. The expression of cotranscription factors is important for SMAD complex-mediated transcriptional regulation. Therefore, modulating these factors, e.g., by using artificial inhibition, is expected to be effective in interfering with TGF-β signaling.

TGF-β signaling is under strict regulation. Several inhibitory regulators can downregulate TGF-β/SMAD signaling. For example, the SMAD6 and SMAD7 (I-SMAD) have similar MH2 domains that can compete with R-SMAD to bind to type I TGF-β receptors [49,50]. Particularly, SMAD7 is more potent in inhibiting TGF-β signaling, as its N-domain can facilitate the interaction of the MH2 domain with the TGF-β receptor and preferentially inhibit BMP signaling [51]. In addition, SMAD7 located in cytoplasm can recruit E3 ligases termed SMAD ubiquitination-related factor 1/2 (SMURF1/2) to the type I TGF-β receptor, promoting ubiquitylation and proteasome-mediated degradation of the TGF-β receptors [52,53]. This inhibitory regulation induced by SMURFs is also counteracted by deubiquitylating enzymes ubiquitin-specific peptidase (USP) 4 and USP15. E3 ligases and USPs work together to maintain the balance of TGF-β/SMAD signaling [54,55]. Notably, the nuclear SMAD7 can also directly interfere with TGF-β signaling with the binding between activated SMAD complex and DNA, and SMAD6 was also found to impact BMP signaling [56]. Additionally, there are other mechanisms that negatively control TGF-β signaling. For instance, Glycogen synthase kinase3-β (Gsk3-β) triggers SMAD3 ubiquitylation and degradation. Ski/SnoN forms an inhibitory complex with SMAD proteins on SBEs of the TGF-β-target gene promoters by recruiting histone deacetylases (HDACs) and additional corepressors to inhibit gene expression [57,58]. The specific inhibitory regulation of TGF-β/SMAD signaling is complex [17,43,59].

SMAD-independent noncanonical pathways

Besides the canonical SMAD signaling pathways, TGF-β can also function through other signal transducers in a context-dependent manner. For instance, TGF-β can activate the mitogen-activated protein kinase (MAPK) family members, including the three main kinases: extracellular-signal regulated kinase (ERK), c-Jun terminal kinase (JNK), and p38 MAPK [60]. Upon TGF-β ligand binding, the activated TβRI with intrinsic tyrosine kinase, as well as serine/threonine kinase activity, can directly phosphorylate the tyrosine and serine on ShcA proteins to promote the formation of ShcA/Grb2/Sos complex, which can activate Ras GTPase, finally leading to sequential activation of Raf, MEK1/2(MAP kinase/ERK kinase), and ERK1/2 [61]. The activated ERK1/2 can phosphorylate AP-1 family members, transcription factors, as well as the linker region of R-SMADs, playing an important role in cell adhesion and migration, inducing EMT and tumor metastasis [62,63].

In addition, TGF-β ligand binding can activate TGF-β-activated kinase 1 (TAK1), also called mitogen-activated protein kinase kinase kinase 7 (MAP3K7), via phosphorylation of TGF-β activated kinase 1 binding protein 1 (TAB1) in epithelial cells [64]. Tumor necrosis factor receptor-associated factor 6 (TRAF6), an E3 ubiquitin ligase, is also activated by TGF-β receptor complexes-mediated autoubiquitylation, and the TRAF6 can associate with TAK1 to promote each other [65,66]. Then, the activated TRAF6 can activate downstream kinases JNK and p38 MAPK, which can further phosphorylate targeted transcription factors [65]. Notably, the

TRAF6-mediated polyubiquitylation of TAK1 can promote the activation of NF-κB signaling, mediating a crosstalk between TGF-β and NF-κB signaling pathway [67,68]. TGF-β-induced JNK and p38 MAPK signaling is essential to promote apoptosis and EMT, as silencing TRAF6 via siRNAs reduced the response of TGF-induced apoptosis [65,69].

Another significant pathway induced by TGF-β is the PI3K-AKT pathway. AKT is a set of three serine/threonine-specific protein kinases, which is also known as protein kinase B (PKB), playing a critical role in a variety of biological processes such as metabolism, cell proliferation, cell migration, and apoptosis [70]. Activation of the TGF-β-mediated AKT/PKB signaling pathway is via phosphoinositide 3-kinase (PI3K) [71]. The regulatory subunit of PI3K, p85, interacts with TβRI in a TGF-β-dependent manner; the inhibition of TβRI can block the TGF-β-induced PI3K/AKT pathway [71]. TRAF6's engagement with TGF-β-induced JNK and p38 MAPK also plays an important role in the TGF-β-induced AKT/PKB pathway by promoting Lys63-linked ubiquitylation and consequent recruitment of AKT to the plasma membrane and its phosphorylation [72]. Upon activation, AKT kinase can induce metabolic reprogramming, EMT, and cell invasion by phosphorylating mammalian targets of rapamycin complex 1/2 (mTORC1/2) and S6 kinase (S6K) [73]. Notably, the mTORC2 can in turn phosphorylate AKT, forming a positive feedback loop [74]. The function of AKT is complex, as it is also involved in RhoA activation, NF-κB pathway, TNF-α-induced EMT via activation of EMT transcription factor SNAILs, and inhibition of proapoptotic protein BCL-2 antagonist of cell death (BAD) [7,70,75,76].

Many other pathways including JAK-STAT pathway and Rho-(like) GTPase pathway, Wnt/β-catenin pathway, and EGFR pathways crosstalk with and are regulated by TGF-β pathways [9,43,47,77,78]. Non-SMAD TGF-β pathways play a critical role in apoptosis, metabolism, cell proliferation, cell differentiation, and cell motility [60]. Therefore, elucidation of mechanism and regulation of non-SMAD TGF-β pathways can provide potential targets for TGF-β-based cancer therapies.

TGF-β and tumor progression

TGF-β plays a dual role in tumor progression, acting as both tumor suppressor and tumor promoter in different contexts. Thus, applying different treatment methods to target specific cells at specific stages can enhance safety and efficacy. To this end, understanding the function of TGF-β in specific cellular context is necessary for drug design to treat cancers.

TGF-β as a tumor suppressor

In premalignant cells, TGF-β functions as a tumor suppressor by inhibiting the cell cycle and promoting apoptosis. TGF-β can induce cell cycle arrest in many types of cells, including epithelial cells, endothelial cells, and hematopoietic cells [79–82]. The main mechanism can be categorized into three modes: promoting cell growth-inhibiting factors, inhibiting cell growth-promoting factors, and regulating expression of apoptosis-related factors.

In normal cells, TGF-β promotes the expression of growth-inhibiting factors such as the cyclin-dependent kinase (CDK) inhibitors $p15^{INK4}$, $p21^{CIP1}$, $p27^{KIP1}$, and $p57^{KIP2}$, which repress the G1 → S transition by binding with CDK2, CDK4, and CDK6 [5,83–86]. For instance,

Fox family members FoxO factors can be recruited by SMAD3, 4 to induce the activation of $p21^{CIP1}$, and this process is reversibly inhibited by PI3K and FoxG1 [87]. The FoxO factors act as nodes to connect these three pathways, mediating cell growth, proliferation, and apoptosis. Also, the SMAD2/3/4 complex can interact with transcription factor Sp1 to induce the expression of $p15^{INK4}$ and $p21^{CIP1}$ [88,89].

In addition, TGF-β can efficiently suppress cell growth by inhibiting the expression of growth-promoting factors including c-Myc and inhibitor of differentiation (Id) proteins [4,90]. For instance, the activated SMAD3/4 complex can cooperate with transcription factors E2F4/5, DP1, and transcriptional repressor p107 to bind with the promoter of c-Myc [90]. Notably, the c-Myc can inhibit the expression of CDK inhibitors $p15^{INK4}$ and $p21^{CIP1}$ through a zinc-finger protein MIZ1 [91,92]. Therefore, the TGF-β-induced suppression of c-Myc can be further enhanced by promoting the expression of growth-inhibiting factors. The Id proteins are small polypeptides with a helix-loop-helix (bHLH) motif, which can not only inhibit cell differentiation but also promote cell proliferation [93]. In epithelial cells, TGF-β can activate the activating transcription factor 3 (ATF3), which functions as an Id1 inhibitor by forming a repressive complex with SMAD3 and SMAD4. Non-SMAD TGF-β signaling pathway can also activate the ATF3 via p38 kinase [94].

The effect of TGF-β on cell apoptosis is complex and dependent on cell types and specific contexts. Although TGF-β exhibits both positive and negative effects on apoptosis, it is more prevalent to promote apoptosis [5]. TGF-β can induce apoptosis in hepatocytes, immature neurons, and lymphocytes [95–97]. However, the mechanism of TGF-β-induced apoptosis is not fully understood. TGF-β was found to induce a zinc-finger transcription factor TGF-β-inducible early response gene 1 (TIEG1), which can suppress intrinsic antiapoptosis regulator Bcl2, consequently inhibiting cell proliferation and inducing apoptosis in various epithelial cell types such as exocrine pancreatic epithelial cells [98,99]. In addition, TGF-β can enhance the expression of the death-associated protein kinase (DAPK) through the canonical SMAD pathway in hepatocytes, consequently, inducing apoptosis [100]. In hematopoietic cells, TGF-β can induce cell apoptosis through a phospholipid-based metabolism. TGF-β mediates the expression of the inositol phosphatase SHIP (Src homology 2 (SH2) domain-containing 5′ inositol phosphatase), consequently blocking the antiapoptosis PI3K/AKT signaling pathway by inhibiting the activation of AKT [101,102]. In pancreatic ductal adenocarcinoma (PDA) cells, TGF-β can induce Sox4 and EMT-TF Snails to promote apoptosis [103].

TGF-β as a tumor promoter

Although TGF-β plays a growth-inhibition role in most premalignant cells, in advanced tumors, the function of TGF-β becomes tumor-promoting. TGF-β plays a critical role in tumor microenvironment (TME), which is featured by remodeling the extracellular matrix (ECM) and promoting Epithelial-mesenchymal transition (EMT) to develop the tumor characteristics such as sustained proliferative capacity, reduced cell death, increased invasion, and metastasis [104,105] (Fig. 3). Additionally, TGF-β exerts a profound influence on tumor metabolic reprogramming, where it modulates important metabolic pathways to accommodate the energetic and biosynthetic demands of rapidly proliferating cancer cells [106]. Tumor cells can be alleviated from TGF-β-mediated suppressive effect via accumulation of mutations or epigenetic modifications such as hypermethylation-induced gene silencing in central components of TGF-β signaling pathways [107–109]. Based on the current human TCGA datasets, mutations or deletions have been found in TGF-β receptor and SMAD coding genes with 25%–50% prevalence in multiple cancers such as gastric colorectal (CRC) and pancreatic adenocarcinomas (PDA) [1]. More specifically, TβRI, TβRII, and SMAD coding genes have 6%–30%, 12%–28%, and 4%–60% prevalence, respectively, in different types of cancers, including head and neck, breast, bladder, prostate, and ovary adenocarcinomas [109]. The role of TGF-β in promoting tumor progression is mainly associated with three complex biological processes, including Epithelial-Mesenchymal Transition (EMT), tumor metabolic reprogramming, and immune evasion, which are discussed later.

Epithelial-mesenchymal transition (EMT)

TME is a process where polarized epithelial cells gain mesenchymal features to be highly mobile. EMT plays a critical role in embryonic development and tumor progression [110,111]. During tumor progression, the high cell motility caused by EMT will promote tumor invasion and metastasis. During the EMT process, the epithelial cells will undergo a transformation and lose cell-to-cell contact that is normally maintained by adhesion molecules and junction complexes localized in the plasma membrane [112]. For instance, the expression of E-cadherin plays a critical role in mechanical stability and intracellular signaling among epithelial cells, and ZO-1 protein, which is the main component of tight conjunctions will decrease significantly [112,113]. On the contrary, expression of mesenchymal markers such as N-cadherin, Vimentin, and Fibronectin will increase to gain high motility [113]. Simultaneously, the reverse process of EMT, termed mesenchymal-epithelial transition (MET), is important in the later stages of tumor metastatic processes, especially metastatic colonization [114]. MET involves the reacquisition of epithelial characteristics, which are essential for metastases [115]. The dynamic equilibrium between EMT and MET collectively facilitates tumor invasion and metastasis. The regulation of EMT involves complex signaling pathways including TGF-β/SMADs, PI3K–AKT, RhoA, and p38 MAPK, and many transcription factors including SLUG, SNAI1, SNAI2, ZEB1, ZEB2, KLF8, and TWIST1, as well as miRNAs [115–120]. Of note, the EMT is also coregulated with apoptosis; tumor cells are resistant to TGF-β-mediated proapoptosis effect during EMT [1].

TGF-β-mediated metabolic reprogramming in the tumor microenvironment

Besides promoting tumor metastasis, TGF-β functions as a pivotal factor in tumor metabolic reprogramming by modulating the cellular metabolism of both tumor cells and cells in the tumor microenvironment [121]. Tumor metabolic microenvironment (TMME) is featured with hypoxia, high acidity, and nutrient deprivation of glucose, lipid, and amino acid [106]. The immunosuppressive TMME favors tumor growth and forms a positive feedback loop to produce more tumor-promoting cytokines and metabolites in the microenvironment.

In tumor cells, TGF-β dramatically influences the metabolic strategy, particularly by enhancing glycolysis while preserving the rate of oxidative phosphorylation (OXPHOS), which is termed as the Warburg effect [122]. One possible explanation of this metabolic shift is that the enhanced glycolysis under aerobic conditions provides a quick ATP supply and generates vital precursors for nucleotides, amino acids, and lipids that are necessary for rapid tumor cell growth and proliferation [123,124]. TGF-β was found to function as an inducer of the overexpression of glucose transporters GLUT1 and GLUT3 to satisfy the high demand for glucose utilization in a variety of cancers such as breast cancer. Pancreatic ductal adenocarcinoma (PDAC), and non-small cell lung cancer (NSCLC) [125–127] (Fig. 2). In addition, this metabolic alteration is positively associated with the EMT process and the expression of EMT regulators such as ZEB1 and Snail [126,127]. These findings suggest there is a close relationship between EMT and tumor metabolic reprogramming. In the glycolysis pathway, TGF-β can regulate the expression of a series of key enzymes such as hexokinase 2 (HK2), 6-phosphofructo-2-kinase/fructose-2,6-bisphosphatase 3 (PFKFB3), and pyruvate kinase M2 (PKM2), which also crosstalk with EMT regulators including SNAI1 and TGIF2 [128–131]. In addition, the glycogen synthase kinase (GSK-3) that can inhibit TGF-β/SMAD signaling is also involved in the mutual regulation between TGF-β signaling and tumor cell metabolism [57]. Interestingly, some studies have found the glycolytic product lactate generates the acidic tumor microenvironment and in turn promotes the expression of TGF-β and EMT [132,133] (Fig. 2).

TGF-β actively participates in lipid and amino metabolism besides glucose metabolism in tumor cells. During lipid metabolism, cholesterol can be either an upstream or downstream component of TGF-β and play a dual role in tumor progression [106]. For instance, cholesterol promotes the endocytosis and degradation of TGF-β receptor and results in TGF-β signaling suppression, while a cholesterol biosynthetic enzyme termed NADPH steroid dehydrogenase-like protein (NSDHL) can inhibit the degradation of TGF-βRII and promote TGF-β-induced EMT and tumor metastasis in triple-negative breast cancer [134,135]. Conversely, TGF-β can inhibit the expression of cholesterol via recruiting EMT-related transcription factor ZEB1 and corepressor CtBP to bind with and repress the expression of a critical cholesterol synthesis regulator called Sterol regulatory element-binding transcription factor 2 (SREBF2) [136]. Furthermore, another important group of lipids, fatty acids, are also upregulated by TGF-β through a multifunctional lipid biosynthesis enzyme FASN [137] (Fig. 2).

During amino acid metabolism, TGF-β can induce the upregulation of specific amino acids to change the cell morphology and promote EMT [138]. For instance, TGF-β profoundly influences glutamine metabolism in tumor cells with the upregulation of glutamine transporter termed glutamine transporter solute carrier family 7 member 5 (SLC7A5) and glutamate synthesis enzyme termed glutaminase 1 (GLS1), and on tryptophan metabolism in immune cells with modulation of rate-limiting enzymes such as IDO-1 (indoleamine 2,3-dioxygenase-1) [138] (Fig. 2). In addition, TGF-β-mediated amino acid metabolism reprogramming can also promote EMT in a variety of cancer types, such as non-small cell lung cancer (NSCLC) [139]. TGF-β's role in tumor cell metabolism is complicated and remains to be further elucidated [106,138].

Another significant aspect of TGF-β-mediated TMME reprogramming is the metabolic shift of tumor stromal cells, especially cancer-associated fibroblasts (CAFs) and immune cells. As an important source of energy and bioprecursors for tumor cells, CAFs also exhibit an enhanced glycolytic and inhibited TCA [106]. Although tumor cells prefer glycolysis, they have intact OXPHOS-related components. Therefore, the glycolytic products of CAFs such as lactate and pyruvate can be uptake by adjacent tumor cells as the fuel of OXPHOS to fulfill the large energy requirement for proliferation, termed reverse Warburg effect (RWE) [140,141]. TGF-β's regulation of metabolic reprogramming for glucose, lipid, and amino acids

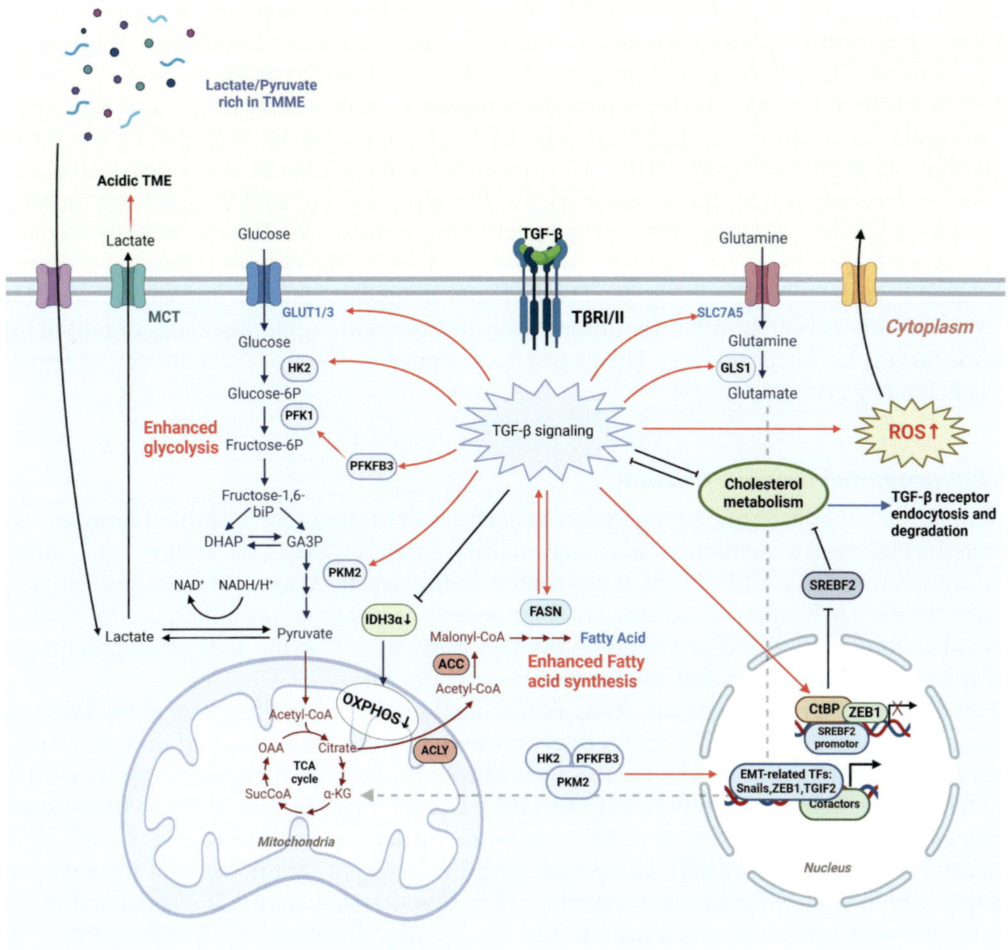

FIG. 2 TGF-β-mediated metabolic reprogramming in the tumor microenvironment. TGF-β signaling mediates various metabolic pathways, including glucose metabolism, lipid metabolism, and amino acid metabolism. For glucose metabolism, it can (1) directly promote glycolysis by upregulating key glycolytic enzymes, such as HK2 and PKM2, and indirectly elevate PFKFB3 to further upregulate PFK1; (2) suppress oxidative phosphorylation (OXPHOS) by down-regulating IDH3α. For fatty acid metabolism, TGF-β, especially the TGF-β1, isoform can increase the expression of fatty acid synthesis key exam FASN, enhancing fatty acid synthesis and accumulation in cancer cells; in turn, FASN enhances TGF-β production, forming a positive feedback loop. For amino acid metabolism, TGF-β signaling differentially regulates the update or synthesis of various amino acids in the cancer cells. For example, it stimulates the uptake of glutamine, serving to follow glutamate synthesis through GLS1. Cholesterol metabolism and TGF-β signaling are mutually regulated. Cholesterol-enriched subdomain-mediated TGF-β receptor endocytosis reduces TGF-β receptor signaling. Cholesterol synthesis enzyme, like NSDHL, inhibits TGF-β1 production. In addition, TGF-β signaling can reduce cholesterol synthesis by inactivating cholesterol synthesis promoter SREBF2. Red arrow: TGF-β1 positively regulated factors or pathways; black arrow: normal flows of various pathways; black bar: being down-regulated factors or pathways. HK2: Hexokinase 2; PFK1: phosphofructokinase 1; PFKFB3: 6-phosphofructo-2-kinase/fructose-2,6-biphosphatase 3; Acetyl-CoA: acetyl coenzyme A; F-1,6-biP: fructose-1,6-bisphosphate; GA3P: glyceraldehyde 3-phosphate; PKM2: pyruvate kinase M2; SREBF2: sterol regulatory element-binding transcription factor 2; FASN: fatty acid synthase; SLC7A5: glutamine transporter solute carrier family 7 member 5; and GLS1: glutaminase 1.

in CAFs largely mirrors that of tumor cells, although additional unique mechanisms are also involved. For example, TGF-β can downregulate TCA key enzymes isocitrate dehydrogenase 3α (IDH3α) and lipid biosynthesis enzyme Caveolin-1 (CAV-1) to promote the RWE [106]. in CAFs. In addition to CAFs, TGF-β can mediate metabolic reprogramming in immune cells to downregulate its cytotoxicity. Recent studies have shown that increasing glucose uptake and glutamine synthesis is important for NK cells and T cells activation and cytotoxicity, while TGF-β can strongly inhibit these processes via the repression of mTOR signaling pathways and some other key enzymes [142]. The specific mechanisms of TGF-β's role in metabolic reprogramming of the tumor microenvironment and its interplay with tumor cells is a topic of intense investigation [106] (Fig. 2). The intricate interplay between TGF-β and metabolic reprogramming in both tumor and tumor microenvironment highlights many potential therapeutic targets in cancer therapy. Drugs can be designed to disrupt this network or regulate the specific key enzymes.

TGF-β-promoted immune evasion

Tumor masses consist of heterogeneous collections of tumor cells, infiltrated immune cells, secreted cytokines, vasculature, and extracellular matrix [143,144]. Tumor cells, tumor-infiltrating immune cells, and cancer-associated fibroblasts (CAFs) are three significant components in the TME of advanced cancers. Tumor cells can secrete multiple types of cytokines to regulate the function and proliferation of other types of cells in the TME, strengthening the ability of tumor metastasis and immune evasion [119,120].

Fibroblasts are a heterogeneous group of mesenchymal cells that function as the main producer of connective tissue ECM and actively participate in tissue repair and immune surveillance [145]. However, like resident leukocytes, fibroblasts remain in a quiescent state with low proliferative capacity and synthetic activity [146]. The activated fibroblasts in TME are known as cancer-associated fibroblasts (CAFs), with elongated morphology and mesenchymal markers such as vimentin [145]. Nevertheless, these markers cannot exclude mesenchymal lineage cells such as pericytes or adipocytes [145]. Research has found TGF-β carried by cancer cell-derived exosomes can promote the normal fibroblasts to differentiate into CAFs through TGF-β/SMAD signaling pathway in bladder cancer [147]. Notably, recent studies have found that the endothelial cells can also differentiate into CAFs through an endothelial-mesenchymal transition (EndMT) mediated by melanoma-derived exosomes [148,149]. The activated CAFs perform an essential role in tumor development, metastasis, and immune evasion. CAFs are main producers of growth factors, chemokines, ECM, EVs, and cytokines, leading to a high level of TGF-β in the TME [149]. For instance, research has found elevated expression of TGF-β, especially TGF-β1 and TGF-β3, caused by CAFs in colorectal cancer (CRC) [150]. The increasing TGF-β secretion can in turn promote the differentiation of resident tissue fibroblasts into CAFs, forming a positive feedback loop [151]. Several other studies also demonstrated that autocrine CAFs can maintain their own state through a TGFβ1/miR-200 s/miR-221/DNA methyltransferase 3 beta (DNMT3B) pathway in breast cancer progression [152]. In addition, CAFs are closely related to EMT, synergistically promoting tumorigenesis and metastasis. The highly expressed IL6 by CAFs can induce EMT in a variety of cancers through IL6/STAT3 pathway, including pancreatic cancer [153], lung cancer [154], gastric cancer [155], and bladder cancer [156].

The excessive TGF-β secreted by CAFs and tumor cells will also impact proliferation and differentiation of immune cells in the TME, promoting immune evasion. The main mechanism of TGF-β-induced immune evasion is to restrain antitumor cells, enhance expression of tumor-promoting cells, and increase tumor tolerance, and both the innate and adaptive immune systems will be impacted [31]. TGF-β can restrain the proliferation of both $CD4^+$ T cells and $CD8^+$ T cells by interfering with the promoter of interleukin-2 (IL-2) gene, mediated by SMAD3 [157,158]. However, the ability of TGF-β to inhibit proliferation is dependent on the extent of T cell differentiation, where TGF-β can only impact naive T cells while having lower effects on activated T cells [159]. One possible mechanism is that the activated T cells have lower expression of type II TGF-β receptor, and IL-10 can restore the sensitivity of activated T cells to the TGF-β by increasing the expression of type II receptor [159]. TGF-β can also influence the differentiation of the $CD4^+$ T cells (Fig. 3). TGF-β can inhibit the

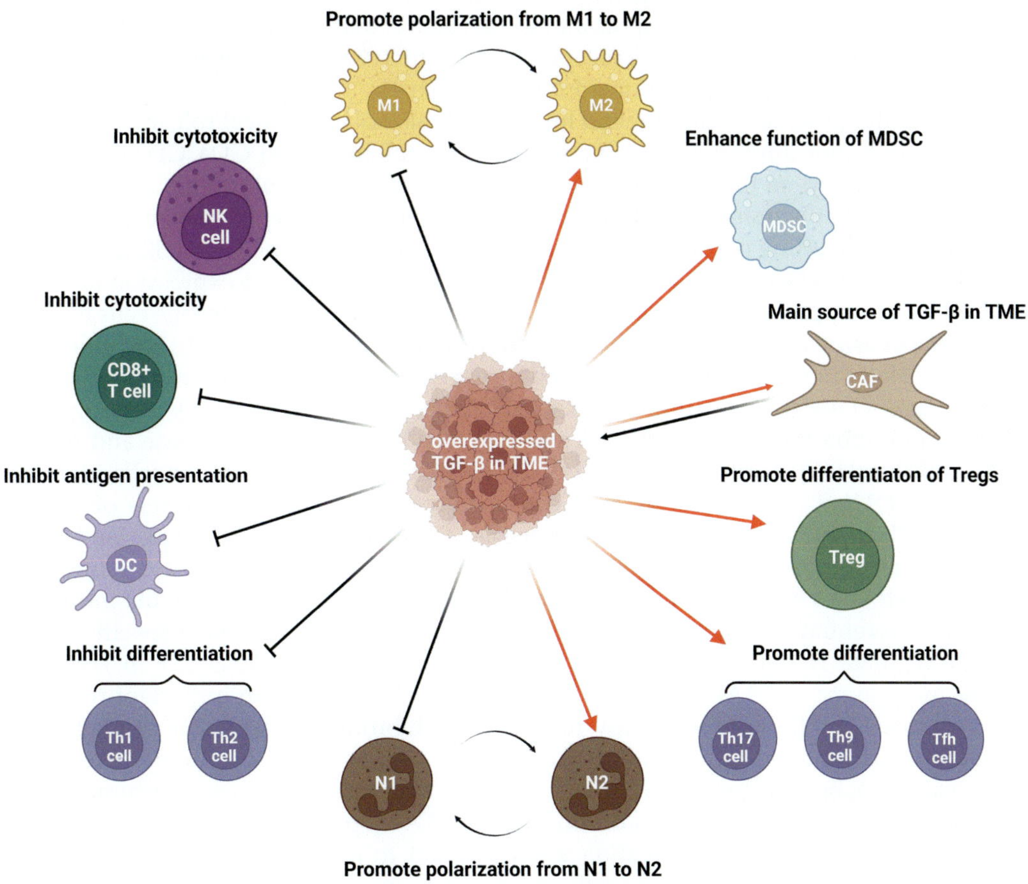

FIG. 3 Schematic overview of tumor-promoting functions of TGF-β on different cells in TME. The overexpressed TGF-β secreted from tumor cells, CAFs, MDSCs, and TAMs, regulates the function and differentiation of immune cells, resulting in a tumor-tolerated environment and immune evasion.

differentiation of Th1 by inhibiting their master differentiation-determining factor T-bet and Stat4, consequently leading to the decreasing expression of interferon γ (IFN-γ) [160,161]. Similarly, TGF-β can inhibit regulatory factor GATA-3 to inhibit the differentiation of Th2 cells [162]. On the contrary, TGF-β will induce the expression of Foxp3, which promotes the differentiation and proliferation of Tregs [163]. Furthermore, TGF-β will promote expression of Th17, Th9, and Tfh (follicular helper T) cells [3] (Fig. 3). Notably, as discussed earlier, Tregs are an important producer and reservoir of TGF-β through GARP, and the release of stored TGF-β is regulated by αvβ8.

Besides $CD4^+$ T cells, TGF-β can directly regulate the expression and function of cytotoxic T lymphocytes (CTLs) ($CD8^+$ T cells) as well [82] (Fig. 3). In CTLs, TGF-β-activated SMADs and transcription factor ATF1 will bind with the promoter regions of granzyme B and IFN-γ and inhibit their expression, leading to lower cytotoxicity [82]. In MC38 and CT26 mouse colon cancer models, blocking of type I TGF-β receptor through small-molecule inhibitor Galunisertib or knockout of tgfbr1 gene specifically in $CD8^+$ T cells can increase its trafficking into the tumor [164]. One of the important mechanisms is that the deficiency of TGFBR1 can abrogate the inhibition of chemokine receptor CXC motif 3 (CXCR3) induced by SMAD2, consequently enhancing the migration toward CXCL10 [164].

Natural killer (NK) cells and myeloid lineage cells also play a critical role in immune surveillance and tumor inhibition. In advanced tumors, the increased $CD4^+CD25^+$ Tregs can release TGF-β stored on their cell membrane, which can directly inhibit NK cells' function [165]. The NK cell functions through several specific activating receptors and coreceptors on its membrane, including natural cytotoxicity receptors (NCR) NKp46, NKp30, and NKp44, as well as NKG2D [11,166]. TGF-β can specifically downregulate the expression of NKp30 and NKG2D to suppress the cytotoxicity of NK cells [11]. The myeloid lineage cells involved in immune evasion can be mainly categorized into immature $Gr\text{-}1^+CD11b^+$ myeloid cells, which are also known as myeloid-derived suppressor cells (MDSCs), tumor-associated macrophages (TAMs), and tumor-associated neutrophils (TANs) [167]. Among them, MDSCs can inhibit the function of B, T, and NK cells by producing arginase and reactive oxygen species [167]. Tumor-associated macrophages (TAMs) are mature cells identified as Mac-1 ($CD11b^+$) and/or $F4/80^+$ and can be categorized into two subsets of M1 and M2 [168,169]. In the early phase of oncogenesis, TAMs tend to be M1 types and play a proinflammatory and tumor-inhibiting effect [169]. However, the hypoxia condition and protumor cytokines such as CSF-1, CCL2, and TGF-β in TME will promote the infiltrating TAMs to differentiate into protumor M2 type [170]. Notably, TAMs have specific mechanisms to regulate TGF-β activation via LRRC33, which functions like GARP in Tregs. The specific mechanism of TAM is complex [169,171]. The TAMs play an important role in tumor metastasis [167], immunotherapy focused on inhibiting M2-TAMs or promoting the expression and function of antitumor M1-TAMs is promising. Interestingly, like TAMs, TANs can also be categorized into N1 and N2 subsets, which are antitumoral and protumoral, respectively [172]. In addition, the N2-TANs play a significant role in tumor progression and metastasis, indicating a good target to develop immunotherapy [173]. The effect of TGF-β on both immune cells and nonimmune cells in the TME is an important topic warranting further investigation [3,10,31] to identify potential therapeutic targets for cancer treatment.

TGF-β-based therapies

Considering the significant role of TGF-β signaling in facilitating tumor progression, targeting TGF-β signaling has become a potentially effective strategy for cancer therapy. The principles of the development of TGF-β-based therapies are to block the TGF-β signaling in specific cell types to enhance the efficacy and reduce the side effects.

Current cancer therapies targeting TGF-β signaling are commonly categorized into the following groups: TGF-β ligand trap, monoclonal neutralizing antibodies against TGF-β, small-molecule inhibitors (SMIs) of TGF-β receptor kinase, antisense oligonucleotides (ASOs) against TGF-β mRNA, and vaccines based on TGF-β signaling [174] (Fig. 4). Among these techniques, the monoclonal neutralizing antibodies against TGF-β and SMIs of TGF-β receptor kinase have been explored for some time. Many drugs have been developed and evaluated in clinical trials, targeting several cancer types.

Traditional strategies and related clinical studies

Utilization of small-molecule inhibitors to suppress TGF-β receptor kinase is one of the most important strategies to target TGF-β signaling. Galunisertib and Vactosertib are two of the best-known and extensively researched. We will use these two examples to describe the strategies of small-molecule inhibitor development.

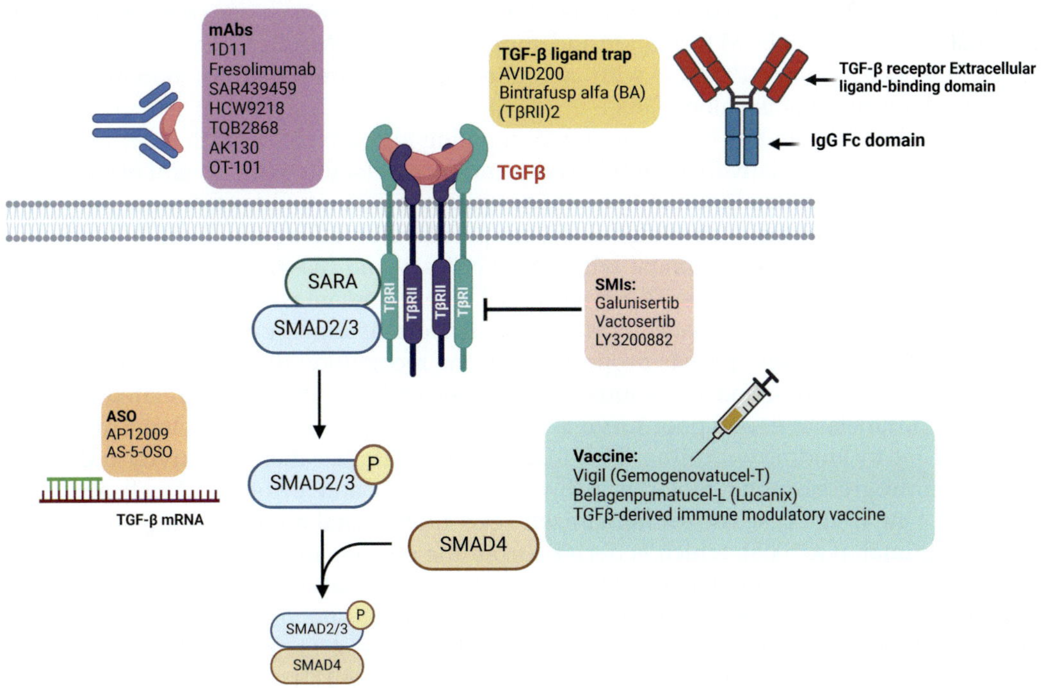

FIG. 4 Recent TGF-β-based antitumor strategies. A variety of TGF-β-based antitumor drugs are under development in preclinical or clinical settings. As shown, the TGF-β signaling pathway provides many therapeutic targets.

Galunisertib (LY2157299 monohydrate) is a first-in-class oral small-molecule that disrupts the TGF-β signaling pathway by inhibiting TGF-β receptor I kinase activity [175]. The TGF-signaling pathway blockade induced by Galunisertib can inhibit EMT and tumor metastasis, repressing tumor growth [176]. And the inhibitory effect of Galunisertib has been found with high selectivity in the preclinical assessment [176]. Several phase I clinical trials have demonstrated the biosafety of Galunisertib. One of the most common adverse effects, cardiac toxicity, was not found in Galunisertib-treated patients [177,178]. Considering that TGF-β is universally expressed and TGF-β signaling is involved in many significant physiological processes, a strict monitoring process needs to be applied to avoid other potential side effects. The selectivity and specificity of the drugs are central. Preclinical studies have found that Galunisertib in combination with other drugs could be more efficient and safer with limited side effects compared to monotherapy. For instance, a recent preclinical trial has demonstrated that Galunisertib combined with gemcitabine will enhance the antitumor efficacy in unresectable pancreatic cancer [179]. A phase I/II study of Galunisertib in combination with nivolumab in patients with advanced refractory solid tumors and non-small cell lung cancer (NSCLC) demonstrated the biosafety and efficacy of the combination therapy [180].

Another small-molecule inhibitor of TGF-β receptor I kinase is Vactosertib (EW-7197/TEW-7197), which can also inhibit ALK-2 and ALK-4 [181]. Vactosertib is commonly used in combination with other drugs to achieve higher safety and efficacy. For example, Vactosertib in combination with pembrolizumab enhances the safety and efficacy in microsatellite stable metastatic colorectal cancer [174,182]. Vactosertib in combination with gemcitabine can inhibit ECM formation and suppress TGF-β/SMAD2 signaling in pancreatic ductal adenocarcinoma (PDAC), with a higher efficacy compared to monotherapy of Vactosertib or gemcitabine alone [183].

Small-molecule inhibitors targeting TGF-β signaling in combination with other antitumor drugs have exhibited relatively high selectivity and bioavailability with several other advantages, including the convenience of oral administration. However, side effects are still one of the most serious concerns, exemplified as dosage problems and drug resistance. Nonetheless, novel drugs are under development with better efficacy and biosafety, e.g., LY3200882, an ATP-competitive inhibitor of the TGF-β receptor I kinase domain with high selectivity [184].

Monoclonal neutralizing antibodies (mAbs) are an important strategy and have been studied for many years. This concept was proposed with the recognition that an overexpression of TGF-β was commonly found in various cancer types and was directly correlated with advanced tumor metastases and poor survival rate of patients [185,186]. Monoclonal antibodies against TGF-β by binding with either TGF-β1, 2, or 3, to neutralize their bioactivity and inhibit the tumor progression and metastasis through NK cells and cytotoxic T cell activation [187–189]. One of the best-known humanized IgG4 monoclonal antibodies targeting TGF-β is Fresolimumab (GC1008), which is derived from the murine antibody 1D11 [187,190,191]. Preclinical studies have found that 1D11 can inhibit human basal-like breast cancer metastasis by increasing infiltration of antitumor cells at the metastatic site, meanwhile enhancing the function of NK cells and T cells in tumors induced by 4 T1 and MDA-MB-231 cells [192,193]. Several clinical trials based on Fresolimumab or 1D11 have been initiated. A phase I trial of Fresolimumab in 28 patients with malignant melanoma and 1 with renal cell carcinoma has found there is no dose-limiting toxicity till 15 mg/kg [194]. In recent years, because monoclonal neutralizing antibodies have shown promising characteristics for cancer treatment, including relatively high efficiency, specificity, and safety, many new monoclonal antibodies targeting TGF-β are being developed and tested in clinical trials, including SAR439459 [195], HCW9218 [196], TQB2868 [197], AK130 [198], and OT-101 [199].

TGF-β ligand trap is a strategy to interfere with TGF-β signaling by blocking the interaction between TGF-β and its receptor. This strategy is developed based on the foundation of TGF-β-targeted mAbs. The mechanism of this technique is to use the human IgG Fc domain to fuse with extracellular ligand-binding domains of the TGF-β receptor, forming a high affinity trap [200]. To increase the efficacy, multiple TGF-β binding domains are added to the same molecule. An engineered single-chain bivalent based on TGF-β type II receptor binding domains has been designed, termed (TβRII)2. It exhibits robust neutralization capacity compared to monovalent ectodomain (TβRII-ED) in vitro, with 100 to 1000-fold higher potency [201]. In addition, the bivalent ligand trap has shown a relatively higher antagonist potency than pan-TGF-β-neutralizing mAb 1D11, significantly inhibiting 4 T1 tumor growth in vivo where the 1D11 is ineffective [201]. Notably, the bivalent TGF-β ligand trap can only antagonize two isoforms of TGF-β (TGF-β1 and TGF-β3). To augment the antagonist potency, a trivalent ligand trap was developed later, showing a stronger inhibition in tumor progression and metastasis in prostate cancer [202]. However, TGF-β2 plays a positive role in hematopoiesis and cardiac function, which is not a desired target in some cases [174,203]. Therefore, the bivalent ligand trap and bivalent trap should be applied based on the specific context. One of the commonly used TGF-β ligand traps is AVID200, which is a traditional bivalent fusion protein of human IgG Fc domain and TGF-β receptor II ectodomains, exhibiting a preferential affinity to TGF-β1 and TGF-β3 [203]. The selective inhibition property of AVID200 makes it a good choice in blood cancer treatment. Benefiting from its small molecular size and ability to target soluble antigens, TGF-β ligand trap in combination with immunotherapies and vaccines shows great therapeutic potential. An important representative combination therapy based on TGF-β ligand trap named bintrafusp alfa will be discussed in the following sections.

Antisense oligonucleotide (ASO) is a genetic approach to specifically knockdown targeted genes by inducing the degradation of mRNA or sterically blocking the translation process [204]. The low cost of this approach and the progress made in human genome sequencing attract much attention and investment in ASO drug development. A variety of ASOs have been developed and applied in cancer settings, especially when other strategies such as SMIs and mAbs are ineffective [205]. Preclinical studies of ASOs targeting TGF-β1 and TGF-β2, respectively, have been applied in many areas. For instance, a TGF-β1 antisense oligos named AS-5-OSO has been designed to inhibit tumor metastasis in fibrosarcomas. One of the best-known ASOs is trabedersen (AP12009), which targets TGF-β2 mRNA expression and has been tested in preclinical studies on pancreatic cancer, malignant gliomas, and malignant melanoma [206,207]. In addition, a phase I/II clinical trial in 33 patients with advanced pancreatic carcinoma, malignant melanoma, or colorectal carcinoma showed AP12009 has an excellent safety profile and antitumor effect [208]. Compared to other strategies, ASOs avoid the general drawbacks such as low specificity [209]. However, there are limitations of ASOs including poor cellular uptake and instability. None of the ASO-based drugs have been formally approved for cancer therapy [209]. To sum up, antisense oligonucleotides remain a promising strategy.

Novel preclinical studies by targeting TGF-β signaling

Recently, many new preclinical studies on TGF-β-based cancer therapy have revealed a variety of promising targets among the components of the TGF-β signaling pathway. Additionally, based on the traditional therapeutic strategies, many improvements and innovations have been developed. TGF-β signaling plays a crucial role in immune regulation and participates in regulating multiple types of gene expression and crosslinks to various signaling pathways in different cell types. Thus, nonspecific targeting of upstream events of TGF-β activation may cause unexpected side effects. Therefore, targeting TGF-β-related factors that are aberrantly regulated specifically in cancers is expected to enhance the specificity of TGF-β blockade and lower the risk of adverse effects.

One promising target is integrins. They are particularly overproduced in TME. In the TGF-β activation phase, integrin heterodimers avβ6 and avβ8 are needed to release active TGF-β from latent complexes. Therefore, inhibition of integrins can block the TGF-β signaling in TME to suppress tumor growth and metastasis. The availability of the subunit integrin β6 (ITGB6) has been found to be a rate-limiting step to form the αvβ6, and anti-ITGB6 treatment can block TGF-β activation, thereby enhancing the infiltration and bioactivity of cytotoxic T cells in the colorectal cancer [210]. Integrin αvβ8 is highly expressed in $CD4^+CD25^+$ Treg cells, macrophages, and dendritic cells. Antisubunit integrin β8 (anti-ITGB8) therapy shows similar effects to anti-ITGB6 treatment, inducing durable immunity in tumors [211]. Additionally, integrin alpha 2 (ITGA2) is overexpressed in tumor cells and can inhibit the nuclear translocation of TFCP2, which is a transcription factor inducing the expression of SMAD2 in pancreatic cancer [212]. Hence, ITGA2 helps to decouple the tumor cells from TGF-β signaling, and silencing ITGA2 may also be a promising therapeutic target to restore the antitumor effect of TGF-β pathway in tumor cells [212].

TGF-β modulators LTBP, GARP, and LRRC33 also play critical roles in regulating TGF-β activation. In the TME, a large amount of TGF-β is stored in the extracellular matrix or on the membrane of Tregs and myeloid lineage cells such as TAMs and MDSCs, which play critical roles in inducing immune evasion. The release of TGF-β of these cells is mediated by GARP and LRRC33. Preclinical experiments have found that the specific inhibition of GARP/TGF-β 1 complexes on the cell membranes of Tregs via monoclonal antibody Clone 58A2 can enhance the cytotoxicity of antitumor T cells and reduce the tumor's resistance to anti-PD-1/PD-L1 immunotherapy, and a relevant phase I trial (NCT03821935) has been initiated [213]. Similarly, LRRC33 in myeloid cells has been found specifically associated with TGF-β1 rather than TGF-β2 or TGF-β3, indicating the LRRC33 axis of TGFβ1 activation might be a good target for immunotherapy [214]. A preclinical study has demonstrated that gene ablation of LRRC33 inhibits TGF-β signaling in B16F10 or MC38 tumors in mouse models and increases the functions of CD8+ T cells and NK cells [214]. Compared to traditional TGF-β1/3 antibody 21D1, gene ablation of LRRC33 shows a relatively weak antitumor effect but significantly improves tumor-specific effects with fewer side effects. Future research may focus on designing antibodies directly against the LRRC33/TGF-β1 complex to augment the efficacy.

SMADs are also potential targets to interfere with the TGF-β signaling. SMAD2/3 (R-SMAD) play central roles in the TGF-β/SMAD pathway. Ubiquitin-mediated degradation is one of the most important ways to regulate the expression level of SMAD3. The ubiquitin-proteasome system involves three enzymes, E1, E2, and E3, while the E3 is the key factor to select the substrates [215]. Cullin ring ubiquitin ligases (CRLs) are multisubunit E3s that build a scaffold based on a cullin subunit, playing crucial roles in cell cycling [215]. Research finds the von Hippel-Lindau tumor suppressor (pVHL) is a member of CRLs to control the degradation of SMAD3 by binding to the conserved LxLxxP motif in the MH2 domain of SMAD3 in mammalian cells [216]. The depletion of pVHL with siRNAs increases the expression of SMAD2/3, while the overexpression of pVHL will reduce the SMAD2/3 level [216]. Therefore, enhancing the expression of pVHL might be a potential way to block the TGF-β/SMAD pathway in human T cells to reduce their inactivation and exhaustion. SMAD6 and SMAD7 are inhibitory SMADs to promote the degradation of TGF-β receptors. As discussed in earlier sections, SMAD7 can recruit the E3 ubiquitin ligase SMURF1/2 and ubiquitin-specific peptidases (USP11, USP15) to induce ubiquitylation and degradation of TGF-β receptors. However, in lung adenocarcinoma (LAD), SMAD7 is transcriptionally suppressed due to DNA hypermethylation [217]. DNA methylation is a suppressive epigenetic modification in eukaryotic cells, which mainly happens in regions rich with CpG. DNA methyltransferase DNMT3B is known as an oncogene in cancer initiation and progression [218]. In LAD patients, a zinc-finger protein called PHF14 recruits DNMT3B to the SMAD7 gene locus, promoting the hypermethylation of the SMAD7 gene and TGF-β-induced LAD metastasis [217]. Therefore, PHF14 and DNMT3B have exhibited potential as targets downstream of TGF-β signaling for translational application. An oncogene protein arginine methyltransferase 5 (PRMT5) plays a crucial role in methylating SMAD4 at R361, promoting the formation of SMAD complexes and subsequent EMT and metastasis in colorectal cancer (CRC) [219]. Hence, post translational modifications (PTMs) are quite meaningful, and a PRMT5 inhibitor GSK3326595 is under phase I clinical trials to treat solid tumors and non-Hodgkin lymphoma [220].

As EMT is an important process that plays a critical role in tumor progression and metastasis, inhibition of EMT via anti-TGF-β drugs has shown therapeutic potential. For instance, a tyrosine kinase inhibitor known as Nintedanib inhibits TGF-β2-induced EMT in ARPE-19 cells, ameliorating relevant cell proliferation and migration [221]. In addition, ST3 beta-galactoside alpha-2,3-sialyltransferase 5 (ST3GAL5) catalyzes the synthesis of a series of gangliosides, which inhibit TGF-β-induced EMT in HEK293T cells and A549 cells by promoting SMAD7/SMURF2-induced ubiquitination and degradation of TβRI [222,223]. Cancer-associated fibroblasts (CAFs), which are the primary source of TGF-β and a significant inducer of EMT in tumors, are also good targets to inhibit tumor growth. Fibroblast activation protein-α (FAPα) is a membrane-bound serine protease with additional endopeptidase activity expressed on the membrane of CAFs as a mesenchymal marker [224,225]. Based on this unique endopeptidase activity of FAP, prodrugs and protoxins targeting TGF-beta can be designed to be activated only upon FAP endopeptidase-mediated cleavage. This approach is expected to limit drugs' effects within the TME and therefore lower systemic effects [226]. For instance, a prodrug generated by conjugating Doxorubicin (Dox) with an anti-CAF FAPα-specific dipeptide (Z-Gly-Pro) shows a similar antitumor efficacy to unmodified Dox, and the treatments haven't shown an obvious cardiotoxic adverse effect [225]. Also, a 26 amino acid protoxin called Melittin can be artificially added with FAP-cleavage sites to be specifically activated by FAPs, which selectively kill FAP-expressing human breast cancer cells such as MCF7 cells [224]. Fibroblast activation protein is a good strategy to increase selectivity and safety. In recent years, many more monoclonal antibodies and other immunotherapies, such as CAR-T, cells have been designed to specifically target FAP on tumor stroma [227,228].

Microtubule targeting agents (MTAs)-based therapy is another recently developed method to specifically modulate microtubule dynamics, cell proliferation, and apoptosis in targeted cells [229,230]. Through combination with other modalities that are highly selective toward tumor cells, e.g., monoclonal antibodies, MTAs can improve the efficacy of antitumor treatment. A variety of preclinical studies have shown the antitumor effect of MTAs in many cancers, such as breast cancer, lung cancer, prostate cancer, and hepatocellular carcinoma [231]. In addition to the antimitotic effects, another function of MTAs is to disrupt the oncogenic signaling by affecting the intracellular transport of factors such as p53, Myc, SMAD2/3, Rb, and Src [232]. MTA-induced disruption of SMAD2/3 can further inhibit the TGF-β signaling pathway and, consequently, repress the TGF-β-induced EMT and tumor metastasis [233]. Among the microtubule targeting agents, eribulin (E7389) is an analog of the marine natural halichondrin B, which functions by triggering cell cycle arrest and apoptosis of cancer cells [234]. Additionally, eribulin can reverse the EMT in triple-negative breast cancer (TNBC) cells both in vitro and in vivo [235,236]. In recent years, researchers have also elucidated the mechanism of anti-EMT effects of eribulin not only based on SMAD2/3 but also relying on the expression of SMAD4. Kaul et al. found that eribulin can inhibit EMT-promoting factor. Snail expression in TNBC cell lines while inducing expression of another factor Slug specifically with a low SMAD4 expression [233]. MTA-based therapy is a relatively new strategy in cancer treatment, providing a novel direction to target the specific molecules in addition to its antimitotic effects. However, to ensure its specificity and safety, other anti-TGF-β drugs should be used in combination.

Synergistic effect of targeting TGF-β and immunotherapies

In recent years, immunotherapy targeting cancers has developed rapidly. However, many solid tumors have a strong resistance to the therapies due to immune evasion induced by overexpression of TGF-β. Anti-TGF-β treatment in combination with other immunotherapies can reduce the resistance and enhance the selectivity of anti-TGF-β drugs at the same time, which will be a promising trend in the future. Anti-PD-1/PD-L1 immune checkpoint blockade and CAR-T cell-mediated therapies are two recent effective strategies in cancer therapy. Besides, vaccines are also a traditional but effective strategy among immunotherapies, which is frequently applied to synergize with other strategies. In this section, we will discuss the advantages of combination therapies.

Cytotoxic T-lymphocyte-associated protein 4 (CTLA-4), programmed death-1 (PD-1), and programmed death-ligand 1 (PD-L1) play crucial roles in tempering T cell function. PD-L1 on cancer cells can bind with PD-1 on tumor-infiltrating T cells (TILs) to activate Src homology region 2 domain-containing phosphatases (SPH2s), resulting in the suppression of the T cell receptor (TCR) pathways, whereby inhibiting T cell activities in many cancers [237]. PD-1/PD-L1 pathway will inhibit non-SMAD TGF-β pathways such as Ras-Raf-MEK-ERK and PI3K-AKT signaling pathways, which play a predominant role in regulating T cell proliferation and differentiation [238].

Immune checkpoint blockade (ICB) based on PD-1/PD-L1 signaling has shown remarkable antitumor effects in many cancers such as lung cancer, bladder cancer, and melanoma

[239]. However, anti-PD-1/PD-L1 therapies still have limitations and only work for specific subsets of cancers; some solid tumors have low responses to this therapy, and some patients are refractory to treatment [240]. One of the most significant influencing factors is that anti-PD-1/PD-L1 treatment will impact and promote the effect of other inhibitory signaling pathways such as T cell immunoglobulin mucin-3 (TIM-3) and TGF-β [241]. There is a significant increase in the expression of TIM-3 on tumor-infiltrating lymphocytes (TILs) in resistant nodules of Kras mice models, which induces the exhaustion of CD8 T cells, while the expression of transcription factor forkhead box P3 (FOXP3) which promote proliferation and differentiation of regulatory T cells (Tregs) shows no significant difference between untreated and resistant samples [241,242]. Similarly, since the Tregs commonly express a higher level of PD-1 than other $CD4^+$ T cells, the inhibition of PD-1/PD-L1 signaling will result in a low $CD4^+$Th1/Tregs ratio, while Tregs will activate the latent TGF-β complex through the specifically expressed GARP/avβ8, in turn stimulating and enhancing the expression of Tregs themselves [243,244]. In non-small cell lung cancer cells (NSCLC), TGF-β can upregulate the expression of PD-L1 by directly enhancing its transcription [245]. SiRNA-based knockdown of SMAD2 in A549 cells results in blockade of TGF-β1-induced upregulation of PD-L1 mRNA, suggesting that the TGF-β-induced enhancement of PD-L1 expression is SMAD2 dependent [245]. In metastatic urothelial cancer (mUC), the increasing expression of TGF-β in cancer-associated fibroblasts will promote EMT and prevent $CD8^+$ T cells from entering the tumor parenchyma, resulting in a large cohort of CD8+ T cells being blocked in the peritumoral stroma and making the tumors resistant to anti-PD-1/PD-L1 treatment [246]. Therefore, there is a substantial need to combine anti-PD-1/PD-L1 therapy with other checkpoint inhibitors to overcome drug resistance. Among the combination therapies, anti-TGF-β's synergy with anti-PD-1/PD-L1 ICB has shown great potential and many advantages. Preclinical studies show that anti-TGF-β treatment can synergize with anti-PD-1/PD-L1 therapy in mUC to induce the tumor extracellular matrix (ECM) remodeling and help $CD8^+$ T cells infiltrate the center of tumors [246], and similar synergistic effects have been found in lung cancer [245], breast cancer [247], colorectal carcinoma [248], and other cancers.

Many anti-TGF-β therapy strategies, such as SMIs and mAbs against TGF-β, can synergize with anti-PD-1/PD-L1 treatment. One of the favorite synergistic anti-TGF-β strategies is TGF-β ligand traps, which are built on TGF-β antibodies and have higher efficacy than traditional mAb against TGF-β. For instance, a novel fusion protein called bifunctional anti-TGF-β/PD-L1 antibody (Bintrafusp alfa/M7824) shows great therapeutic potential [245]. Bintrafusp alfa (BA) is a recently developed first-in-class bifunctional fusion protein composed of the extracellular domain of transforming growth factor TGF-βRII (TGF-β ligand trap) and a human IgG1 or IgG4 mAb blocking PD-L1 [247,249,250]. Preclinical studies have found the BA (M7824) can significantly reduce the expression of TGF-β and TGF-β-induced EMT in tumor microenvironment and enhance the antitumor activity of NK cells and $CD8^+$ T cells in EMT6 tumors [239]. Also, M7824 exhibits an inhibitory effect on Tregs, which doesn't occur in anti-PD-1/PD-L1 therapy [247]. A phase I dose-escalation clinical trial (NCT02517398) of M7824 in patients with heavily pretreated advanced solid tumors has demonstrated the safety profile is manageable [251]. However, four patients in this clinical trial have shown Grade ≥3 treatment-related adverse events (AEs), including bullous pemphigoid with subsequent skin infection, asymptomatic lipase increase with keratoacanthoma, colitis leading to anemia and rectal hemorrhage, and hypokalemia accompanied by gastroparesis [251]. The expansion cohort of Phase I Trial (NCT02517398) in 80 patients with platinum-experienced NSCLC has revealed a similar result that M7824 has a manageable tolerability and encouraging efficacy, especially in patients with PD-L1–high tumors [250]. Compared to traditional monotherapies, the fusion protein exhibits an enhanced ability to localize to the targeted tumor areas, thus having a higher efficacy and safety than separately applying a combination of anti-TGF-β and anti-PD-1/PD-L1 therapy in MC38 mouse tumor model [252]. ^{89}Zr-radiolabelling of bintrafusp alfa has found that the localization of the drugs mainly relies on the PD-1/PD-L1 antibody portion and is concentrated in tumors, while only small amount of drug will be uptake in TGF-β-rich areas including lung, spleen, liver, and kidney [253]. Also, M7824 exhibits a long-term antitumor effect like PD-1/PD-L1 antibodies and can work in combination with radiotherapy or chemotherapy [254]. Furthermore, the fusion protein is easier to clear after binding with TGF-β in the TME, where the drug/TGF-β complexes will be internalized by PD-L1$^+$ cells and degraded in the lysosomes, significantly lowering the drug/TGF-β complex concentration in the circulation [252]. However, bintrafusp alfa has limitations. As described in the phase I clinical trial (NCT02517398), adverse events have been found in some patients. In addition, a recent phase II clinical trial (NCT03833661) of bintrafusp alfa has been initiated in patients with locally advanced/metastatic biliary tract cancers, while the results show an improved efficacy of BA compared to anti-PD-1/PD-L1, monotherapy doesn't reach the expected threshold [255]. Future exploring directions of bintrafusp alfa can be on synergizing with other therapies, as a preclinical study has found bintrafusp alfa synergizes with localized radiotherapy (RT) to enhance the infiltration of antitumor cells into the tumor and reduce the immune evasion and corresponding drug resistance [252].

Chimeric antigen receptor (CAR) T cell therapy is an immunotherapy that recently made breakthroughs in tumor treatment. The basic mechanism is artificially adding engineered receptors, which consist of an extracellular binding domain and a transmembrane domain linked to intracellular signaling domains on T cells, to specifically target tumor cells that express ligands recognized by the engineered receptors [256]. Notably, the first developed CAR-T cells are all applied in the treatment of blood cancers, including lymphomas, leukemia, and multiple myeloma, by targeting cell surface antigen CD19 and B cell maturation antigen (BCMA) [257–259]. Nevertheless, other types of tumors, especially solid tumors, show strong resistance to CAR-T cell therapy via immune evasion induced by high concentrations of tumor-promoting cytokines in TME [260]. Therefore, the overexpressed TGF-β in the tumor microenvironment may play a critical role in tumor-evading CAR-T-mediated killing. To reduce the TGF-β-induced resistance in CAR-T cell therapy, two main strategies have been developed. One is to generate TGF-β-resistant CAR-T cells; the other is to develop CAR-T cells directly targeting TGF-β. For the first strategy, early exploration focused on pretreating CAR-T cells with anti-TGF-β drugs before introducing them into the hosts. An in vitro study has found that pretreating CAR-T cells with ATP-competitive TGF-βRI kinase inhibitor SD-208 can augment the antitumor activity of the receptor tyrosine kinase-like orphan receptor 1 (ROR1) antigen-specific CAR-T cells which against triple-negative breast cancer [261]. Additionally, more advanced preclinical studies attempted to genetically engineer CAR-T cells to become refractory to TGF-β signaling. For instance, preclinical studies have developed CAR-T cells that coexpress the dominant negative (DN) TGFBRII and target BCMA or prostate-specific membrane antigen (PSMA) via lentiviral transduction [262,263]. And the armored CAR-T cells have shown a long-term augmented tumor infiltration and TGF-β resistance, consequently exhibiting high antitumor efficacy [262,263]. Furthermore, CRISPR/Cas9 is also frequently used to modify the CAR-T cells. CRISPR/Cas9-mediated TGFβRII knockout CAR-T cells have been found with higher TGF-β resistance and antitumor efficacy both in vitro and in vivo [260,264]. A phase I clinical trial has demonstrated the feasibility and safety of CRISPR/Cas9-mediated TGF-β-resistant CAR-T cells that target PSMA in patients with refractory metastatic castration-resistant prostate cancer (mCRPC) [265]. For the second strategy, researchers have made CAR-T cells directly target the soluble ligands in addition to cell surface antigens, creating the opportunity to synergize with anti-TGF-β therapies [266,267]. Preclinical studies have demonstrated that engineered T cells with chimeric antigen receptor (CAR) targeting the soluble TGF-β ligands successfully enhance the activity of other effective T cells and suppress the differentiation of Tregs [267].

Nonetheless, limitations exist. It is hard to control the expansion of activated CAR-T cells, which may lead to neurotoxicity and cytokine release syndrome [268,269]. Additionally, CAR-T cells may also attack the normal tissues that express the targeted antigens as well [270]. Furthermore, strong selection pressure may be exhibited on the tumors by monovalent CAR-T cell therapy, resulting in a loss of targeted tumor surface antigen and resistance to the therapy [271]. Therefore, advanced CAR-T cell technology should be developed to achieve high efficacy and safety, and multitargeting CAR-T cell therapy could be a promising venue to explore. Recently, a novel adapter CAR-T system (AdCAR) has been generated to improve the controllability, flexibility, and selectivity, which can target multiple antigens at the same time [256]. By fusing with an adapter molecule (AMs), which usually are antigens targeting tumor-secreting cytokines like TGF-β, the engineered CAR-T cells can target different tumor-associated antigens (TAAs) at the same time, which significantly increases the efficacy and reduces the immune evasion of the tumor [256]. Preclinical studies based on the AdCAR system to target soluble LAP as well as latent TGF-β both in vitro and in vivo have been initiated, showing a faster response and higher antitumor efficacy in a controllable TGF-β concentration-dependent manner [259].

Vaccine is a traditional but effective immunotherapy to strengthen patients' immunity, and one of the most used types of vaccine in cancer therapy is the CTL inducer [189]. Preclinical studies have found the inhibition of TGF-β with mAb 1D11 can enhance the efficacy of CTL-inducing vaccines like glioma-associated antigen peptide vaccines [272]. Another strategy is to transfect cancer cells with anti-TGF-β elements to directly repress the TGF-β expression. There are two mature vaccines following this strategy, termed Vigil and Belagenpumatucel-L. Vigil (Gemogenovatucel-T) is developed from a patient's own tumor cells obtained during surgery and then externally modified with a plasmid encoding GM-CSF and shRNA [174]. This combination not only boosts immune stimulation but also specifically inhibits TGF-beta-activating protease Furin to reduce the production of TGF-β1 and TGF-β2 [174]. Similarly, Belagenpumatucel-L (Lucanix) is generated by transfecting NSCLC cells with a TGF-β2 antisense transgene to suppress the expression of TGF-β2 and enhance the immune response against the tumor within the TME [174]. Recently, a novel form of vaccine called immune modulatory vaccines (IMVs) has been designed based on a newly found T cell subset termed antiregulatory T cells (anti-Tregs). Anti-Tregs are a set of self-reactive, antigen-specific, and proinflammatory T cells that inhibit or kill immune-suppressive cells including Tregs, dendritic cells, MDSCs, TAMs, and CAFs [273,274]. Anti-Tregs can specifically recognize human leukocyte antigens (HLA)-restricted epitopes derived from proteins expressed by these immune-suppressive cells, including indoleamine 2,3-dioxygenase (IDO), tryptophan 2,6-dioxygenase (TDO), PD-L1, Foxp3, and TGFβ [273,274]. The effects of anti-Tregs indicate that there might be specific short peptide epitopes within TGF-β to stimulate the anti-Tregs. A recent study indeed revealed preexisting TGF-β-specific T cell immunity elicited by TGFβ-specific $CD4^+$ T cells and $CD8^+$ T cells (CTLs) [275,276]. These T cells can specifically recognize and respond to a TGFβ-derived epitope TGFβ141–160 (REAVPEPVLLSRAELRLLRL, or TGFβ-15) and thereby termed TGFβ-15-specific T cells [276]. Recently, a TGFβ-derived peptide vaccination designed based on murine TGFβ1-derived T cell epitopes was found to significantly reduce TGF-β-induced immune evasion and resistance in Pan02, a syngeneic murine model of PDAC [277]. Also, the TGF-β-derived IMVs can synergize with other cancer therapies, such as anti-PD-1/PD-L1 ICB, with a greater therapeutic potential [277].

The improvements made by combining TGF-β-based therapy with immunotherapies including ICI, CAR-T, and vaccines suggest a great therapeutic potential for targeting TGF-β pathways. Compared to the traditional monotherapies, combination therapy significantly reduces the drug resistance caused by TGF-β-induced tumor immune evasion, providing a promising future exploring direction. Nonetheless, obstacles remain for the development of safe and effective antitumor therapeutics by targeting TGF-β signaling. Considering TGF-β plays quite diverse roles during physiological functions in healthy tissues, inhibition of TGF-β signaling could cause detrimental effects on normal tissue homeostasis. In addition, current TGF-β inhibitors are not as selective as they are hoped to be. This issue may cause unforeseen off-target effects. Finally, it is necessary to examine the feasibility of newly developed anti-TGF-β therapeutic strategy in clinical trials, and long-term follow-up is required to assess their effect on tumor development.

Conclusions

TGF-β plays a dual role in different phases of tumor progression, and it is one of the most important factors to induce immune evasion in advanced tumors. Here, we discussed how TGF-β regulates immune responses via SMAD and non-SMAD pathways and its role in immune cells and tumor cells. The overexpressed TGF-β in TME can significantly suppress the activities of antitumor cells, e.g., NK cells and $CD8^+$ T cells, meanwhile promoting differentiation and proliferation of tumor-promoting cells, e.g., Tregs, MDSCs, and TAMs. Considering that TGF-β exhibited obvious procancer properties, therapeutics targeting TGF-β signaling would be a feasible approach and have shown potential in preclinical and clinical studies. Notably, enhanced efficacy and safety were observed when suppression of TGF-β signaling was synergized with several currently developed approaches, including anti-PD-1/PD-L1, CAR-T cells, and vaccine-based therapies. Such combination approaches will be future therapeutic directions worth pursuing.

While targeting TGF-β signaling is a prominent strategy in cancer therapy, it's important to note that its activation is not exclusive to cancer cells, contrasting with more specific targets like the PD-1/PD-L1 pathway. This lack of specificity has led to unforeseen challenges in both theoretical and clinical spheres, and patients who accepted pan-TGF-β inhibition treatment may face high risks of off-target cardiotoxic adverse effects. Thus, systematic suppression of TGF-β signaling may not be an efficient approach for cancer treatment. To overcome this obstacle, targeting downstream pathways of TGF-β signaling that are more specifically featured in cancer cells or subtypes of cell types or combination therapy synergized with other immunotherapies would be expected for the future studies, as to reduce serious side effects and enhance efficacy. For example, a recently developed strategy that selectively inhibited Th cell TGF-β signaling in tumor-draining lymph nodes can cause reorganization of tumor vasculature and cancer cell death [278].

Furthermore, various cancer type-dependent activities have been discovered recently, which may promote the development of more precise therapy by selectively interacting anti-TGF-β signaling to cancer type-specific activities. For instance, the emerging understanding of the role of TGF-β receptor II exosomes in modulating CD8+ T cell exhaustion highlights another layer of regulatory mechanisms that could impact therapeutic advancements [279]. Therefore, a more profound and comprehensive exploration of TGF-β mechanisms is needed to devise innovative and effective strategies to mitigate its impacts on cancer progression and drug resistance.

Acknowledgments

This work was supported by the NIH (R01 AI160774, R01 AI123193, and R56 AG071256) and the National Multiple Sclerosis Society (RG-1802-304830) to YYW. The figures were created using BioRender.com. We apologize to colleagues whose work we were unable to discuss and cite due to space limitations.

References

[1] Batlle E, Massagué J. Transforming growth factor-β signaling in immunity and Cancer. Immunity 2019;50: 924–40. https://doi.org/10.1016/j.immuni.2019.03.024.

[2] Derynck R, Budi EH. Specificity, versatility, and control of TGF-β family signaling. Sci Signal 2019;12. https://doi.org/10.1126/scisignal.aav5183.

[3] Sanjabi S, Oh SA, Li MO. Regulation of the immune response by TGF-β: from conception to autoimmunity and infection. Cold Spring Harb Perspect Biol 2017;9. https://doi.org/10.1101/cshperspect.a022236.

[4] Siegel PM, Massagué J. Cytostatic and apoptotic actions of TGF-beta in homeostasis and cancer. Nat Rev Cancer 2003;3:807–21. https://doi.org/10.1038/nrc1208.

[5] Zhang Y, Alexander PB, Wang XF. TGF-β family signaling in the control of cell proliferation and survival. Cold Spring Harb Perspect Biol 2017;9. https://doi.org/10.1101/cshperspect.a022145.

[6] Mullen AC, Wrana JL. TGF-β family signaling in embryonic and somatic stem-cell renewal and differentiation. Cold Spring Harb Perspect Biol 2017;9. https://doi.org/10.1101/cshperspect.a022186.

[7] Wu Y, Deng J, Rychahou PG, Qiu S, Evers BM, Zhou BP. Stabilization of snail by NF-kappaB is required for inflammation-induced cell migration and invasion. Cancer Cell 2009;15:416–28. https://doi.org/10.1016/j.ccr.2009.03.016.

[8] Massagué J. TGFβ signalling in context. Nat Rev Mol Cell Biol 2012;13:616–30. https://doi.org/10.1038/nrm3434.

[9] Meng XM, Nikolic-Paterson DJ, Lan HY. TGF-β: the master regulator of fibrosis. Nat Rev Nephrol 2016;12: 325–38. https://doi.org/10.1038/nrneph.2016.48.

[10] Seoane J, Gomis RR. TGF-β family signaling in tumor suppression and cancer progression. Cold Spring Harb Perspect Biol 2017;9. https://doi.org/10.1101/cshperspect.a022277.

[11] Castriconi R, Cantoni C, Della Chiesa M, Vitale M, Marcenaro E, Conte R, Biassoni R, Bottino C, Moretta L, Moretta A. Transforming growth factor beta 1 inhibits expression of NKp30 and NKG2D receptors: consequences for the NK-mediated killing of dendritic cells. Proc Natl Acad Sci USA 2003;100:4120–5. https://doi.org/10.1073/pnas.0730640100.

[12] Laouar Y, Sutterwala FS, Gorelik L, Flavell RA. Transforming growth factor-beta controls T helper type 1 cell development through regulation of natural killer cell interferon-gamma. Nat Immunol 2005;6:600–7. https://doi.org/10.1038/ni1197.

[13] Chen B, Mu C, Zhang Z, He X, Liu X. The love-hate relationship between TGF-β signaling and the immune system during development and tumorigenesis. Front Immunol 2022;13, 891268. https://doi.org/10.3389/fimmu.2022.891268.

[14] Hinck AP. Structural studies of the TGF-βs and their receptors – insights into evolution of the TGF-β superfamily. FEBS Lett 2012;586:1860–70. https://doi.org/10.1016/j.febslet.2012.05.028.

[15] Hinck AP, Mueller TD, Springer TA. Structural biology and evolution of the TGF-β family. Cold Spring Harb Perspect Biol 2016;8. https://doi.org/10.1101/cshperspect.a022103.

[16] Heldin CH, Moustakas A. Signaling receptors for TGF-β family members. Cold Spring Harb Perspect Biol 2016;8. https://doi.org/10.1101/cshperspect.a022053.

[17] David CJ, Massagué J. Contextual determinants of TGFβ action in development, immunity and cancer. Nat Rev Mol Cell Biol 2018;19:419–35. https://doi.org/10.1038/s41580-018-0007-0.

[18] Travis MA, Sheppard D. TGF-β activation and function in immunity. Annu Rev Immunol 2014;32:51–82. https://doi.org/10.1146/annurev-immunol-032713-120257.

[19] Robertson IB, Rifkin DB. Regulation of the bioavailability of TGF-β and TGF-β-related proteins. Cold Spring Harb Perspect Biol 2016;8. https://doi.org/10.1101/cshperspect.a021907.

[20] Robertson IB, Horiguchi M, Zilberberg L, Dabovic B, Hadjiolova K, Rifkin DB. Latent TGF-β-binding proteins. Matrix Biol 2015;47:44–53. https://doi.org/10.1016/j.matbio.2015.05.005.

[21] Shi M, Zhu J, Wang R, Chen X, Mi L, Walz T, Springer TA. Latent TGF-β structure and activation. Nature 2011;474:343–9. https://doi.org/10.1038/nature10152.

[22] Qin Y, Garrison BS, Ma W, Wang R, Jiang A, Li J, Mistry M, Bronson RT, Santoro D, Franco C, et al. A milieu molecule for TGF-β required for microglia function in the nervous system. Cell 2018;174:156–171.e116. https://doi.org/10.1016/j.cell.2018.05.027.

[23] Stockis J, Colau D, Coulie PG, Lucas S. Membrane protein GARP is a receptor for latent TGF-beta on the surface of activated human Treg. Eur J Immunol 2009;39:3315–22. https://doi.org/10.1002/eji.200939684.

[24] Wang R, Zhu J, Dong X, Shi M, Lu C, Springer TA. GARP regulates the bioavailability and activation of TGFβ. Mol Biol Cell 2012;23:1129–39. https://doi.org/10.1091/mbc.E11-12-1018.

[25] Tran DQ, Andersson J, Wang R, Ramsey H, Unutmaz D, Shevach EM. GARP (LRRC32) is essential for the surface expression of latent TGF-beta on platelets and activated FOXP3+ regulatory T cells. Proc Natl Acad Sci USA 2009;106:13445–50. https://doi.org/10.1073/pnas.0901944106.

[26] Duan Z, Lin X, Wang L, Zhen Q, Jiang Y, Chen C, Yang J, Lee CH, Qin Y, Li Y, et al. Specificity of TGF-β1 signal designated by LRRC33 and integrin α(V)β(8). Nat Commun 2022;13:4988. https://doi.org/10.1038/s41467-022-32655-9.

[27] Lyons RM, Gentry LE, Purchio AF, Moses HL. Mechanism of activation of latent recombinant transforming growth factor beta 1 by plasmin. J Cell Biol 1990;110:1361–7. https://doi.org/10.1083/jcb.110.4.1361.

[28] Wang M, Zhao D, Spinetti G, Zhang J, Jiang LQ, Pintus G, Monticone R, Lakatta EG. Matrix metalloproteinase 2 activation of transforming growth factor-beta1 (TGF-beta1) and TGF-beta1-type II receptor signaling within the aged arterial wall. Arterioscler Thromb Vasc Biol 2006;26:1503–9. https://doi.org/10.1161/01.ATV.0000225777.58488.f2.

[29] Kobayashi T, Kim H, Liu X, Sugiura H, Kohyama T, Fang Q, Wen FQ, Abe S, Wang X, Atkinson JJ, et al. Matrix metalloproteinase-9 activates TGF-β and stimulates fibroblast contraction of collagen gels. Am J Phys Lung Cell Mol Phys 2014;306:L1006–15. https://doi.org/10.1152/ajplung.00015.2014.

[30] Illman SA, Lehti K, Keski-Oja J, Lohi J. Epilysin (MMP-28) induces TGF-beta mediated epithelial to mesenchymal transition in lung carcinoma cells. J Cell Sci 2006;119:3856–65. https://doi.org/10.1242/jcs.03157.

[31] Tauriello DVF, Sancho E, Batlle E. Overcoming TGFβ-mediated immune evasion in cancer. Nat Rev Cancer 2022;22:25–44. https://doi.org/10.1038/s41568-021-00413-6.

[32] Yang Z, Mu Z, Dabovic B, Jurukovski V, Yu D, Sung J, Xiong X, Munger JS. Absence of integrin-mediated TGFbeta1 activation in vivo recapitulates the phenotype of TGFbeta1-null mice. J Cell Biol 2007;176:787–93. https://doi.org/10.1083/jcb.200611044.

[33] Edwards JP, Thornton AM, Shevach EM. Release of active TGF-β1 from the latent TGF-β1/GARP complex on T regulatory cells is mediated by integrin β8. J Immunol 2014;193:2843–9. https://doi.org/10.4049/jimmunol.1401102.

[34] Metelli A, Wu BX, Riesenberg B, Guglietta S, Huck JD, Mills C, Li A, Rachidi S, Krieg C, Rubinstein MP, et al. Thrombin contributes to cancer immune evasion via proteolysis of platelet-bound GARP to activate LTGF-β. Sci Transl Med 2020;12. https://doi.org/10.1126/scitranslmed.aay4860.

[35] Bates RC, Bellovin DI, Brown C, Maynard E, Wu B, Kawakatsu H, Sheppard D, Oettgen P, Mercurio AM. Transcriptional activation of integrin beta6 during the epithelial-mesenchymal transition defines a novel prognostic indicator of aggressive colon carcinoma. J Clin Invest 2005;115:339–47. https://doi.org/10.1172/jci23183.

[36] Metelli A, Wu BX, Fugle CW, Rachidi S, Sun S, Zhang Y, Wu J, Tomlinson S, Howe PH, Yang Y, et al. Surface expression of TGFβ docking receptor GARP promotes oncogenesis and immune tolerance in breast Cancer. Cancer Res 2016;76:7106–17. https://doi.org/10.1158/0008-5472.Can-16-1456.

[37] ten Dijke P, Heldin C-H. The SMAD family. In: Smad signal transduction, vol. 5. Springer; 2006. p. 1–13.

[38] Lawler S, Feng XH, Chen RH, Maruoka EM, Turck CW, Griswold-Prenner I, Derynck R. The type II transforming growth factor-beta receptor autophosphorylates not only on serine and threonine but also on tyrosine residues. J Biol Chem 1997;272:14850–9. https://doi.org/10.1074/jbc.272.23.14850.

[39] Moustakas A, Heldin CH. The regulation of TGFbeta signal transduction. Development 2009;136:3699–714. https://doi.org/10.1242/dev.030338.

[40] Shi Y, Massagué J. Mechanisms of TGF-beta signaling from cell membrane to the nucleus. Cell 2003;113:685–700. https://doi.org/10.1016/s0092-8674(03)00432-x.

[41] Chaikuad A, Bullock AN. Structural basis of intracellular TGF-β signaling: receptors and Smads. Cold Spring Harb Perspect Biol 2016;8. https://doi.org/10.1101/cshperspect.a022111.

[42] Souchelnytskyi S, Tamaki K, Engström U, Wernstedt C, ten Dijke P, Heldin CH. Phosphorylation of Ser465 and Ser467 in the C terminus of Smad2 mediates interaction with Smad4 and is required for transforming growth factor-beta signaling. J Biol Chem 1997;272:28107–15. https://doi.org/10.1074/jbc.272.44.28107.

[43] Tzavlaki K, Moustakas A. TGF-β signaling. Biomolecules 2020;10. https://doi.org/10.3390/biom10030487.

[44] Shi Y, Wang YF, Jayaraman L, Yang H, Massagué J, Pavletich NP. Crystal structure of a Smad MH1 domain bound to DNA: insights on DNA binding in TGF-beta signaling. Cell 1998;94:585–94. https://doi.org/10.1016/s0092-8674(00)81600-1.

[45] Baburajendran N, Jauch R, Tan CY, Narasimhan K, Kolatkar PR. Structural basis for the cooperative DNA recognition by Smad4 MH1 dimers. Nucleic Acids Res 2011;39:8213–22. https://doi.org/10.1093/nar/gkr500.

[46] Dennler S, Huet S, Gauthier JM. A short amino-acid sequence in MH1 domain is responsible for functional differences between Smad2 and Smad3. Oncogene 1999;18:1643–8. https://doi.org/10.1038/sj.onc.1202729.

[47] Zhang YE. Non-Smad signaling pathways of the TGF-β family. Cold Spring Harb Perspect Biol 2017;9. https://doi.org/10.1101/cshperspect.a022129.

[48] Simonsson M, Kanduri M, Grönroos E, Heldin CH, Ericsson J. The DNA binding activities of Smad2 and Smad3 are regulated by coactivator-mediated acetylation. J Biol Chem 2006;281:39870–80. https://doi.org/10.1074/jbc.M607868200.

[49] Nakao A, Afrakhte M, Morén A, Nakayama T, Christian JL, Heuchel R, Itoh S, Kawabata M, Heldin NE, Heldin CH, ten Dijke P. Identification of Smad7, a TGFbeta-inducible antagonist of TGF-beta signalling. Nature 1997;389:631–5. https://doi.org/10.1038/39369.

[50] Mochizuki T, Miyazaki H, Hara T, Furuya T, Imamura T, Watabe T, Miyazono K. Roles for the MH2 domain of Smad7 in the specific inhibition of transforming growth factor-beta superfamily signaling. J Biol Chem 2004;279:31568–74. https://doi.org/10.1074/jbc.M313977200.

[51] Hanyu A, Ishidou Y, Ebisawa T, Shimanuki T, Imamura T, Miyazono K. The N domain of Smad7 is essential for specific inhibition of transforming growth factor-beta signaling. J Cell Biol 2001;155:1017–27. https://doi.org/10.1083/jcb.200106023.

[52] Kavsak P, Rasmussen RK, Causing CG, Bonni S, Zhu H, Thomsen GH, Wrana JL. Smad7 binds to Smurf2 to form an E3 ubiquitin ligase that targets the TGF beta receptor for degradation. Mol Cell 2000;6:1365–75. https://doi.org/10.1016/s1097-2765(00)00134-9.

[53] Ebisawa T, Fukuchi M, Murakami G, Chiba T, Tanaka K, Imamura T, Miyazono K. Smurf1 interacts with transforming growth factor-beta type I receptor through Smad7 and induces receptor degradation. J Biol Chem 2001;276:12477–80. https://doi.org/10.1074/jbc.C100008200.

[54] Inui M, Manfrin A, Mamidi A, Martello G, Morsut L, Soligo S, Enzo E, Moro S, Polo S, Dupont S, et al. USP15 is a deubiquitylating enzyme for receptor-activated SMADs. Nat Cell Biol 2011;13:1368–75. https://doi.org/10.1038/ncb2346.

[55] Zhang L, Zhou F, Drabsch Y, Gao R, Snaar-Jagalska BE, Mickanin C, Huang H, Sheppard KA, Porter JA, Lu CX, ten Dijke P. USP4 is regulated by AKT phosphorylation and directly deubiquitylates TGF-β type I receptor. Nat Cell Biol 2012;14:717–26. https://doi.org/10.1038/ncb2522.

[56] Zhang S, Fei T, Zhang L, Zhang R, Chen F, Ning Y, Han Y, Feng XH, Meng A, Chen YG. Smad7 antagonizes transforming growth factor beta signaling in the nucleus by interfering with functional Smad-DNA complex formation. Mol Cell Biol 2007;27:4488–99. https://doi.org/10.1128/mcb.01636-06.

[57] Guo X, Ramirez A, Waddell DS, Li Z, Liu X, Wang XF. Axin and GSK3- control Smad3 protein stability and modulate TGF- signaling. Genes Dev 2008;22:106–20. https://doi.org/10.1101/gad.1590908.

[58] Tecalco-Cruz AC, Ríos-López DG, Vázquez-Victorio G, Rosales-Alvarez RE, Macías-Silva M. Transcriptional cofactors Ski and SnoN are major regulators of the TGF-β/Smad signaling pathway in health and disease. Signal Transduct Target Ther 2018;3:15. https://doi.org/10.1038/s41392-018-0015-8.

[59] Hata A, Chen YG. TGF-β signaling from receptors to Smads. Cold Spring Harb Perspect Biol 2016;8. https://doi.org/10.1101/cshperspect.a022061.

[60] Moustakas A, Heldin CH. Non-Smad TGF-beta signals. J Cell Sci 2005;118:3573–84. https://doi.org/10.1242/jcs.02554.

[61] Lee MK, Pardoux C, Hall MC, Lee PS, Warburton D, Qing J, Smith SM, Derynck R. TGF-beta activates Erk MAP kinase signalling through direct phosphorylation of ShcA. EMBO J 2007;26:3957–67. https://doi.org/10.1038/sj.emboj.7601818.

[62] Zavadil J, Bitzer M, Liang D, Yang YC, Massimi A, Kneitz S, Piek E, Bottinger EP. Genetic programs of epithelial cell plasticity directed by transforming growth factor-beta. Proc Natl Acad Sci USA 2001;98:6686–91. https://doi.org/10.1073/pnas.111614398.

[63] Xie L, Law BK, Chytil AM, Brown KA, Aakre ME, Moses HL. Activation of the Erk pathway is required for TGF-beta1-induced EMT in vitro. Neoplasia 2004;6:603–10. https://doi.org/10.1593/neo.04241.

[64] Yamaguchi K, Shirakabe K, Shibuya H, Irie K, Oishi I, Ueno N, Taniguchi T, Nishida E, Matsumoto K. Identification of a member of the MAPKKK family as a potential mediator of TGF-beta signal transduction. Science 1995;270:2008–11. https://doi.org/10.1126/science.270.5244.2008.

[65] Yamashita M, Fatyol K, Jin C, Wang X, Liu Z, Zhang YE. TRAF6 mediates Smad-independent activation of JNK and p38 by TGF-beta. Mol Cell 2008;31:918–24. https://doi.org/10.1016/j.molcel.2008.09.002.

[66] Kim SI, Kwak JH, Na HJ, Kim JK, Ding Y, Choi ME. Transforming growth factor-beta (TGF-beta1) activates TAK1 via TAB1-mediated autophosphorylation, independent of TGF-beta receptor kinase activity in mesangial cells. J Biol Chem 2009;284:22285–96. https://doi.org/10.1074/jbc.M109.007146.

[67] Broglie P, Matsumoto K, Akira S, Brautigan DL, Ninomiya-Tsuji J. Transforming growth factor beta-activated kinase 1 (TAK1) kinase adaptor, TAK1-binding protein 2, plays dual roles in TAK1 signaling by recruiting both an activator and an inhibitor of TAK1 kinase in tumor necrosis factor signaling pathway. J Biol Chem 2010;285:2333–9. https://doi.org/10.1074/jbc.M109.090522.

[68] Hamidi A, von Bulow V, Hamidi R, Winssinger N, Barluenga S, Heldin CH, Landström M. Polyubiquitination of transforming growth factor β (TGFβ)-associated kinase 1 mediates nuclear factor-κB activation in response to different inflammatory stimuli. J Biol Chem 2012;287:123–33. https://doi.org/10.1074/jbc.M111.285122.

[69] Sorrentino A, Thakur N, Grimsby S, Marcusson A, von Bulow V, Schuster N, Zhang S, Heldin CH, Landström M. The type I TGF-beta receptor engages TRAF6 to activate TAK1 in a receptor kinase-independent manner. Nat Cell Biol 2008;10:1199–207. https://doi.org/10.1038/ncb1780.

[70] Fruman DA, Rommel C. PI3K and cancer: lessons, challenges and opportunities. Nat Rev Drug Discov 2014;13:140–56. https://doi.org/10.1038/nrd4204.

[71] Bakin AV, Tomlinson AK, Bhowmick NA, Moses HL, Arteaga CL. Phosphatidylinositol 3-kinase function is required for transforming growth factor beta-mediated epithelial to mesenchymal transition and cell migration. J Biol Chem 2000;275:36803–10. https://doi.org/10.1074/jbc.M005912200.

[72] Yang WL, Wang J, Chan CH, Lee SW, Campos AD, Lamothe B, Hur L, Grabiner BC, Lin X, Darnay BG, Lin HK. The E3 ligase TRAF6 regulates Akt ubiquitination and activation. Science 2009;325:1134–8. https://doi.org/10.1126/science.1175065.

[73] Lamouille S, Derynck R. Cell size and invasion in TGF-beta-induced epithelial to mesenchymal transition is regulated by activation of the mTOR pathway. J Cell Biol 2007;178:437–51. https://doi.org/10.1083/jcb.200611146.

[74] Lamouille S, Connolly E, Smyth JW, Akhurst RJ, Derynck R. TGF-β-induced activation of mTOR complex 2 drives epithelial-mesenchymal transition and cell invasion. J Cell Sci 2012;125:1259–73. https://doi.org/10.1242/jcs.095299.

[75] Zhou BP, Deng J, Xia W, Xu J, Li YM, Gunduz M, Hung MC. Dual regulation of Snail by GSK-3beta-mediated phosphorylation in control of epithelial-mesenchymal transition. Nat Cell Biol 2004;6:931–40. https://doi.org/10.1038/ncb1173.

[76] Wang H, Fang R, Wang XF, Zhang F, Chen DY, Zhou B, Wang HS, Cai SH, Du J. Stabilization of Snail through AKT/GSK-3β signaling pathway is required for TNF-α-induced epithelial-mesenchymal transition in prostate cancer PC3 cells. Eur J Pharmacol 2013;714:48–55. https://doi.org/10.1016/j.ejphar.2013.05.046.

[77] Bhowmick NA, Ghiassi M, Bakin A, Aakre M, Lundquist CA, Engel ME, Arteaga CL, Moses HL. Transforming growth factor-beta1 mediates epithelial to mesenchymal transdifferentiation through a RhoA-dependent mechanism. Mol Biol Cell 2001;12:27–36. https://doi.org/10.1091/mbc.12.1.27.

[78] Tang LY, Heller M, Meng Z, Yu LR, Tang Y, Zhou M, Zhang YE. Transforming growth factor-β (TGF-β) directly activates the JAK1-STAT3 Axis to induce hepatic fibrosis in coordination with the SMAD pathway. J Biol Chem 2017;292:4302–12. https://doi.org/10.1074/jbc.M116.773085.

[79] Pierce Jr DF, Johnson MD, Matsui Y, Robinson SD, Gold LI, Purchio AF, Daniel CW, Hogan BL, Moses HL. Inhibition of mammary duct development but not alveolar outgrowth during pregnancy in transgenic mice expressing active TGF-beta 1. Genes Dev 1993;7:2308–17. https://doi.org/10.1101/gad.7.12a.2308.

[80] Hartsough MT, Mulder KM. Transforming growth factor-beta signaling in epithelial cells. Pharmacol Ther 1997;75:21–41. https://doi.org/10.1016/s0163-7258(97)00020-x.

[81] Lebrin F, Deckers M, Bertolino P, Ten Dijke P. TGF-beta receptor function in the endothelium. Cardiovasc Res 2005;65:599–608. https://doi.org/10.1016/j.cardiores.2004.10.036.

[82] Thomas DA, Massagué J. TGF-beta directly targets cytotoxic T cell functions during tumor evasion of immune surveillance. Cancer Cell 2005;8:369–80. https://doi.org/10.1016/j.ccr.2005.10.012.

[83] Hannon GJ, Beach D. p15INK4B is a potential effector of TGF-beta-induced cell cycle arrest. Nature 1994;371:257–61. https://doi.org/10.1038/371257a0.

[84] Datto MB, Li Y, Panus JF, Howe DJ, Xiong Y, Wang XF. Transforming growth factor beta induces the cyclin-dependent kinase inhibitor p21 through a p53-independent mechanism. Proc Natl Acad Sci USA 1995;92: 5545–9. https://doi.org/10.1073/pnas.92.12.5545.

[85] Reynisdóttir I, Polyak K, Iavarone A, Massagué J. Kip/Cip and Ink4 Cdk inhibitors cooperate to induce cell cycle arrest in response to TGF-beta. Genes Dev 1995;9:1831–45. https://doi.org/10.1101/gad.9.15.1831.

[86] Scandura JM, Boccuni P, Massagué J, Nimer SD. Transforming growth factor beta-induced cell cycle arrest of human hematopoietic cells requires p57KIP2 up-regulation. Proc Natl Acad Sci USA 2004;101:15231–6. https://doi.org/10.1073/pnas.0406771101.

[87] Seoane J, Le HV, Shen L, Anderson SA, Massagué J. Integration of Smad and forkhead pathways in the control of neuroepithelial and glioblastoma cell proliferation. Cell 2004;117:211–23. https://doi.org/10.1016/s0092-8674(04)00298-3.

[88] Feng XH, Lin X, Derynck R. Smad2, Smad3 and Smad4 cooperate with Sp1 to induce p15(Ink4B) transcription in response to TGF-beta. EMBO J 2000;19:5178–93. https://doi.org/10.1093/emboj/19.19.5178.

[89] Pardali K, Kurisaki A, Morén A, ten Dijke P, Kardassis D, Moustakas A. Role of Smad proteins and transcription factor Sp1 in p21(Waf1/Cip1) regulation by transforming growth factor-beta. J Biol Chem 2000;275:29244–56. https://doi.org/10.1074/jbc.M909467199.

[90] Chen CR, Kang Y, Siegel PM, Massagué J. E2F4/5 and p107 as Smad cofactors linking the TGFbeta receptor to c-myc repression. Cell 2002;110:19–32. https://doi.org/10.1016/s0092-8674(02)00801-2.

[91] Staller P, Peukert K, Kiermaier A, Seoane J, Lukas J, Karsunky H, Möröy T, Bartek J, Massagué J, Hänel F, Eilers M. Repression of p15INK4b expression by Myc through association with Miz-1. Nat Cell Biol 2001;3:392–9. https://doi.org/10.1038/35070076.

[92] Seoane J, Le HV, Massagué J. Myc suppression of the p21(Cip1) Cdk inhibitor influences the outcome of the p53 response to DNA damage. Nature 2002;419:729–34. https://doi.org/10.1038/nature01119.

[93] Roschger C, Cabrele C. The Id-protein family in developmental and cancer-associated pathways. Cell Commun Signal 2017;15:7. https://doi.org/10.1186/s12964-016-0161-y.

[94] Kang Y, Chen CR, Massagué J. A self-enabling TGFbeta response coupled to stress signaling: Smad engages stress response factor ATF3 for Id1 repression in epithelial cells. Mol Cell 2003;11:915–26. https://doi.org/10.1016/s1097-2765(03)00109-6.

[95] Oberhammer FA, Pavelka M, Sharma S, Tiefenbacher R, Purchio AF, Bursch W, Schulte-Hermann R. Induction of apoptosis in cultured hepatocytes and in regressing liver by transforming growth factor beta 1. Proc Natl Acad Sci USA 1992;89:5408–12. https://doi.org/10.1073/pnas.89.12.5408.

[96] Schuster N, Krieglstein K. Mechanisms of TGF-beta-mediated apoptosis. Cell Tissue Res 2002;307:1–14. https://doi.org/10.1007/s00441-001-0479-6.

[97] Spender LC, O'Brien DI, Simpson D, Dutt D, Gregory CD, Allday MJ, Clark LJ, Inman GJ. TGF-beta induces apoptosis in human B cells by transcriptional regulation of BIK and BCL-XL. Cell Death Differ 2009;16: 593–602. https://doi.org/10.1038/cdd.2008.183.

[98] Tachibana I, Imoto M, Adjei PN, Gores GJ, Subramaniam M, Spelsberg TC, Urrutia R. Overexpression of the TGFbeta-regulated zinc finger encoding gene, TIEG, induces apoptosis in pancreatic epithelial cells. J Clin Invest 1997;99:2365–74. https://doi.org/10.1172/jci119418.

[99] Chalaux E, López-Rovira T, Rosa JL, Pons G, Boxer LM, Bartrons R, Ventura F. A zinc-finger transcription factor induced by TGF-beta promotes apoptotic cell death in epithelial Mv1Lu cells. FEBS Lett 1999;457:478–82. https://doi.org/10.1016/s0014-5793(99)01051-0.

[100] Jang CW, Chen CH, Chen CC, Chen JY, Su YH, Chen RH. TGF-beta induces apoptosis through Smad-mediated expression of DAP-kinase. Nat Cell Biol 2002;4:51–8. https://doi.org/10.1038/ncb731.

[101] Aman MJ, Lamkin TD, Okada H, Kurosaki T, Ravichandran KS. The inositol phosphatase SHIP inhibits Akt/PKB activation in B cells. J Biol Chem 1998;273:33922–8. https://doi.org/10.1074/jbc.273.51.33922.

[102] Valderrama-Carvajal H, Cocolakis E, Lacerte A, Lee EH, Krystal G, Ali S, Lebrun JJ. Activin/TGF-beta induce apoptosis through Smad-dependent expression of the lipid phosphatase SHIP. Nat Cell Biol 2002;4:963–9. https://doi.org/10.1038/ncb885.

[103] David CJ, Huang YH, Chen M, Su J, Zou Y, Bardeesy N, Iacobuzio-Donahue CA, Massagué J. TGF-β tumor suppression through a lethal EMT. Cell 2016;164:1015–30. https://doi.org/10.1016/j.cell.2016.01.009.

[104] Hanahan D, Weinberg RA. Hallmarks of cancer: the next generation. Cell 2011;144:646–74. https://doi.org/10.1016/j.cell.2011.02.013.

[105] Pastushenko I, Blanpain C. EMT transition states during tumor progression and metastasis. Trends Cell Biol 2019;29:212–26. https://doi.org/10.1016/j.tcb.2018.12.001.

[106] Shi X, Yang J, Deng S, Xu H, Wu D, Zeng Q, Wang S, Hu T, Wu F, Zhou H. TGF-β signaling in the tumor metabolic microenvironment and targeted therapies. J Hematol Oncol 2022;15:135. https://doi.org/10.1186/s13045-022-01349-6.

[107] Kang SH, Bang YJ, Im YH, Yang HK, Lee DA, Lee HY, Lee HS, Kim NK, Kim SJ. Transcriptional repression of the transforming growth factor-beta type I receptor gene by DNA methylation results in the development of TGF-beta resistance in human gastric cancer. Oncogene 1999;18:7280–6. https://doi.org/10.1038/sj.onc.1203146.

[108] Pinto M, Oliveira C, Cirnes L, Carlos Machado J, Ramires M, Nogueira A, Carneiro F, Seruca R. Promoter methylation of TGFbeta receptor I and mutation of TGFbeta receptor II are frequent events in MSI sporadic gastric carcinomas. J Pathol 2003;200:32–8. https://doi.org/10.1002/path.1327.

[109] Goulet CR, Pouliot F. TGFβ signaling in the tumor microenvironment. Adv Exp Med Biol 2021;1270:89–105. https://doi.org/10.1007/978-3-030-47189-7_6.

[110] Huber MA, Kraut N, Beug H. Molecular requirements for epithelial-mesenchymal transition during tumor progression. Curr Opin Cell Biol 2005;17:548–58. https://doi.org/10.1016/j.ceb.2005.08.001.

[111] Thiery JP, Acloque H, Huang RY, Nieto MA. Epithelial-mesenchymal transitions in development and disease. Cell 2009;139:871–90. https://doi.org/10.1016/j.cell.2009.11.007.

[112] Lindley LE, Briegel KJ. Molecular characterization of TGFbeta-induced epithelial-mesenchymal transition in normal finite lifespan human mammary epithelial cells. Biochem Biophys Res Commun 2010;399:659–64. https://doi.org/10.1016/j.bbrc.2010.07.138.

[113] Mani SA, Guo W, Liao MJ, Eaton EN, Ayyanan A, Zhou AY, Brooks M, Reinhard F, Zhang CC, Shipitsin M, et al. The epithelial-mesenchymal transition generates cells with properties of stem cells. Cell 2008;133:704–15. https://doi.org/10.1016/j.cell.2008.03.027.

[114] Ocaña OH, Córcoles R, Fabra A, Moreno-Bueno G, Acloque H, Vega S, Barrallo-Gimeno A, Cano A, Nieto MA. Metastatic colonization requires the repression of the epithelial-mesenchymal transition inducer Prrx1. Cancer Cell 2012;22:709–24. https://doi.org/10.1016/j.ccr.2012.10.012.

[115] Lamouille S, Xu J, Derynck R. Molecular mechanisms of epithelial-mesenchymal transition. Nat Rev Mol Cell Biol 2014;15:178–96. https://doi.org/10.1038/nrm3758.

[116] Zohn IE, Li Y, Skolnik EY, Anderson KV, Han J, Niswander L. p38 and a p38-interacting protein are critical for downregulation of E-cadherin during mouse gastrulation. Cell 2006;125:957–69. https://doi.org/10.1016/j.cell.2006.03.048.

[117] Peinado H, Olmeda D, Cano A. Snail, Zeb and bHLH factors in tumour progression: an alliance against the epithelial phenotype? Nat Rev Cancer 2007;7:415–28. https://doi.org/10.1038/nrc2131.

[118] Wang X, Zheng M, Liu G, Xia W, McKeown-Longo PJ, Hung MC, Zhao J. Krüppel-like factor 8 induces epithelial to mesenchymal transition and epithelial cell invasion. Cancer Res 2007;67:7184–93. https://doi.org/10.1158/0008-5472.Can-06-4729.

[119] Nieto MA, Huang RY, Jackson RA, Thiery JP. EMT: 2016. Cell 2016;166:21–45. https://doi.org/10.1016/j.cell.2016.06.028.

[120] Brabletz T, Kalluri R, Nieto MA, Weinberg RA. EMT in cancer. Nat Rev Cancer 2018;18:128–34. https://doi.org/10.1038/nrc.2017.118.

[121] Guido C, Whitaker-Menezes D, Capparelli C, Balliet R, Lin Z, Pestell RG, Howell A, Aquila S, Andò S, Martinez-Outschoorn U, et al. Metabolic reprogramming of cancer-associated fibroblasts by TGF-β drives tumor growth: connecting TGF-β signaling with "Warburg-like" cancer metabolism and L-lactate production. Cell Cycle 2012;11:3019–35. https://doi.org/10.4161/cc.21384.

[122] Weinhouse S. Glycolysis, respiration, and anomalous gene expression in experimental hepatomas: G.H.A. Clowes memorial lecture. Cancer Res 1972;32:2007–16.

[123] Vander Heiden MG, Cantley LC, Thompson CB. Understanding the Warburg effect: the metabolic requirements of cell proliferation. Science 2009;324:1029–33. https://doi.org/10.1126/science.1160809.

[124] Chae HS, Hong ST. Overview of cancer metabolism and signaling transduction. Int J Mol Sci 2022;24. https://doi.org/10.3390/ijms24010012.

[125] Li W, Wei Z, Liu Y, Li H, Ren R, Tang Y. Increased 18F-FDG uptake and expression of Glut1 in the EMT transformed breast cancer cells induced by TGF-beta. Neoplasma 2010;57:234–40. https://doi.org/10.4149/neo_2010_03_234.

[126] Masin M, Vazquez J, Rossi S, Groeneveld S, Samson N, Schwalie PC, Deplancke B, Frawley LE, Gouttenoire J, Moradpour D, et al. GLUT3 is induced during epithelial-mesenchymal transition and promotes tumor cell proliferation in non-small cell lung cancer. Cancer Metab 2014;2:11. https://doi.org/10.1186/2049-3002-2-11.

[127] Liu M, Hancock SE, Sultani G, Wilkins BP, Ding E, Osborne B, Quek LE, Turner N. Snail-overexpression induces epithelial-mesenchymal transition and metabolic reprogramming in human pancreatic ductal adenocarcinoma and non-tumorigenic ductal cells. J Clin Med 2019;8. https://doi.org/10.3390/jcm8060822.

[128] Hamabe A, Konno M, Tanuma N, Shima H, Tsunekuni K, Kawamoto K, Nishida N, Koseki J, Mimori K, Gotoh N, et al. Role of pyruvate kinase M2 in transcriptional regulation leading to epithelial-mesenchymal transition. Proc Natl Acad Sci USA 2014;111:15526–31. https://doi.org/10.1073/pnas.1407717111.

[129] Rodríguez-García A, Samsó P, Fontova P, Simon-Molas H, Manzano A, Castaño E, Rosa JL, Martinez-Outshoorn U, Ventura F, Navarro-Sabaté À, Bartrons R. TGF-β1 targets Smad, p38 MAPK, and PI3K/Akt signaling pathways to induce PFKFB3 gene expression and glycolysis in glioblastoma cells. FEBS J 2017;284:3437–54. https://doi.org/10.1111/febs.14201.

[130] Yin X, Choudhury M, Kang JH, Schaefbauer KJ, Jung MY, Andrianifahanana M, Hernandez DM, Leof EB. Hexokinase 2 couples glycolysis with the profibrotic actions of TGF-β. Sci Signal 2019;12. https://doi.org/10.1126/scisignal.aax4067.

[131] Zahra K, Dey T, Ashish, Mishra SP, Pandey U. Pyruvate kinase M2 and cancer: the role of PKM2 in promoting tumorigenesis. Front Oncol 2020;10:159. https://doi.org/10.3389/fonc.2020.00159.

[132] Baumann F, Leukel P, Doerfelt A, Beier CP, Dettmer K, Oefner PJ, Kastenberger M, Kreutz M, Nickl-Jockschat T, Bogdahn U, et al. Lactate promotes glioma migration by TGF-beta2-dependent regulation of matrix metalloproteinase-2. Neuro-Oncology 2009;11:368–80. https://doi.org/10.1215/15228517-2008-106.

[133] Li X, Zhang Z, Zhang Y, Cao Y, Wei H, Wu Z. Upregulation of lactate-inducible snail protein suppresses oncogene-mediated senescence through p16(INK4a) inactivation. J Exp Clin Cancer Res 2018;37:39. https://doi.org/10.1186/s13046-018-0701-y.

[134] Chen YG. Endocytic regulation of TGF-beta signaling. Cell Res 2009;19:58–70. https://doi.org/10.1038/cr.2008.315.

[135] Chen M, Zhao Y, Yang X, Zhao Y, Liu Q, Liu Y, Hou Y, Sun H, Jin W. NSDHL promotes triple-negative breast cancer metastasis through the TGFβ signaling pathway and cholesterol biosynthesis. Breast Cancer Res Treat 2021;187:349–62. https://doi.org/10.1007/s10549-021-06213-8.

[136] Zhao Z, Hao D, Wang L, Li J, Meng Y, Li P, Wang Y, Zhang C, Zhou H, Gardner K, Di LJ. CtBP promotes metastasis of breast cancer through repressing cholesterol and activating TGF-β signaling. Oncogene 2019;38:2076–91. https://doi.org/10.1038/s41388-018-0570-z.

[137] Mullen GE, Yet L. Progress in the development of fatty acid synthase inhibitors as anticancer targets. Bioorg Med Chem Lett 2015;25:4363–9. https://doi.org/10.1016/j.bmcl.2015.08.087.
[138] Liu H, Chen YG. The interplay between TGF-β signaling and cell metabolism. Front Cell Dev Biol 2022;10, 846723. https://doi.org/10.3389/fcell.2022.846723.
[139] Nakasuka F, Tabata S, Sakamoto T, Hirayama A, Ebi H, Yamada T, Umetsu K, Ohishi M, Ueno A, Goto H, et al. TGF-β-dependent reprogramming of amino acid metabolism induces epithelial-mesenchymal transition in non-small cell lung cancers. Commun Biol 2021;4:782. https://doi.org/10.1038/s42003-021-02323-7.
[140] Pavlides S, Whitaker-Menezes D, Castello-Cros R, Flomenberg N, Witkiewicz AK, Frank PG, Casimiro MC, Wang C, Fortina P, Addya S, et al. The reverse Warburg effect: aerobic glycolysis in cancer associated fibroblasts and the tumor stroma. Cell Cycle 2009;8:3984–4001. https://doi.org/10.4161/cc.8.23.10238.
[141] Avagliano A, Granato G, Ruocco MR, Romano V, Belviso I, Carfora A, Montagnani S, Arcucci A. Metabolic reprogramming of cancer associated fibroblasts: the slavery of stromal fibroblasts. Biomed Res Int 2018;2018:6075403. https://doi.org/10.1155/2018/6075403.
[142] Cong J. Metabolism of natural killer cells and other innate lymphoid cells. Front Immunol 2020;11:1989. https://doi.org/10.3389/fimmu.2020.01989.
[143] Anderson NM, Simon MC. The tumor microenvironment. Curr Biol 2020;30:R921–r925. https://doi.org/10.1016/j.cub.2020.06.081.
[144] Barrett RL, Puré E. Cancer-associated fibroblasts and their influence on tumor immunity and immunotherapy. elife 2020;9. https://doi.org/10.7554/eLife.57243.
[145] Sahai E, Astsaturov I, Cukierman E, DeNardo DG, Egeblad M, Evans RM, Fearon D, Greten FR, Hingorani SR, Hunter T, et al. A framework for advancing our understanding of cancer-associated fibroblasts. Nat Rev Cancer 2020;20:174–86. https://doi.org/10.1038/s41568-019-0238-1.
[146] Darby IA, Hewitson TD. Fibroblast differentiation in wound healing and fibrosis. Int Rev Cytol 2007;257: 143–79. https://doi.org/10.1016/s0074-7696(07)57004-x.
[147] Ringuette Goulet C, Bernard G, Tremblay S, Chabaud S, Bolduc S, Pouliot F. Exosomes induce fibroblast differentiation into cancer-associated fibroblasts through TGFβ signaling. Mol Cancer Res 2018;16:1196–204. https://doi.org/10.1158/1541-7786.Mcr-17-0784.
[148] Yeon JH, Jeong HE, Seo H, Cho S, Kim K, Na D, Chung S, Park J, Choi N, Kang JY. Cancer-derived exosomes trigger endothelial to mesenchymal transition followed by the induction of cancer-associated fibroblasts. Acta Biomater 2018;76:146–53. https://doi.org/10.1016/j.actbio.2018.07.001.
[149] Ping Q, Yan R, Cheng X, Wang W, Zhong Y, Hou Z, Shi Y, Wang C, Li R. Cancer-associated fibroblasts: overview, progress, challenges, and directions. Cancer Gene Ther 2021;28:984–99. https://doi.org/10.1038/s41417-021-00318-4.
[150] Calon A, Lonardo E, Berenguer-Llergo A, Espinet E, Hernando-Momblona X, Iglesias M, Sevillano M, Palomo-Ponce S, Tauriello DV, Byrom D, et al. Stromal gene expression defines poor-prognosis subtypes in colorectal cancer. Nat Genet 2015;47:320–9. https://doi.org/10.1038/ng.3225.
[151] Kojima Y, Acar A, Eaton EN, Mellody KT, Scheel C, Ben-Porath I, Onder TT, Wang ZC, Richardson AL, Weinberg RA, Orimo A. Autocrine TGF-beta and stromal cell-derived factor-1 (SDF-1) signaling drives the evolution of tumor-promoting mammary stromal myofibroblasts. Proc Natl Acad Sci USA 2010;107:20009–14. https://doi.org/10.1073/pnas.1013805107.
[152] Tang X, Tu G, Yang G, Wang X, Kang L, Yang L, Zeng H, Wan X, Qiao Y, Cui X, et al. Autocrine TGF-β1/miR-200s/miR-221/DNMT3B regulatory loop maintains CAF status to fuel breast cancer cell proliferation. Cancer Lett 2019;452:79–89. https://doi.org/10.1016/j.canlet.2019.02.044.
[153] Wu YS, Chung I, Wong WF, Masamune A, Sim MS, Looi CY. Paracrine IL-6 signaling mediates the effects of pancreatic stellate cells on epithelial-mesenchymal transition via Stat3/Nrf2 pathway in pancreatic cancer cells. Biochim Biophys Acta Gen Subj 2017;1861:296–306. https://doi.org/10.1016/j.bbagen.2016.10.006.
[154] Wang L, Cao L, Wang H, Liu B, Zhang Q, Meng Z, Wu X, Zhou Q, Xu K. Cancer-associated fibroblasts enhance metastatic potential of lung cancer cells through IL-6/STAT3 signaling pathway. Oncotarget 2017;8:76116–28. https://doi.org/10.18632/oncotarget.18814.
[155] Wu X, Tao P, Zhou Q, Li J, Yu Z, Wang X, Li J, Li C, Yan M, Zhu Z, et al. IL-6 secreted by cancer-associated fibroblasts promotes epithelial-mesenchymal transition and metastasis of gastric cancer via JAK2/STAT3 signaling pathway. Oncotarget 2017;8:20741–50. https://doi.org/10.18632/oncotarget.15119.
[156] Goulet CR, Champagne A, Bernard G, Vandal D, Chabaud S, Pouliot F, Bolduc S. Cancer-associated fibroblasts induce epithelial-mesenchymal transition of bladder cancer cells through paracrine IL-6 signalling. BMC Cancer 2019;19:137. https://doi.org/10.1186/s12885-019-5353-6.

[157] Brabletz T, Pfeuffer I, Schorr E, Siebelt F, Wirth T, Serfling E. Transforming growth factor beta and cyclosporin A inhibit the inducible activity of the interleukin-2 gene in T cells through a noncanonical octamer-binding site. Mol Cell Biol 1993;13:1155–62. https://doi.org/10.1128/mcb.13.2.1155-1162.1993.

[158] McKarns SC, Schwartz RH, Kaminski NE. Smad3 is essential for TGF-beta 1 to suppress IL-2 production and TCR-induced proliferation, but not IL-2-induced proliferation. J Immunol 2004;172:4275–84. https://doi.org/10.4049/jimmunol.172.7.4275.

[159] Cottrez F, Groux H. Regulation of TGF-beta response during T cell activation is modulated by IL-10. J Immunol 2001;167:773–8. https://doi.org/10.4049/jimmunol.167.2.773.

[160] Gorelik L, Constant S, Flavell RA. Mechanism of transforming growth factor beta-induced inhibition of T helper type 1 differentiation. J Exp Med 2002;195:1499–505. https://doi.org/10.1084/jem.20012076.

[161] Lin JT, Martin SL, Xia L, Gorham JD. TGF-beta 1 uses distinct mechanisms to inhibit IFN-gamma expression in CD4+ T cells at priming and at recall: differential involvement of Stat4 and T-bet. J Immunol 2005;174:5950–8. https://doi.org/10.4049/jimmunol.174.10.5950.

[162] Gorelik L, Fields PE, Flavell RA. Cutting edge: TGF-beta inhibits Th type 2 development through inhibition of GATA-3 expression. J Immunol 2000;165:4773–7. https://doi.org/10.4049/jimmunol.165.9.4773.

[163] Nakamura K, Kitani A, Strober W. Cell contact-dependent immunosuppression by CD4(+)CD25(+) regulatory T cells is mediated by cell surface-bound transforming growth factor beta. J Exp Med 2001;194:629–44. https://doi.org/10.1084/jem.194.5.629.

[164] Gunderson AJ, Yamazaki T, McCarty K, Fox N, Phillips M, Alice A, Blair T, Whiteford M, O'Brien D, Ahmad R, et al. TGFβ suppresses CD8(+) T cell expression of CXCR3 and tumor trafficking. Nat Commun 2020;11:1749. https://doi.org/10.1038/s41467-020-15404-8.

[165] Ghiringhelli F, Ménard C, Terme M, Flament C, Taieb J, Chaput N, Puig PE, Novault S, Escudier B, Vivier E, et al. CD4+CD25+ regulatory T cells inhibit natural killer cell functions in a transforming growth factor-beta-dependent manner. J Exp Med 2005;202:1075–85. https://doi.org/10.1084/jem.20051511.

[166] Bauer S, Groh V, Wu J, Steinle A, Phillips JH, Lanier LL, Spies T. Activation of NK cells and T cells by NKG2D, a receptor for stress-inducible MICA. Science 1999;285:727–9. https://doi.org/10.1126/science.285.5428.727.

[167] Yang L, Pang Y, Moses HL. TGF-beta and immune cells: an important regulatory axis in the tumor microenvironment and progression. Trends Immunol 2010;31:220–7. https://doi.org/10.1016/j.it.2010.04.002.

[168] Condeelis J, Pollard JW. Macrophages: obligate partners for tumor cell migration, invasion, and metastasis. Cell 2006;124:263–6. https://doi.org/10.1016/j.cell.2006.01.007.

[169] He Z, Zhang S. Tumor-associated macrophages and their functional transformation in the hypoxic tumor microenvironment. Front Immunol 2021;12, 741305. https://doi.org/10.3389/fimmu.2021.741305.

[170] Park JE, Dutta B, Tse SW, Gupta N, Tan CF, Low JK, Yeoh KW, Kon OL, Tam JP, Sze SK. Hypoxia-induced tumor exosomes promote M2-like macrophage polarization of infiltrating myeloid cells and microRNA-mediated metabolic shift. Oncogene 2019;38:5158–73. https://doi.org/10.1038/s41388-019-0782-x.

[171] Noy R, Pollard JW. Tumor-associated macrophages: from mechanisms to therapy. Immunity 2014;41:49–61. https://doi.org/10.1016/j.immuni.2014.06.010.

[172] Masucci MT, Minopoli M, Carriero MV. Tumor associated neutrophils. Their role in tumorigenesis, metastasis, prognosis and therapy. Front. Oncologia 2019;9:1146. https://doi.org/10.3389/fonc.2019.01146.

[173] Gregory AD, Houghton AM. Tumor-associated neutrophils: new targets for cancer therapy. Cancer Res 2011;71:2411–6. https://doi.org/10.1158/0008-5472.Can-10-2583.

[174] Kim BG, Malek E, Choi SH, Ignatz-Hoover JJ, Driscoll JJ. Novel therapies emerging in oncology to target the TGF-β pathway. J Hematol Oncol 2021;14:55. https://doi.org/10.1186/s13045-021-01053-x.

[175] Herbertz S, Sawyer JS, Stauber AJ, Gueorguieva I, Driscoll KE, Estrem ST, Cleverly AL, Desaiah D, Guba SC, Benhadji KA, et al. Clinical development of galunisertib (LY2157299 monohydrate), a small molecule inhibitor of transforming growth factor-beta signaling pathway. Drug Des Devel Ther 2015;9:4479–99. https://doi.org/10.2147/dddt.S86621.

[176] Yingling JM, McMillen WT, Yan L, Huang H, Sawyer JS, Graff J, Clawson DK, Britt KS, Anderson BD, Beight DW, et al. Preclinical assessment of galunisertib (LY2157299 monohydrate), a first-in-class transforming growth factor-β receptor type I inhibitor. Oncotarget 2018;9:6659–77. https://doi.org/10.18632/oncotarget.23795.

[177] Anderton MJ, Mellor HR, Bell A, Sadler C, Pass M, Powell S, Steele SJ, Roberts RR, Heier A. Induction of heart valve lesions by small-molecule ALK5 inhibitors. Toxicol Pathol 2011;39:916–24. https://doi.org/10.1177/0192623311416259.

[178] Kovacs RJ, Maldonado G, Azaro A, Fernández MS, Romero FL, Sepulveda-Sánchez JM, Corretti M, Carducci M, Dolan M, Gueorguieva I, et al. Cardiac safety of TGF-β receptor I kinase inhibitor LY2157299 monohydrate in

cancer patients in a first-in-human dose study. Cardiovasc Toxicol 2015;15:309–23. https://doi.org/10.1007/s12012-014-9297-4.

[179] Melisi D, Garcia-Carbonero R, Macarulla T, Pezet D, Deplanque G, Fuchs M, Trojan J, Oettle H, Kozloff M, Cleverly A, et al. Galunisertib plus gemcitabine vs. gemcitabine for first-line treatment of patients with unresectable pancreatic cancer. Br J Cancer 2018;119:1208–14. https://doi.org/10.1038/s41416-018-0246-z.

[180] Nadal E, Saleh M, Aix SP, Ochoa-de-Olza M, Patel SP, Antonia S, Zhao Y, Gueorguieva I, Man M, Estrem ST, et al. A phase Ib/II study of galunisertib in combination with nivolumab in solid tumors and non-small cell lung cancer. BMC Cancer 2023;23:708. https://doi.org/10.1186/s12885-023-11153-1.

[181] Son JY, Park SY, Kim SJ, Lee SJ, Park SA, Kim MJ, Kim SW, Kim DK, Nam JS, Sheen YY. EW-7197, a novel ALK-5 kinase inhibitor, potently inhibits breast to lung metastasis. Mol Cancer Ther 2014;13:1704–16. https://doi.org/10.1158/1535-7163.Mct-13-0903.

[182] Jung SY, Hwang S, Clarke JM, Bauer TM, Keedy VL, Lee H, Park N, Kim SJ, Lee JI. Pharmacokinetic characteristics of vactosertib, a new activin receptor-like kinase 5 inhibitor, in patients with advanced solid tumors in a first-in-human phase 1 study. Investig New Drugs 2020;38:812–20. https://doi.org/10.1007/s10637-019-00835-y.

[183] Lee JE, Lee P, Yoon YC, Han BS, Ko S, Park MS, Lee YJ, Kim SE, Cho YJ, Lim JH, et al. Vactosertib, TGF-β receptor I inhibitor, augments the sensitization of the anti-cancer activity of gemcitabine in pancreatic cancer. Biomed Pharmacother 2023;162, 114716. https://doi.org/10.1016/j.biopha.2023.114716.

[184] Pei H, Parthasarathy S, Joseph S, McMillen W, Xu X, Castaneda S, Inigo I, Britt K, Anderson B, Zhao G. LY3200882, a novel, highly selective TGFβRI small molecule inhibitor. Cancer Res 2017;77:955.

[185] Wojtowicz-Praga S. Reversal of tumor-induced immunosuppression by TGF-beta inhibitors. Investig New Drugs 2003;21:21–32. https://doi.org/10.1023/a:1022951824806.

[186] Ivanović V, Demajo M, Krtolica K, Krajnović M, Konstantinović M, Baltić V, Prtenjak G, Stojiljković B, Breberina M, Nesković-Konstantinović Z, et al. Elevated plasma TGF-beta1 levels correlate with decreased survival of metastatic breast cancer patients. Clin Chim Acta 2006;371:191–3. https://doi.org/10.1016/j.cca.2006.02.027.

[187] Dasch JR, Pace DR, Waegell W, Inenaga D, Ellingsworth L. Monoclonal antibodies recognizing transforming growth factor-beta. Bioactivity neutralization and transforming growth factor beta 2 affinity purification. J Immunol 1989;142:1536–41.

[188] Arteaga CL, Hurd SD, Winnier AR, Johnson MD, Fendly BM, Forbes JT. Anti-transforming growth factor (TGF)-beta antibodies inhibit breast cancer cell tumorigenicity and increase mouse spleen natural killer cell activity. Implications for a possible role of tumor cell/host TGF-beta interactions in human breast cancer progression. J Clin Invest 1993;92:2569–76. https://doi.org/10.1172/jci116871.

[189] Terabe M, Ambrosino E, Takaku S, O'Konek JJ, Venzon D, Lonning S, McPherson JM, Berzofsky JA. Synergistic enhancement of CD8+ T cell-mediated tumor vaccine efficacy by an anti-transforming growth factor-beta monoclonal antibody. Clin Cancer Res 2009;15:6560–9. https://doi.org/10.1158/1078-0432.Ccr-09-1066.

[190] Trachtman H, Fervenza FC, Gipson DS, Heering P, Jayne DR, Peters H, Rota S, Remuzzi G, Rump LC, Sellin LK, et al. A phase 1, single-dose study of fresolimumab, an anti-TGF-β antibody, in treatment-resistant primary focal segmental glomerulosclerosis. Kidney Int 2011;79:1236–43. https://doi.org/10.1038/ki.2011.33.

[191] Bedinger D, Lao L, Khan S, Lee S, Takeuchi T, Mirza AM. Development and characterization of human monoclonal antibodies that neutralize multiple TGFβ isoforms. MAbs 2016;8:389–404. https://doi.org/10.1080/19420862.2015.1115166.

[192] Nam JS, Terabe M, Mamura M, Kang MJ, Chae H, Stuelten C, Kohn E, Tang B, Sabzevari H, Anver MR, et al. An anti-transforming growth factor beta antibody suppresses metastasis via cooperative effects on multiple cell compartments. Cancer Res 2008;68:3835–43. https://doi.org/10.1158/0008-5472.Can-08-0215.

[193] Ganapathy V, Ge R, Grazioli A, Xie W, Banach-Petrosky W, Kang Y, Lonning S, McPherson J, Yingling JM, Biswas S, et al. Targeting the transforming growth factor-beta pathway inhibits human basal-like breast cancer metastasis. Mol Cancer 2010;9:122. https://doi.org/10.1186/1476-4598-9-122.

[194] Morris JC, Tan AR, Olencki TE, Shapiro GI, Dezube BJ, Reiss M, Hsu FJ, Berzofsky JA, Lawrence DP. Phase I study of GC1008 (fresolimumab): a human anti-transforming growth factor-beta (TGFβ) monoclonal antibody in patients with advanced malignant melanoma or renal cell carcinoma. PLoS One 2014;9, e90353. https://doi.org/10.1371/journal.pone.0090353.

[195] Williamson SK, Hodi FS, Johnson ML, Barve MA, Juric D, Baranda JC, Schneider RE, Bauer TM, Lin TT, Wang R. Safety, pharmacokinetic and pharmacodynamic results from dose escalation of SAR439459, a TGFβ inhibitor, as monotherapy or in combination with cemiplimab in a phase 1/1b study. Wolters Kluwer Health; 2021.

[196] Chaturvedi P, George V, Shrestha N, Wang M, Dee MJ, Zhu X, Liu B, Egan J, D'Eramo F, Spanoudis C, et al. Immunotherapeutic HCW9218 augments anti-tumor activity of chemotherapy via NK cell-mediated reduction of therapy-induced senescent cells. Mol Ther 2022;30:1171–87. https://doi.org/10.1016/j.ymthe.2022.01.025.

[197] Li L, Wen Q, Ding R. Therapeutic targeting of VEGF and/or TGF-β to enhance anti-PD-(L)1 therapy: the evidence from clinical trials. Front Oncol 2022;12, 905520. https://doi.org/10.3389/fonc.2022.905520.

[198] Min J, Huang Z, Pang X, Zhong T, Jin C, Chen N, Xia D, Zhang P, Wang Z, Xia Y. 486P AK130, a first-in-class Fc-mutant anti-TIGIT antibody fused with TGF-βRII protein, elicits potent anti-tumor efficacy in pre-clinical studies. Ann Oncol 2022;33:S762.

[199] Hwang L, Ng K, Wang W, Trieu VN. OT-101: an anti-TGF-beta-2 antisense-primed tumors to subsequent chemotherapies. American Society of Clinical Oncology; 2016.

[200] Mourskaia AA, Dong Z, Ng S, Banville M, Zwaagstra JC, O'Connor-McCourt MD, Siegel PM. Transforming growth factor-beta1 is the predominant isoform required for breast cancer cell outgrowth in bone. Oncogene 2009;28:1005–15. https://doi.org/10.1038/onc.2008.454.

[201] Zwaagstra JC, Sulea T, Baardsnes J, Lenferink AE, Collins C, Cantin C, Paul-Roc B, Grothe S, Hossain S, Richer LP, et al. Engineering and therapeutic application of single-chain bivalent TGF-β family traps. Mol Cancer Ther 2012;11:1477–87. https://doi.org/10.1158/1535-7163.Mct-12-0060.

[202] Qin T, Barron L, Xia L, Huang H, Villarreal MM, Zwaagstra J, Collins C, Yang J, Zwieb C, Kodali R, et al. A novel highly potent trivalent TGF-β receptor trap inhibits early-stage tumorigenesis and tumor cell invasion in murine Pten-deficient prostate glands. Oncotarget 2016;7:86087–102. https://doi.org/10.18632/oncotarget.13343.

[203] Rodríguez A, Yang C, Furutani E, García de Teresa B, Velázquez M, Filiatrault J, Sambel LA, Phan T, Flores-Guzmán P, Sánchez S, et al. Inhibition of TGFβ1 and TGFβ3 promotes hematopoiesis in Fanconi anemia. Exp Hematol 2021;93:70–84.e74. https://doi.org/10.1016/j.exphem.2020.11.002.

[204] Dias N, Stein CA. Antisense oligonucleotides: basic concepts and mechanisms. Mol Cancer Ther 2002;1:347–55.

[205] Gleave ME, Monia BP. Antisense therapy for cancer. Nat Rev Cancer 2005;5:468–79. https://doi.org/10.1038/nrc1631.

[206] Schlingensiepen KH, Schlingensiepen R, Steinbrecher A, Hau P, Bogdahn U, Fischer-Blass B, Jachimczak P. Targeted tumor therapy with the TGF-beta 2 antisense compound AP 12009. Cytokine Growth Factor Rev 2006;17:129–39. https://doi.org/10.1016/j.cytogfr.2005.09.002.

[207] Schlingensiepen KH, Jaschinski F, Lang SA, Moser C, Geissler EK, Schlitt HJ, Kielmanowicz M, Schneider A. Transforming growth factor-beta 2 gene silencing with trabedersen (AP 12009) in pancreatic cancer. Cancer Sci 2011;102:1193–200. https://doi.org/10.1111/j.1349-7006.2011.01917.x.

[208] Oettle H, Hilbig A, Seufferlein T, Tsianakas A, Luger T, Schmid R, Von Wichert G, Endlicher E, Garbe C, Kaehler K. Phase I/II study with trabedersen (AP 12009) monotherapy for the treatment of patients with advanced pancreatic cancer, malignant melanoma, and colorectal carcinoma. J Clin Oncol 2011;29:2513.

[209] Le BT, Raguraman P, Kosbar TR, Fletcher S, Wilton SD, Veedu RN. Antisense oligonucleotides targeting angiogenic factors as potential cancer therapeutics. Mol Ther Nucleic Acids 2019;14:142–57. https://doi.org/10.1016/j.omtn.2018.11.007.

[210] Busenhart P, Montalban-Arques A, Katkeviciute E, Morsy Y, Van Passen C, Hering L, Atrott K, Lang S, Garzon JFG, Naschberger E, et al. Inhibition of integrin αvβ6 sparks T-cell antitumor response and enhances immune checkpoint blockade therapy in colorectal cancer. J Immunother Cancer 2022;10. https://doi.org/10.1136/jitc-2021-003465.

[211] Dodagatta-Marri E, Ma HY, Liang B, Li J, Meyer DS, Chen SY, Sun KH, Ren X, Zivak B, Rosenblum MD, et al. Integrin αvβ8 on T cells suppresses anti-tumor immunity in multiple models and is a promising target for tumor immunotherapy. Cell Rep 2021;36, 109309. https://doi.org/10.1016/j.celrep.2021.109309.

[212] Cai H, Guo F, Wen S, Jin X, Wu H, Ren D. Overexpressed integrin alpha 2 inhibits the activation of the transforming growth factor β pathway in pancreatic cancer via the TFCP2-SMAD2 axis. J Exp Clin Cancer Res 2022;41:73. https://doi.org/10.1186/s13046-022-02286-5.

[213] de Streel G, Bertrand C, Chalon N, Liénart S, Bricard O, Lecomte S, Devreux J, Gaignage M, De Boeck G, Mariën L, et al. Selective inhibition of TGF-β1 produced by GARP-expressing Tregs overcomes resistance to PD-1/PD-L1 blockade in cancer. Nat Commun 2020;11:4545. https://doi.org/10.1038/s41467-020-17811-3.

[214] Jiang A, Qin Y, Springer TA. Loss of LRRC33-dependent TGFβ1 activation enhances antitumor immunity and checkpoint blockade therapy. Cancer Immunol Res 2022;10:453–67. https://doi.org/10.1158/2326-6066.Cir-21-0593.

[215] Fouad S, Wells OS, Hill MA, D'Angiolella V. Cullin ring ubiquitin ligases (CRLs) in cancer: responses to ionizing radiation (IR) treatment. Front Physiol 2019;10:1144. https://doi.org/10.3389/fphys.2019.01144.

[216] Zhou J, Dabiri Y, Gama-Brambila RA, Ghafoory S, Altinbay M, Mehrabi A, Golriz M, Blagojevic B, Reuter S, Han K, et al. pVHL-mediated SMAD3 degradation suppresses TGF-β signaling. J Cell Biol 2022;221. https://doi.org/10.1083/jcb.202012097.
[217] Tian H, Liu C, Yu J, Han J, Du J, Liang S, Wang W, Liu Q, Lian R, Zhu T, et al. PHF14 enhances DNA methylation of SMAD7 gene to promote TGF-β-driven lung adenocarcinoma metastasis. Cell Discov 2023;9:41. https://doi.org/10.1038/s41421-023-00528-0.
[218] Lin Z, Huang W, He Q, Li D, Wang Z, Feng Y, Liu D, Zhang T, Wang Y, Xie M, et al. FOXC1 promotes HCC proliferation and metastasis by upregulating DNMT3B to induce DNA hypermethylation of CTH promoter. J Exp Clin Cancer Res 2021;40:50. https://doi.org/10.1186/s13046-021-01829-6.
[219] Liu A, Yu C, Qiu C, Wu Q, Huang C, Li X, She X, Wan K, Liu L, Li M, et al. PRMT5 methylating SMAD4 activates TGF-β signaling and promotes colorectal cancer metastasis. Oncogene 2023;42:1572–84. https://doi.org/10.1038/s41388-023-02674-x.
[220] Rasco D, Tolcher A, Siu LL, Heinhuis K, Postel-Vinay S, Barbash O, Egger JL, Gorman S, Horner T, Dhar A. Abstract CT038: a phase I, open-label, dose-escalation study to investigate the safety, pharmacokinetics, pharmacodynamics, and clinical activity of GSK3326595 in subjects with solid tumors and non-Hodgkin's lymphoma. Cancer Res 2017;77, CT038.
[221] Yin Y, Liu S, Pu L, Luo J, Liu H, Wu W. Nintedanib prevents TGF-β2-induced epithelial-mesenchymal transition in retinal pigment epithelial cells. Biomed Pharmacother 2023;161, 114543. https://doi.org/10.1016/j.biopha.2023.114543.
[222] Watanabe S, Lei M, Nakagawa E, Takeshita E, Inamori KI, Shishido F, Sasaki M, Mitsuhashi S, Matsumoto N, Kimura Y, et al. Neurological insights on two siblings with GM3 synthase deficiency due to novel compound heterozygous ST3GAL5 variants. Brain Dev 2023;45:270–7. https://doi.org/10.1016/j.braindev.2023.01.002.
[223] Zhang J, van der Zon G, Ma J, Mei H, Cabukusta B, Agaser CC, Madunić K, Wuhrer M, Zhang T, Ten Dijke P. ST3GAL5-catalyzed gangliosides inhibit TGF-β-induced epithelial-mesenchymal transition via TβRI degradation. EMBO J 2023;42, e110553. https://doi.org/10.15252/embj.2021110553.
[224] LeBeau AM, Brennen WN, Aggarwal S, Denmeade SR. Targeting the cancer stroma with a fibroblast activation protein-activated promelittin protoxin. Mol Cancer Ther 2009;8:1378–86. https://doi.org/10.1158/1535-7163.Mct-08-1170.
[225] Huang S, Fang R, Xu J, Qiu S, Zhang H, Du J, Cai S. Evaluation of the tumor targeting of a FAPα-based doxorubicin prodrug. J Drug Target 2011;19:487–96. https://doi.org/10.3109/1061186x.2010.511225.
[226] Brennen WN, Isaacs JT, Denmeade SR. Rationale behind targeting fibroblast activation protein-expressing carcinoma-associated fibroblasts as a novel chemotherapeutic strategy. Mol Cancer Ther 2012;11:257–66. https://doi.org/10.1158/1535-7163.Mct-11-0340.
[227] Bughda R, Dimou P, D'Souza RR, Klampatsa A. Fibroblast activation protein (FAP)-targeted CAR-T cells: launching an attack on tumor stroma. Immunotargets Ther 2021;10:313–23. https://doi.org/10.2147/itt.S291767.
[228] Shahvali S, Rahiman N, Jaafari MR, Arabi L. Targeting fibroblast activation protein (FAP): advances in CAR-T cell, antibody, and vaccine in cancer immunotherapy. Drug Deliv Transl Res 2023;13:2041–56. https://doi.org/10.1007/s13346-023-01308-9.
[229] Dumontet C, Jordan MA. Microtubule-binding agents: a dynamic field of cancer therapeutics. Nat Rev Drug Discov 2010;9:790–803. https://doi.org/10.1038/nrd3253.
[230] Wang X, Gigant B, Zheng X, Chen Q. Microtubule-targeting agents for cancer treatment: seven binding sites and three strategies. MedComm Oncol 2023;2, e46.
[231] Loong HH, Yeo W. Microtubule-targeting agents in oncology and therapeutic potential in hepatocellular carcinoma. Onco Targets Ther 2014;7:575–85. https://doi.org/10.2147/ott.S46019.
[232] Komlodi-Pasztor E, Sackett D, Wilkerson J, Fojo T. Mitosis is not a key target of microtubule agents in patient tumors. Nat Rev Clin Oncol 2011;8:244–50. https://doi.org/10.1038/nrclinonc.2010.228.
[233] Kaul R, Risinger AL, Mooberry SL. Eribulin rapidly inhibits TGF-β-induced snail expression and can induce Slug expression in a Smad4-dependent manner. Br J Cancer 2019;121:611–21. https://doi.org/10.1038/s41416-019-0556-9.
[234] Kuznetsov G, Towle MJ, Cheng H, Kawamura T, TenDyke K, Liu D, Kishi Y, Yu MJ, Littlefield BA. Induction of morphological and biochemical apoptosis following prolonged mitotic blockage by halichondrin B macrocyclic ketone analog E7389. Cancer Res 2004;64:5760–6. https://doi.org/10.1158/0008-5472.Can-04-1169.
[235] Yoshida T, Ozawa Y, Kimura T, Sato Y, Kuznetsov G, Xu S, Uesugi M, Agoulnik S, Taylor N, Funahashi Y, Matsui J. Eribulin mesilate suppresses experimental metastasis of breast cancer cells by reversing phenotype

from epithelial-mesenchymal transition (EMT) to mesenchymal-epithelial transition (MET) states. Br J Cancer 2014;110:1497–505. https://doi.org/10.1038/bjc.2014.80.

[236] Horimoto Y, Tokuda E, Murakami F, Uomori T, Himuro T, Nakai K, Orihata G, Iijima K, Togo S, Shimizu H, Saito M. Analysis of circulating tumour cell and the epithelial mesenchymal transition (EMT) status during eribulin-based treatment in 22 patients with metastatic breast cancer: a pilot study. J Transl Med 2018;16:287. https://doi.org/10.1186/s12967-018-1663-8.

[237] Cha JH, Chan LC, Li CW, Hsu JL, Hung MC. Mechanisms controlling PD-L1 expression in cancer. Mol Cell 2019;76:359–70. https://doi.org/10.1016/j.molcel.2019.09.030.

[238] Patsoukis N, Brown J, Petkova V, Liu F, Li L, Boussiotis VA. Selective effects of PD-1 on Akt and Ras pathways regulate molecular components of the cell cycle and inhibit T cell proliferation. Sci Signal 2012;5:ra46. https://doi.org/10.1126/scisignal.2002796.

[239] Knudson KM, Hicks KC, Luo X, Chen JQ, Schlom J, Gameiro SR. M7824, a novel bifunctional anti-PD-L1/TGFβ trap fusion protein, promotes anti-tumor efficacy as monotherapy and in combination with vaccine. Onco Targets Ther 2018;7, e1426519. https://doi.org/10.1080/2162402x.2018.1426519.

[240] Seidel JA, Otsuka A, Kabashima K. Anti-PD-1 and anti-CTLA-4 therapies in cancer: mechanisms of action, efficacy, and limitations. Front Oncol 2018;8:86. https://doi.org/10.3389/fonc.2018.00086.

[241] Koyama S, Akbay EA, Li YY, Herter-Sprie GS, Buczkowski KA, Richards WG, Gandhi L, Redig AJ, Rodig SJ, Asahina H, et al. Adaptive resistance to therapeutic PD-1 blockade is associated with upregulation of alternative immune checkpoints. Nat Commun 2016;7:10501. https://doi.org/10.1038/ncomms10501.

[242] Jia K, He Y, Dziadziuszko R, Zhao S, Zhang X, Deng J, Wang H, Hirsch FR, Zhou C, Yu H, Zhang L. T cell immunoglobulin and mucin-domain containing-3 in non-small cell lung cancer. Transl Lung Cancer Res 2019;8:895–906. https://doi.org/10.21037/tlcr.2019.11.17.

[243] Chen W, Jin W, Hardegen N, Lei KJ, Li L, Marinos N, McGrady G, Wahl SM. Conversion of peripheral CD4+CD25- naive T cells to CD4+CD25+ regulatory T cells by TGF-beta induction of transcription factor Foxp3. J Exp Med 2003;198:1875–86. https://doi.org/10.1084/jem.20030152.

[244] Chen SY, Mamai O, Akhurst RJ. TGFβ: signaling blockade for cancer immunotherapy. Annu Rev Cancer Biol 2022;6:123–46. https://doi.org/10.1146/annurev-cancerbio-070620-103554.

[245] David JM, Dominguez C, McCampbell KK, Gulley JL, Schlom J, Palena C. A novel bifunctional anti-PD-L1/TGF-β trap fusion protein (M7824) efficiently reverts mesenchymalization of human lung cancer cells. Onco Targets Ther 2017;6, e1349589. https://doi.org/10.1080/2162402x.2017.1349589.

[246] Mariathasan S, Turley SJ, Nickles D, Castiglioni A, Yuen K, Wang Y, Kadel III EE, Koeppen H, Astarita JL, Cubas R, et al. TGFβ attenuates tumour response to PD-L1 blockade by contributing to exclusion of T cells. Nature 2018;554:544–8. https://doi.org/10.1038/nature25501.

[247] Jochems C, Tritsch SR, Pellom ST, Su Z, Soon-Shiong P, Wong HC, Gulley JL, Schlom J. Analyses of functions of an anti-PD-L1/TGFβR2 bispecific fusion protein (M7824). Oncotarget 2017;8:75217–31. https://doi.org/10.18632/oncotarget.20680.

[248] Khalili-Tanha G, Fiuji H, Gharib M, Moghbeli M, Khalili-Tanha N, Rahmani F, Shakour N, Maftooh M, Hassanian SM, Asgharzadeh F, et al. Dual targeting of TGF-β and PD-L1 inhibits tumor growth in TGF-β/PD-L1-driven colorectal carcinoma. Life Sci 2023;328, 121865. https://doi.org/10.1016/j.lfs.2023.121865.

[249] Strauss J, Gatti-Mays ME, Cho BC, Hill A, Salas S, McClay E, Redman JM, Sater HA, Donahue RN, Jochems C, et al. Bintrafusp alfa, a bifunctional fusion protein targeting TGF-β and PD-L1, in patients with human papillomavirus-associated malignancies. J Immunother Cancer 2020;8. https://doi.org/10.1136/jitc-2020-001395.

[250] Paz-Ares L, Kim TM, Vicente D, Felip E, Lee DH, Lee KH, Lin CC, Flor MJ, Di Nicola M, Alvarez RM, et al. Bintrafusp alfa, a bifunctional fusion protein targeting TGF-β and PD-L1, in second-line treatment of patients with NSCLC: results from an expansion cohort of a phase 1 trial. J Thorac Oncol 2020;15:1210–22. https://doi.org/10.1016/j.jtho.2020.03.003.

[251] Strauss J, Heery CR, Schlom J, Madan RA, Cao L, Kang Z, Lamping E, Marté JL, Donahue RN, Grenga I, et al. Phase I trial of M7824 (MSB0011359C), a bifunctional fusion protein targeting PD-L1 and TGFβ, in advanced solid tumors. Clin Cancer Res 2018;24:1287–95. https://doi.org/10.1158/1078-0432.Ccr-17-2653.

[252] Lan Y, Moustafa M, Knoll M, Xu C, Furkel J, Lazorchak A, Yeung TL, Hasheminasab SM, Jenkins MH, Meister S, et al. Simultaneous targeting of TGF-β/PD-L1 synergizes with radiotherapy by reprogramming the tumor microenvironment to overcome immune evasion. Cancer Cell 2021;39:1388–1403.e1310. https://doi.org/10.1016/j.ccell.2021.08.008.

[253] Burvenich IJG, Goh YW, Guo N, Gan HK, Rigopoulos A, Cao D, Liu Z, Ackermann U, Wichmann CW, McDonald AF, et al. Radiolabelling and preclinical characterization of (89)Zr-Df-radiolabelled bispecific anti-PD-L1/TGF-βRII fusion protein bintrafusp alfa. Eur J Nucl Med Mol Imaging 2021;48:3075–88. https://doi.org/10.1007/s00259-021-05251-0.

[254] Lan Y, Zhang D, Xu C, Hance KW, Marelli B, Qi J, Yu H, Qin G, Sircar A, Hernández VM, et al. Enhanced preclinical antitumor activity of M7824, a bifunctional fusion protein simultaneously targeting PD-L1 and TGF-β. Sci Transl Med 2018;10. https://doi.org/10.1126/scitranslmed.aan5488.

[255] Yoo C, Javle MM, Verdaguer Mata H, de Braud F, Trojan J, Raoul JL, Kim JW, Ueno M, Lee CK, Hijioka S, et al. Phase 2 trial of bintrafusp alfa as second-line therapy for patients with locally advanced/metastatic biliary tract cancers. Hepatology 2023;78:758–70. https://doi.org/10.1097/hep.0000000000000365.

[256] Seitz CM, Mittelstaet J, Atar D, Hau J, Reiter S, Illi C, Kieble V, Engert F, Drees B, Bender G, et al. Novel adapter CAR-T cell technology for precisely controllable multiplex cancer targeting. Onco Targets Ther 2021;10, 2003532. https://doi.org/10.1080/2162402x.2021.2003532.

[257] Cohen AD, Garfall AL, Stadtmauer EA, Melenhorst JJ, Lacey SF, Lancaster E, Vogl DT, Weiss BM, Dengel K, Nelson A, et al. B cell maturation antigen-specific CAR T cells are clinically active in multiple myeloma. J Clin Invest 2019;129:2210–21. https://doi.org/10.1172/jci126397.

[258] Ying Z, Huang XF, Xiang X, Liu Y, Kang X, Song Y, Guo X, Liu H, Ding N, Zhang T, et al. A safe and potent anti-CD19 CAR T cell therapy. Nat Med 2019;25:947–53. https://doi.org/10.1038/s41591-019-0421-7.

[259] Werchau N, Kotter B, Criado-Moronati E, Gosselink A, Cordes N, Lock D, Lennartz S, Kolbe C, Winter N, Teppert K, et al. Combined targeting of soluble latent TGF-ß and a solid tumor-associated antigen with adapter CAR T cells. Onco Targets Ther 2022;11:2140534. https://doi.org/10.1080/2162402x.2022.2140534.

[260] Alishah K, Birtel M, Masoumi E, Jafarzadeh L, Mirzaee HR, Hadjati J, Voss RH, Diken M, Asad S. CRISPR/Cas9-mediated TGFβRII disruption enhances anti-tumor efficacy of human chimeric antigen receptor T cells in vitro. J Transl Med 2021;19:482. https://doi.org/10.1186/s12967-021-03146-0.

[261] Stüber T, Monjezi R, Wallstabe L, Kühnemundt J, Nietzer SL, Dandekar G, Wöckel A, Einsele H, Wischhusen J, Hudecek M. Inhibition of TGF-β-receptor signaling augments the antitumor function of ROR1-specific CAR T-cells against triple-negative breast cancer. J Immunother Cancer 2020;8. https://doi.org/10.1136/jitc-2020-000676.

[262] Kloss CC, Lee J, Zhang A, Chen F, Melenhorst JJ, Lacey SF, Maus MV, Fraietta JA, Zhao Y, June CH. Dominant-negative TGF-β receptor enhances PSMA-targeted human CAR T cell proliferation and augments prostate cancer eradication. Mol Ther 2018;26:1855–66. https://doi.org/10.1016/j.ymthe.2018.05.003.

[263] Alabanza LM, Xiong Y, Vu B, Webster B, Wu D, Hu P, Zhu Z, Dropulic B, Dash P, Schneider D. Armored BCMA CAR T cells eliminate multiple myeloma and are resistant to the suppressive effects of TGF-β. Front Immunol 2022;13, 832645. https://doi.org/10.3389/fimmu.2022.832645.

[264] Tang N, Cheng C, Zhang X, Qiao M, Li N, Mu W, Wei XF, Han W, Wang H. TGF-β inhibition via CRISPR promotes the long-term efficacy of CAR T cells against solid tumors. JCI Insight 2020;5. https://doi.org/10.1172/jci.insight.133977.

[265] Narayan V, Barber-Rotenberg JS, Jung IY, Lacey SF, Rech AJ, Davis MM, Hwang WT, Lal P, Carpenter EL, Maude SL, et al. PSMA-targeting TGFβ-insensitive armored CAR T cells in metastatic castration-resistant prostate cancer: a phase 1 trial. Nat Med 2022;28:724–34. https://doi.org/10.1038/s41591-022-01726-1.

[266] Chang ZL, Lorenzini MH, Chen X, Tran U, Bangayan NJ, Chen YY. Rewiring T-cell responses to soluble factors with chimeric antigen receptors. Nat Chem Biol 2018;14:317–24. https://doi.org/10.1038/nchembio.2565.

[267] Hou AJ, Chang ZL, Lorenzini MH, Zah E, Chen YY. TGF-β-responsive CAR-T cells promote anti-tumor immune function. Bioeng Transl Med 2018;3:75–86. https://doi.org/10.1002/btm2.10097.

[268] Lee DW, Gardner R, Porter DL, Louis CU, Ahmed N, Jensen M, Grupp SA, Mackall CL. Current concepts in the diagnosis and management of cytokine release syndrome. Blood 2014;124:188–95. https://doi.org/10.1182/blood-2014-05-552729.

[269] Gauthier J, Turtle CJ. Insights into cytokine release syndrome and neurotoxicity after CD19-specific CAR-T cell therapy. Curr Res Transl Med 2018;66:50–2. https://doi.org/10.1016/j.retram.2018.03.003.

[270] Brudno JN, Kochenderfer JN. Toxicities of chimeric antigen receptor T cells: recognition and management. Blood 2016;127:3321–30. https://doi.org/10.1182/blood-2016-04-703751.

[271] Shalabi H, Kraft IL, Wang HW, Yuan CM, Yates B, Delbrook C, Zimbelman JD, Giller R, Stetler-Stevenson M, Jaffe ES, et al. Sequential loss of tumor surface antigens following chimeric antigen receptor T-cell

therapies in diffuse large B-cell lymphoma. Haematologica 2018;103:e215–8. https://doi.org/10.3324/haematol.2017.183459.

[272] Ueda R, Fujita M, Zhu X, Sasaki K, Kastenhuber ER, Kohanbash G, McDonald HA, Harper J, Lonning S, Okada H. Systemic inhibition of transforming growth factor-beta in glioma-bearing mice improves the therapeutic efficacy of glioma-associated antigen peptide vaccines. Clin Cancer Res 2009;15:6551–9. https://doi.org/10.1158/1078-0432.Ccr-09-1067.

[273] Andersen MH. Immune regulation by self-recognition: novel possibilities for anticancer immunotherapy. J Natl Cancer Inst 2015;107. https://doi.org/10.1093/jnci/djv154.

[274] Andersen MH. Novel immune modulatory vaccines targeting TGFβ. Cell Mol Immunol 2023;20:551–3. https://doi.org/10.1038/s41423-023-01000-5.

[275] Holmström MO, Mortensen REJ, Pavlidis AM, Martinenaite E, Weis-Banke SE, Aaboe-Jørgensen M, Bendtsen SK, Met Ö, Pedersen AW, Donia M, et al. Cytotoxic T cells isolated from healthy donors and cancer patients kill TGFβ-expressing cancer cells in a TGFβ-dependent manner. Cell Mol Immunol 2021;18:415–26. https://doi.org/10.1038/s41423-020-00593-5.

[276] Mortensen REJ, Holmström MO, Andersen MH. Characterization of TGFβ-specific CD4(+)T cells through the modulation of TGFβ expression in malignant myeloid cells. Cell Mol Immunol 2021;18:2575–7. https://doi.org/10.1038/s41423-021-00770-0.

[277] Perez-Penco M, Weis-Banke SE, Schina A, Siersbæk M, Hübbe ML, Jørgensen MA, Lecoq I, Lara de la Torre L, Bendtsen SK, Martinenaite E, et al. TGFβ-derived immune modulatory vaccine: targeting the immunosuppressive and fibrotic tumor microenvironment in a murine model of pancreatic cancer. J Immunother Cancer 2022;10. https://doi.org/10.1136/jitc-2022-005491.

[278] Li S, Liu M, Do MH, Chou C, Stamatiades EG, Nixon BG, Shi W, Zhang X, Li P, Gao S, et al. Cancer immunotherapy via targeted TGF-β signalling blockade in T(H) cells. Nature 2020;587:121–5. https://doi.org/10.1038/s41586-020-2850-3.

[279] Xie F, Zhou X, Su P, Li H, Tu Y, Du J, Pan C, Wei X, Zheng M, Jin K, et al. Breast cancer cell-derived extracellular vesicles promote CD8(+)T cell exhaustion via TGF-β type II receptor signaling. Nat Commun 2022;13:4461. https://doi.org/10.1038/s41467-022-31250-2.

CHAPTER

5

Targeting immunometabolism to improve cancer therapies

Darby J. Ballard[a], *Liqing Wang*[a,b], *Jugal Kishore Das*[a], *Anil Kumar*[a], *Yijie Ren*[a], *and Jianxun Song*[a]

[a]Department of Microbial Pathogenesis and Immunology, Texas A&M University Health Science Center, Bryan, TX, United States [b]Department of Biochemistry and Biophysics, Texas A&M University, College Station, TX, United States

Abstract

Immunotherapy presents a promising approach to cancer treatment, yet its effectiveness is often constrained by several challenges. A novel strategy to overcome these limitations involves targeting the metabolic interplay within the tumor microenvironment (TME). By modulating metabolic pathways, researchers aim to bolster T cell function and improve antitumor immune responses. This approach includes enhancing T cell metabolism while simultaneously inhibiting the metabolic processes of tumor cells. Such metabolic interventions could potentially amplify the efficacy of existing immunotherapies. Advances in comprehending the intricate metabolic interactions between tumors and the immune system are crucial for unlocking new therapeutic possibilities in this burgeoning field of immunometabolism. Continued exploration in this area holds significant promise for advancing cancer treatment strategies and improving patient outcomes.

Abbreviations

ALL acute lymphoblastic leukemia
APC antigen-presenting cell
ATP adenosine triphosphate
BiTE bispecific T cell-engager
CAR chimeric antigen receptor
COX-2 cyclooxygenase-2
CD39 cluster of differentiation 39
CD4 cluster of differentiation 4
CD73 cluster of differentiation 73
CD8 cluster of differentiation 8
CRS cytokine release syndrome
CTLA-4 cytotoxic T-lymphocyte-associated protein 4

https://doi.org/10.1016/B978-0-443-13827-0.00018-4

CVID	common variable immunodeficiency
FAO	fatty acid oxidation
IBD	inflammatory bowel disease
ICI	immune checkpoint inhibitor
IDO	indoleamine 2,3-dioxygenase
IL-12	interleukin-12
IL-15	interleukin-15
IL-17	interleukin-17
IL-2	interleukin-2
IL-21	interleukin-21
IL-27	interleukin-27
IMR	immunometabolic regulation
irAE	immune-related adverse events
LDH	lactate dehydrogenase
MDSC	myeloid-derived suppressor cells
MS	multiple sclerosis
MSMD	Mendelian susceptibility to mycobacterial disease
NF-κB	nuclear factor-kappa B
NK	natural killer
NSCLC	non-small cell lung carcinoma
OXPHOS	oxidative phosphorylation
PD-1	programmed cell death protein-1
PD-L1	programmed death-ligand 1
RCC	renal cell carcinoma
ROS	reactive oxygen species
SLE	systemic lupus erythematosus
TCR	T cell-receptor
TGF-β	transforming growth factor-beta
Th1	T helper 1
Th17	T helper 17
TIL	tumor-infiltrating lymphocyte
TMB	tumor mutational burden
TME	tumor microenvironment
TNF-α	tumor necrosis factor-alpha
Treg	regulatory T cell

Conflict of interest

All authors have declared that no conflict of interest exists.

Cytokines in immunometabolism

Cytokines are signaling molecules that play a crucial role in immune system function. In recent years, there has been increasing interest in the role of these key regulators in immunometabolism and their impact on cancer immunotherapy. Some cytokines that have been implicated in immunometabolism and may play a role in improving cancer immunotherapy include:

Interleukin-2 (IL-2): Interleukin-2 (IL-2), a multifaceted cytokine crucial for immune homeostasis, controls many cellular processes, including immune cell activation, proliferation, and differentiation. At the core of its functionality lies the modulation of immunometabolism, in which IL-2 affects cellular bioenergetics, metabolic pathways, and immune cell function [1,2].

IL-2 impacts immunometabolism through several mechanisms. First, IL-2 signaling promotes the activation of critical metabolic pathways to meet the heightened bioenergetic demands of activated immune cells. This augmentation of metabolic flux facilitates ATP generation, providing the necessary energy substrates to fuel immune cell effector functions such as cytokine production and cytotoxicity [3,4]. Additionally, IL-2 influences nutrient uptake and utilization in immune cells, enhancing glucose uptake to fuel glycolysis, the preferential metabolic pathway for rapidly proliferating T cells [5,6]. Moreover, IL-2 signaling promotes the uptake of amino acids, essential for protein synthesis and cellular proliferation [7,8].

Furthermore, IL-2-mediated metabolic rewiring extends beyond energy production to impact cellular redox status and biosynthetic processes. IL-2 signaling modulates the intracellular redox balance, regulating reactive oxygen species (ROS) levels to ensure optimal immune cell function and survival. Additionally, IL-2 promotes lipid biosynthesis and mitochondrial biogenesis, facilitating the generation of membrane components and organelles essential for cell growth and division [9,10].

Overall, IL-2 is a pivotal regulator of immunometabolism, fine-tuning metabolic pathways to support immune cell activation, proliferation, and effector function. By enhancing energy production, nutrient uptake, and biosynthetic processes, IL-2 optimizes immune cell metabolism and elicits robust immune responses against pathogens and tumors [11–13]. IL-2 contributes to immune homeostasis by regulating T cell development, tolerance, and function. Defects in IL-2 signaling can lead to immune dysregulation, contributing to autoimmune diseases such as type 1 diabetes and multiple sclerosis [13,14]. Conversely, excessive IL-2 signaling is associated with immune hyperactivation and inflammatory disorders [15].

Therapeutically, IL-2 has been utilized to treat certain cancers, particularly metastatic melanoma and renal cell carcinoma [16,17]. High-dose IL-2 therapy aims to enhance antitumor immunity by promoting the expansion and activation of cytotoxic T cells and NK cells. However, IL-2 therapy can be limited by systemic toxicity and adverse effects. Researchers are exploring novel IL-2-based immunotherapies to mitigate these challenges, including engineered IL-2 variants with improved pharmacokinetics and reduced toxicity [18,19].

Interleukin-12 (IL-12): Interleukin-12 (IL-12) is a critical cytokine that plays a key role in regulating immune cell metabolism, particularly impacting T cells and natural killer (NK) cells. IL-12 is primarily produced by antigen-presenting cells, such as dendritic cells and macrophages, in response to microbial infections or inflammatory stimuli [20]. Upon binding to its receptor on immune cells, IL-12 triggers signaling pathways that promote T cell differentiation into effector T cells, specifically Th1 cells, and enhances NK cell cytotoxicity [21].

IL-12 influences immunometabolism by stimulating glycolysis and oxidative phosphorylation in activated T cells and NK cells. This metabolic shift supports the energy demands for immune cell activation, proliferation, and cytokine production during immune responses. IL-12-mediated metabolic reprogramming enhances effector functions of T cells and NK cells, contributing to effective pathogen clearance and antitumor immunity [22–24].

In health, IL-12 promotes protective immune responses against intracellular pathogens such as bacteria and viruses. By driving Th1 cell differentiation and enhancing NK cell activity, IL-12 eliminates infected cells and controls microbial infections. Conversely, dysregulated IL-12 signaling is implicated in several inflammatory and autoimmune diseases [21,25]. Excessive IL-12 production can lead to chronic inflammation and tissue damage, contributing to autoimmune conditions like rheumatoid arthritis, multiple sclerosis, and inflammatory bowel disease [26–29].

IL-12-based therapies have been explored for their potential in cancer immunotherapy. Recombinant IL-12 and IL-12-based cytokine complexes are investigated as immunomodulatory agents to enhance antitumor immune responses. These therapies activate cytotoxic T cells and NK cells within the tumor microenvironment, promoting tumor cell killing and immune surveillance [20]. Clinical trials evaluating IL-12 agonists and IL-12-based combination therapies demonstrate promising results in various cancers, including melanoma, renal cell carcinoma, and lung cancer [24,30,31]. The goal is to harness IL-12's immunomodulatory effects to overcome immune suppression and improve patient responses to immunotherapy.

Interleukin-15 (IL-15): Interleukin-15 (IL-15) plays a pivotal role in immune cell metabolism, particularly influencing the behavior of memory CD8+ T cells and natural killer (NK) cells within the immune system. IL-15 is essential for sustaining antitumor immunity by modulating the activation, proliferation, and longevity of memory CD8+ T cells, crucial for effective immune responses against cancer [32]. Additionally, IL-15 supports FAO in memory T cells, ensuring a sustainable energy supply essential for their functions in immunological memory.

IL-15 regulates immune cell functions and controls metabolic pathways, including glycolysis and mitochondrial respiration. This cytokine enhances glucose uptake and ATP production in immune cells, elevating their functionality and antitumor immune responses [33,34]. Moreover, dysregulation of IL-15 signaling contributes to disease pathology. Elevated IL-15 levels are observed in autoimmune disorders like rheumatoid arthritis, inflammatory bowel disease, and psoriasis, perpetuating immune-mediated tissue damage and inflammation [35,36]. Conversely, reduced IL-15 levels compromise immune cell functions, impairing responses against viral infections or tumors [37].

Given its pivotal role in immune cell metabolism and function, IL-15 has become a target for therapeutic interventions across various diseases. In cancer immunotherapy, IL-15-based treatments, including IL-15 agonists and cytokine complexes, are explored to enhance antitumor immune responses [18,36,38]. IL-15 therapies offer promising avenues to augment immune-mediated tumor clearance by stimulating T cell functions.

Furthermore, targeting IL-15 signaling holds potential in autoimmune diseases. Modulating IL-15 receptors or downstream pathways could help mitigate excessive immune activation and inflammation observed in autoimmune conditions, providing relief to patients with autoimmune disorders [39]. IL-15-based therapies are also investigated in infectious diseases to bolster immune responses against viral infections, particularly in cases where NK cell or memory T cell functions are compromised [40].

Interleukin-17 (IL-17): Interleukin-17 (IL-17) is a pro-inflammatory cytokine produced primarily by T helper 17 (Th17) cells, a subset of CD4+ T cells. IL-17 plays a critical role in immune responses against extracellular pathogens and the pathogenesis of autoimmune diseases [41,42]. Beyond its immunological functions, emerging evidence suggests that IL-17 also influences cellular metabolism, contributing to the metabolic reprogramming of immune cells in various disease contexts [43].

IL-17 exerts its effects on immune cell metabolism by modulating several key pathways. One prominent mechanism involves the activation of the NF-κB pathway in target cells, which promotes the expression of genes involved in glucose metabolism and glycolysis [43,44]. IL-17 can enhance glycolysis in neutrophils and macrophages, enabling these cells to meet the increased energy demands associated with inflammatory responses. Moreover, IL-17 can influence lipid metabolism by promoting gene expression in fatty acid synthesis and lipid storage, contributing to immune cells' activation and effector functions [42,45].

Dysregulation of IL-17 signaling is associated with various pathological conditions. When IL-17 is upregulated, excessive inflammation can occur, leading to tissue damage and autoimmune diseases such as rheumatoid arthritis, psoriasis, and IBD [46–48]. IL-17-driven inflammation contributes to joint destruction in rheumatoid arthritis by promoting osteoclast differentiation and bone resorption. In psoriasis, IL-17 induces keratinocyte proliferation and the production of inflammatory mediators, resulting in skin lesions characteristic of the disease [49]. Additionally, IL-17 is implicated in the pathogenesis of inflammatory bowel disease, promoting intestinal inflammation and tissue damage.

Conversely, downregulation of IL-17 signaling can impair immune responses against pathogens and increase susceptibility to infections. Deficiencies in IL-17 signaling have been associated with recurrent fungal infections, particularly mucocutaneous candidiasis, highlighting the importance of IL-17 in host defense against fungal pathogens [50,51].

Therapeutically targeting the IL-17 pathway has emerged as an effective strategy for treating autoimmune diseases characterized by IL-17-mediated inflammation. Monoclonal antibodies targeting IL-17 or its receptor have been developed and approved for clinical use in conditions such as psoriasis and psoriatic arthritis. These biologic therapies effectively suppress IL-17-driven inflammation, significantly helping patients with these chronic autoimmune disorders [39,42].

IL-17 plays a multifaceted role in immunometabolism, influencing cellular metabolism in immune cells and contributing to the pathogenesis of autoimmune diseases. Therapeutic interventions targeting the IL-17 pathway have shown promise in managing IL-17-mediated inflammatory conditions, highlighting the therapeutic potential of modulating immunometabolic pathways in disease treatment.

Interleukin-21 (IL-21): Interleukin-21 (IL-21) is a pivotal cytokine produced by activated T cells and natural killer T cells, influencing immune cell activation and differentiation, particularly within CD4+ and CD8+ T cell subsets. IL-21 extends its impact beyond canonical functions by regulating metabolic pathways in immune cells, notably glycolysis and mitochondrial respiration. IL-21 signaling enhances glucose uptake and ATP production in T cells, bolstering their functionality and fortifying antitumor immune responses [52–54].

In cancer immunotherapy, preclinical investigations have highlighted IL-21's potential to augment the efficacy of antitumor T cells, promoting their activation and proliferation. Additionally, IL-21 synergistically enhances the effectiveness of other immunotherapies, such as immune checkpoint inhibitors, in combating cancer [30,55,56]. However, the clinical translation of IL-21-based therapies is challenged by associated toxicities. Consequently, ongoing research explores alternative strategies targeting IL-21 signaling, such as IL-21 fusion proteins or combination therapies, to maximize therapeutic benefit while minimizing adverse effects [57,58].

Beyond cancer, dysregulated IL-21 signaling contributes to various disease states, including autoimmune disorders like systemic lupus erythematosus (SLE) and inflammatory bowel disease (IBD) [59–62]. Aberrant IL-21 activity exacerbates immune dysregulation and chronic inflammation, driving disease pathology. Conversely, deficiencies in IL-21 signaling are linked to immunodeficiencies such as common variable immunodeficiency (CVID), compromising immune responses against pathogens [63,64].

Understanding the mechanisms underlying IL-21-mediated IMR holds promise for advancing therapeutic interventions to elevate antitumor immune responses and mitigate immune dysregulation in autoimmune disorders. By optimizing immune cell metabolism, IL-21-based therapies offer potential avenues for combating cancer and ameliorating autoimmune pathology, representing a significant stride towards precision medicine approaches to disease management [65,66]. Ongoing research continues to elucidate IL-21's role in immunometabolism and its therapeutic implications, shaping the landscape of immunotherapy and autoimmune disease management.

Interleukin-27 (IL-27): Interleukin-27 (IL-27) is a cytokine with diverse roles in immune regulation, inflammation, and antimicrobial defense. Emerging evidence suggests that IL-27 also plays a critical role in modulating immune cell metabolism, impacting immune responses, and contributing to the pathogenesis of various diseases [67,68].

A critical mechanism through which IL-27 regulates immune cell metabolism involves regulating T cell function and differentiation. IL-27 can suppress glycolysis in effector T cells, promoting a shift towards OXPHOS and a more quiescent metabolic state [69–71]. This metabolic reprogramming is associated with reduced pro-inflammatory cytokine production and effector T cell activity, favoring immune tolerance and regulatory T cell (Treg) development [71].

Dysregulation of IL-27 signaling is associated with autoimmune diseases and chronic inflammation. Reduced IL-27 production or responsiveness has been observed in autoimmune conditions such as multiple sclerosis (MS) and IBD, contributing to aberrant T cell activation and tissue damage [69,72]. Conversely, excessive IL-27 signaling is implicated in immune suppression and tolerance induction, promoting immune evasion in certain cancers and chronic infections [67,73–75].

IL-27 also influences macrophage polarization and function, impacting metabolic pathways in these innate immune cells. IL-27 can skew macrophages towards an anti-inflammatory, M2-like phenotype characterized by increased oxidative metabolism and fatty acid oxidation (FAO) [76,77]. This metabolic shift is associated with tissue repair and resolution of inflammation but can also contribute to immune suppression and disease progression in specific contexts.

Therapeutic targeting of IL-27 signaling holds promise for treating immune-related disorders and cancer. Strategies aimed at modulating IL-27 activity include the development of IL-27 agonists or antagonists to modulate immune responses selectively. Clinical trials evaluating IL-27-based therapies in autoimmune diseases and cancer are underway, aiming to harness the immunomodulatory properties of IL-27 for therapeutic benefit [69,78].

Transforming growth factor-beta (TGF-β): Transforming growth factor-β (TGF-β) is a multifunctional cytokine that is essential to immune regulation and tissue homeostasis. Beyond its well-characterized functions in cell growth, differentiation, and wound healing, TGF-β also influences immune cell metabolism, contributing to the modulation of immune responses and the pathogenesis of various diseases.

One key mechanism through which TGF-β modifies immunometabolism is by regulating cellular bioenergetics. TGF-β signaling can suppress glycolysis in effector T cells, reducing proliferation and cytokine production. Instead, TGF-β promotes OXPHOS, favoring a more quiescent and regulatory phenotype in T cells. This metabolic shift is critical for developing and maintaining Tregs, which play a central role in immune tolerance and suppression of excessive immune responses.

Dysregulation of TGF-β signaling is associated with a spectrum of pathological conditions. Enhanced TGF-β signaling contributes to immune suppression and facilitates immune evasion in cancer [79]. Tumor-derived TGF-β promotes the generation of immunosuppressive

Tregs and myeloid-derived suppressor cells (MDSCs), inhibiting effector T cell function and promoting tumor growth [80,81].

Moreover, excessive TGF-β signaling is implicated in the pathogenesis of fibrotic diseases, including pulmonary and liver fibrosis [82]. TGF-β drives the differentiation of fibroblasts into myofibroblasts, leading to excessive extracellular matrix deposition and tissue remodeling. Conversely, insufficient TGF-β signaling can lead to immune dysregulation and autoimmune diseases. Defects in TGF-β signaling pathways have been implicated in autoimmune disorders such as SLE and IBD, resulting in aberrant activation of immune cells and tissue damage [83,84].

Therapeutic targeting of TGF-β signaling holds promise for the treatment of immune-related disorders. Strategies aimed at modulating TGF-β activity include the development of small molecule inhibitors, monoclonal antibodies targeting TGF-β receptors, and gene therapy approaches to restore or suppress TGF-β signaling selectively in diseased tissues [85,86]. Clinical trials evaluating TGF-β inhibitors in cancer and fibrotic diseases have shown encouraging results, highlighting the therapeutic potential of targeting TGF-β to modulate immunometabolism and immune responses [87].

Tumor necrosis factor-alpha (TNF-α): Tumor necrosis factor-alpha (TNF-α) is a potent pro-inflammatory cytokine that is crucial in regulating immunometabolism. TNF-α is primarily produced by activated macrophages, T cells, and other immune cells in response to infection, inflammation, or tissue injury. This cytokine exerts diverse effects on immune cell metabolism, influencing cellular energy production and immune cell function [88].

Regarding immunometabolism, TNF-α promotes glycolysis, a key metabolic pathway that provides energy and intermediates for immune cell activation and effector functions. By enhancing glycolysis, TNF-α stimulates the production of ATP and promotes the synthesis of cytokines and chemokines necessary for mounting effective immune responses against pathogens [89].

TNF-α's role in health and disease is complex. In healthy individuals, TNF-α contributes to immune surveillance and host defense by promoting the recruitment and activation of immune cells at sites of infection or injury [90–92]. However, dysregulated TNF-α production is associated with various inflammatory and autoimmune diseases, including rheumatoid arthritis, inflammatory bowel disease, and psoriasis [93–95]. Excessive TNF-α can lead to chronic inflammation, tissue damage, and disease progression.

Therapeutically, TNF-α inhibitors have revolutionized the treatment of inflammatory disorders. Drugs like infliximab, adalimumab, and etanercept block TNF-α signaling and effectively suppress inflammation, providing relief and improving the quality of life for patients with rheumatoid arthritis and Crohn's disease [96,97]. These biological therapies have demonstrated efficacy in reducing symptoms, preventing joint damage, and inducing remission in many patients.

Conclusion

In summary, IL-2, IL-12, IL-15, IL-17, IL-21, IL-27, TGF-beta, and TNF-α impact immune cell metabolism through diverse mechanisms, influencing nutrient utilization, metabolic pathways, and cellular functions during immune responses. Understanding the metabolic regulation by these cytokines is critical for developing targeted immunotherapies and treatments for immune-related diseases. Each cytokine uniquely modulates immunometabolism to maintain immune homeostasis or promote specific immune responses in health and disease. The role of cytokines in immunometabolism and cancer immunotherapy is an active area of research, and further studies are needed to fully understand the mechanisms underlying these processes.

Targeting immunometabolism to improve cancer immunotherapy

Immunotherapy has catalyzed a paradigm shift in cancer treatment, heralding a new era of therapeutic possibilities. By leveraging the inherent capabilities of the immune system to combat cancer, immunotherapy has yielded unprecedented outcomes across various malignancies characterized by durable responses and prolonged survival for patients. This introduction offers a glimpse into the evolving landscape of immunotherapy, delineating recent breakthroughs, persistent hurdles, and burgeoning strategies aimed at refining and enhancing cancer immunotherapy.

Current immunotherapies

Immunotherapy has emerged as a cornerstone in the management of diverse cancer types, spanning melanoma, lung carcinoma, and bladder malignancies [98,99]. The current armamentarium of immunotherapeutic modalities encompasses immune checkpoint inhibitors, chimeric antigen receptor (CAR)-T cell therapy, and cancer vaccines [100–102]. Immune checkpoint inhibitors, exemplified by anti-PD-1 and anti-CTLA-4 antibodies, disrupt inhibitory signaling pathways, thereby unleashing T cell-mediated cytotoxicity against tumor cells. CAR-T cell therapy involves the genetic engineering of T cells to express chimeric receptors targeting tumor-specific antigens, facilitating targeted tumor cell eradication. Concurrently, cancer vaccines harness the immune system's capacity to recognize and eliminate malignant cells by presenting tumor antigens, thereby instigating an antitumor immune response. Despite the notable strides made in immunotherapy, some patients fail to respond to treatment, while others develop acquired resistance over time [103].

Recent advances in immunometabolic therapies

In recent years, immunotherapy has witnessed unprecedented progress, reshaping the paradigm of cancer treatment and offering new hope for patients with various malignancies. These advancements span diverse therapeutic modalities, each with distinct mechanisms of action and therapeutic applications.

Expansion of approved immune checkpoint inhibitors (ICIs): Immune checkpoint inhibitors (ICIs) have emerged as a cornerstone of cancer immunotherapy, revolutionizing treatment across multiple cancer types. FDA approvals for ICIs, including anti-PD-1 (programmed cell death protein 1) and anti-CTLA-4 (cytotoxic T-lymphocyte-associated protein 4) antibodies, have significantly expanded the therapeutic arsenal against cancers such as melanoma, non-small cell lung carcinoma (NSCLC), and renal cell carcinoma (RCC) [104–106]. Since the initial approval of ipilimumab (anti-CTLA-4) in 2011, the landscape of cancer care has evolved rapidly, with subsequent approvals of pembrolizumab, nivolumab, and atezolizumab targeting PD-1/PD-L1 interactions [107–109]. Mechanistically, ICIs disrupt inhibitory signaling pathways exploited by tumors to evade immune surveillance, unleashing the cytotoxic potential of T cells against cancer cells. These antibodies restore T cell activity by blocking immune checkpoints, such as PD-1/PD-L1 and CTLA-4, thereby promoting antitumor immune responses. Over the past decade, the expansion of FDA approvals underscores the remarkable efficacy and growing acceptance of ICIs as standard-of-care treatments across diverse cancer types [110–112]. However, challenges persist, including identifying predictive biomarkers for patient response, managing immune-related adverse events, overcoming resistance mechanisms, and optimizing treatment strategies to maximize long-term clinical outcomes [113,114].

Advances in chimeric antigen receptor (CAR) T cell-therapy: Chimeric antigen receptor (CAR) T cell-therapy represents a transformative approach to cancer treatment, particularly for hematological malignancies [115]. Since the first FDA approval of tisagenlecleucel (Kymriah) for pediatric acute lymphoblastic leukemia (ALL) in 2017, CAR T cell-therapy has demonstrated remarkable efficacy in patients with refractory or relapsed disease [116]. This innovative therapy involves genetically modifying patients' T cells to express chimeric receptors targeting tumor-specific antigens, thereby redirecting T cells to recognize and eliminate cancer cells. Mechanistically, CAR T cells engage with tumor cells via their chimeric receptors, leading to T cell activation, cytokine release, and subsequent tumor cell destruction [117]. While initial successes have been observed primarily in hematologic malignancies, ongoing research aims to extend the applicability of CAR T cell-therapy to solid tumors [118]. Challenges such as cytokine release syndrome (CRS), neurotoxicity, persistence of CAR T cells, and tumor heterogeneity underscore the need for further optimization and refinement of this groundbreaking therapeutic strategy. Moreover, efforts are underway to enhance the safety profile and increase the scalability of CAR T cell-therapy, including developing next-generation CAR constructs and novel manufacturing techniques.

Enhanced cancer vaccines: Personalized cancer vaccines represent a promising approach to cancer immunotherapy, offering tailored treatment options based on individual patient's tumor characteristics [119]. Initiated in the mid-1990s and gaining traction in the 2010s, personalized cancer vaccines aim to stimulate robust and specific immune responses against tumor cells [120]. By leveraging the unique mutational landscape of each patient's tumor, these vaccines deliver tumor-specific antigens to antigen-presenting cells (APCs), activating antigen-specific T cells and priming them for targeted tumor recognition and elimination. Mechanistically, personalized cancer vaccines aim to overcome immune tolerance and induce durable antitumor immunity by presenting tumor-specific antigens to the immune system. While challenges such as identifying optimal antigen targets, optimizing vaccine delivery methods, and overcoming immunosuppressive tumor microenvironments (TME) persist, personalized cancer vaccines hold immense promise for improving clinical outcomes and advancing the field of cancer immunotherapy [121].

Novel immunotherapeutic modalities: In addition to established immunotherapies, such as ICIs and CAR T cell-therapy, novel immunotherapeutic strategies are emerging as promising avenues for cancer treatment. Bispecific T cell-engagers (BiTEs) and T cell-receptor (TCR)-engineered T cells represent innovative approaches that harness the immune system's cytotoxicity against tumor cells [122–124]. BiTEs facilitate T cell-mediated killing of tumor cells by bridging T cells and cancer cells via bispecific antibodies targeting specific antigens, promoting direct tumor cell lysis [125]. Similarly, TCR-engineered T cells enable the recognition and destruction of cancer cells expressing specific antigens through the engineered T cell receptor, leading to targeted tumor cell elimination [126]. While these novel immunotherapies are in the early stages of development, they hold great promise for expanding the therapeutic armamentarium against cancer and improving patient treatment outcomes [127,128]. Continued research and clinical investigation are essential to further characterize their mechanisms of action, optimize their efficacy, and overcome potential challenges in clinical translation. Additionally, ongoing efforts focus on enhancing the safety and specificity of these modalities to minimize off-target effects and maximize therapeutic benefits in diverse cancer settings.

Ongoing challenges

Despite the remarkable progress achieved in cancer immunotherapy, several formidable challenges persist, necessitating continued research and innovation to optimize treatment outcomes and expand patient access to these transformative therapies:

Patient heterogeneity: One of the foremost challenges in cancer immunotherapy is the inherent variability in patient responses. While some patients experience robust and durable responses to immunotherapy, others derive minimal or transient benefits. Patient heterogeneity encompasses many factors, including tumor biology, host immune status, and the TME [129,130]. Comprehensive characterization of these factors is essential for stratifying patients based on their likelihood of responding to specific immunotherapeutic interventions. Biomarkers such as tumor mutational burden (TMB), programmed death-ligand 1 (PD-L1) expression, and immune cell infiltration have shown promise in predicting patient responses to immunotherapy [131,132]. However, further refinement and validation of predictive biomarkers are needed to enhance their clinical utility and enable personalized treatment approaches. Additionally, exploring novel biomarkers and integrating multi-omic analyses may provide deeper insights into the complex interplay between tumors and the immune system, ultimately improving patient stratification and treatment outcomes [133,134].

Acquired resistance: While many patients initially benefit from immunotherapy, a significant proportion eventually develop acquired resistance, leading to disease progression and therapeutic failure. The mechanisms underlying acquired resistance to immunotherapy are multifaceted and incompletely understood. Tumor immune evasion strategies contribute to acquired resistance, including upregulation of alternative immune checkpoints, loss of tumor antigenicity, and emergence of immunosuppressive cell populations [135,136]. Additionally, adaptive changes in the TME, such as increased fibrosis and immune exclusion, further diminish treatment efficacy [137,138]. Unraveling the molecular and cellular mechanisms driving acquired resistance is essential for developing strategies to overcome or prevent resistance and prolong treatment responses. Preclinical models and longitudinal patient studies offer valuable insights into the dynamic evolution of resistance mechanisms throughout treatment [139–142]. Combinatorial approaches targeting multiple resistance pathways and adaptive therapeutic strategies informed by real-time monitoring of tumor dynamics hold promise for mitigating acquired resistance and improving long-term treatment outcomes.

Immune-related adverse events (irAEs): While immunotherapy has revolutionized cancer treatment, it is accompanied by a unique spectrum of immune-related adverse events (irAEs) that can affect virtually any organ system. These irAEs result from dysregulated immune activation and can manifest as autoimmune-like phenomena, including dermatitis, colitis, pneumonitis, hepatitis, and endocrinopathies [143,144]. The severity of irAEs can range from mild, self-limiting symptoms to life-threatening complications requiring immediate intervention. Prompt recognition and management of irAEs are paramount to preventing immunotherapy-related morbidity and mortality [145,146]. Treatment guidelines recommend early detection through regular monitoring of symptoms and laboratory parameters and prompt initiation of corticosteroids or other immunosuppressive agents for moderate to severe irAEs [147–149]. Additionally, interdisciplinary collaboration between oncologists, immunologists, and other specialists is essential for the comprehensive management of irAEs and optimization of patient outcomes. Further research is needed to define the underlying mechanisms of irAE development and identify predictive biomarkers to stratify patients at risk of developing irAEs and guide personalized management strategies.

High costs: While immunotherapy offers tremendous promise in cancer treatment, the high costs associated with these therapies' present significant challenges to patient access and healthcare sustainability. The cost of immunotherapy encompasses various components, including drug acquisition costs, administration fees, supportive care expenses, and monitoring costs [150–152]. Additionally, the high cost of cancer immunotherapy is compounded by the need for prolonged treatment duration and the potential for combination therapies, further exacerbating financial burdens for patients and healthcare systems. Addressing the affordability of immunotherapy is essential to ensure equitable access for all patients, regardless of socioeconomic status or geographic location [153–155]. Strategies to mitigate costs may include value-based pricing models, innovative reimbursement strategies, and collaborative efforts between stakeholders to streamline drug development and regulatory processes. Moreover, investments in research and development to improve treatment efficacy, durability, and safety may yield long-term cost savings by reducing the need for expensive supportive care interventions and hospitalizations [156]. Achieving sustainable immunotherapy access requires a multifaceted approach that balances innovation with affordability, prioritizing patient outcomes and societal benefit.

Potential solutions

To overcome the formidable obstacles encountered in cancer immunotherapy and enhance treatment efficacy, targeting immunometabolism has emerged as a promising avenue. The relationship between the metabolism of tumors and immune cells underscores the potential to modulate metabolic pathways within the TME to bolster antitumor immune responses. Several strategies for targeting immunometabolism have garnered significant attention:

First, promoting glycolysis and augmenting glucose availability in the TME can potentiate T cell effector functions [157–159]. The TME is characterized by glucose deprivation, leading to impaired glycolytic metabolism and diminished cytotoxic activity of tumor-infiltrating lymphocytes (TILs). The administration of glycolytic inhibitors that selectively target tumor cells while sparing lymphocytes presents a promising approach to alleviate glucose restriction for T cells. Additionally, supplementing exogenous glycolytic substrates such as pyruvate or acetyl-CoA can potentially enhance T cell metabolism and activity [160–162].

Second, targeting amino acid metabolism offers another avenue to modulate the TME. Depletion of essential amino acids like tryptophan and arginine, or accumulation of their metabolites, exerts inhibitory effects on T cell activation and function [163–165]. Strategies aimed at increasing amino acid availability or transport and blocking their catabolism in the TME may sustain T cell responses and augment immunotherapy efficacy. Numerous clinical trials are currently investigating agents targeting amino acid metabolism in combination with immunotherapies to optimize treatment outcomes [166,167].

Third, promoting OXPHOS in T cells can enhance their persistence and long-term survival within the TME. While glycolysis fuels initial T cell activation, OXPHOS is crucial in supporting sustained T cell responses and memory formation [167,168]. Selectively inhibiting OXPHOS in Tregs while preserving effector T cells may confer a survival advantage to the latter, thereby bolstering antitumor immunity. Targeting the transcription factors governing OXPHOS and mitochondrial biogenesis presents a promising approach to modulate T cell metabolism and enhance immunotherapy efficacy selectively [161,169].

Fourth, blocking the accumulation of immunosuppressive metabolic byproducts can reinvigorate antitumor immunity. For instance, lactate, produced by tumor cells and M2-like tumor-associated macrophages, suppresses T cell and natural killer cell activity [170,171]. Clinical trials are underway to evaluate the inhibition of lactate dehydrogenase (LDH) to mitigate lactate production. Similarly, strategies aimed at lowering prostaglandin E2, adenosine, and kynurenine levels in the TME through inhibition of cyclooxygenase-2 (COX-2), CD39/CD73, and indoleamine 2,3-dioxygenase (IDO), respectively, represent promising avenues to alleviate immunosuppression and enhance immunotherapy efficacy [172–174].

Fifth, restoring cholesterol metabolism in T cells may augment their proliferative and cytotoxic capacity within the TME. Cholesterol depletion in the TME inhibits T cell responses while replenishing cholesterol rescues T cell activity [175,176]. Approaches such as blocking cholesterol esterification or enhancing cholesterol transport in T cells hold promise for supporting their function in the TME and enhancing immunotherapy efficacy.

Finally, promoting mitochondrial biogenesis and health in T cells is crucial for sustaining their antitumor effects. Natural compounds like sulforaphane can activate transcription factors involved in mitochondrial biogenesis and antioxidant responses [177,178]. Restoring mitochondrial content and reducing excess reactive oxygen species in T cells through these approaches may enhance their metabolic and effector functions, boosting immunotherapy efficacy.

In conclusion, while immunotherapy holds promise as a cancer treatment modality, challenges persist that limit its effectiveness. Targeting the metabolic interplay within the TME offers innovative solutions to augment antitumor immune responses and enhance the efficacy of immunotherapies. Modulating metabolic pathways to support T cell function while inhibiting tumor cell metabolism represents a novel approach to cancer treatment. Continued advancements in understanding the complex metabolic dynamics between tumors and the immune system will pave the way for further progress in this emerging field of immunometabolism.

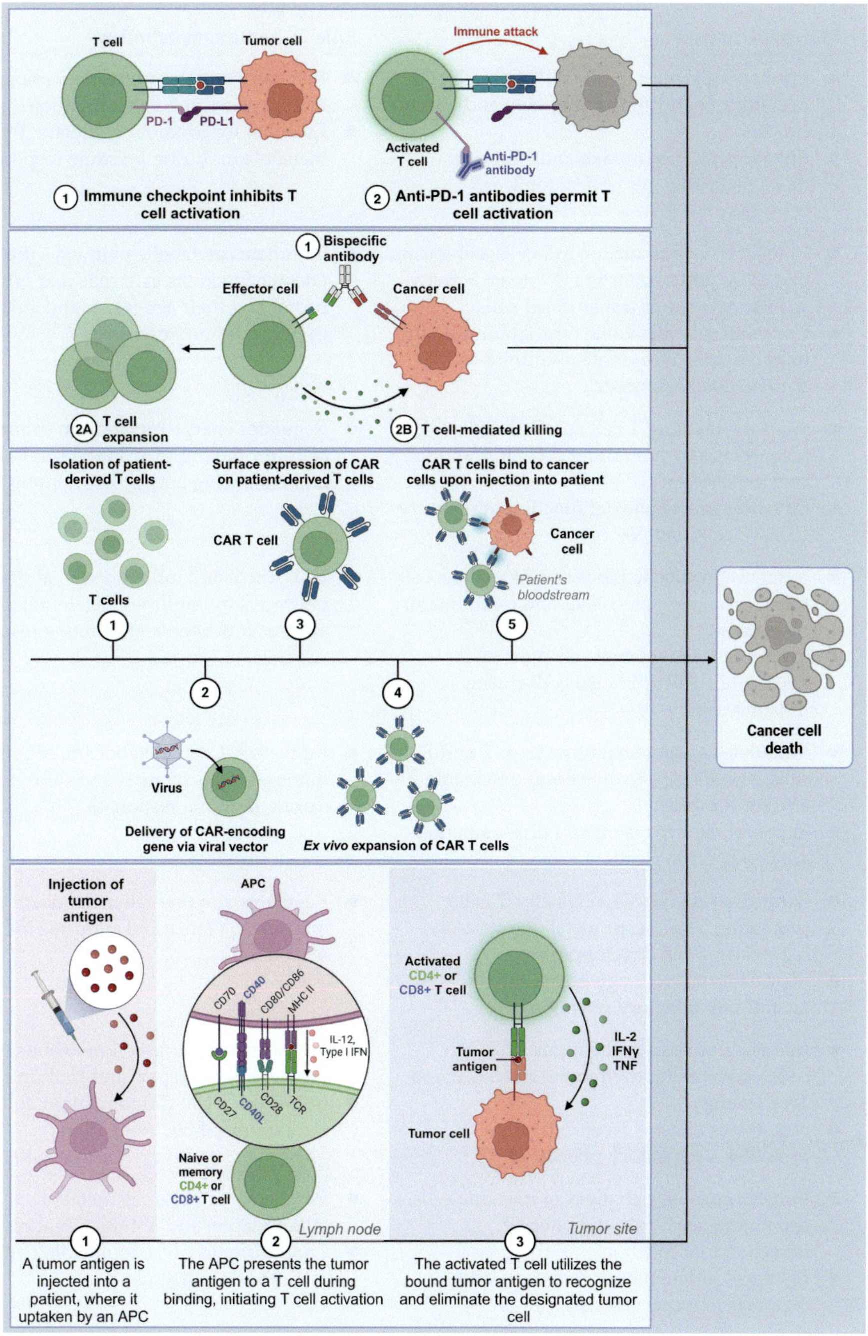
T cell
Tumor cell
PD-1
PD-L1
Immune attack
Activated T cell
Anti-PD-1 antibody
1 Immune checkpoint inhibits T cell activation
2 Anti-PD-1 antibodies permit T cell activation
1 Bispecific antibody
Effector cell
Cancer cell
2A T cell expansion
2B T cell-mediated killing
Isolation of patient-derived T cells
Surface expression of CAR on patient-derived T cells
CAR T cells bind to cancer cells upon injection into patient
T cells
CAR T cell
Cancer cell
Patient's bloodstream
1
2
3
4
5
Virus
Delivery of CAR-encoding gene via viral vector
Ex vivo expansion of CAR T cells
Cancer cell death
Injection of tumor antigen
APC
CD70
CD40
CD80/CD86
MHC II
IL-12, Type I IFN
CD27
CD40L
CD28
TCR
Naive or memory CD4+ or CD8+ T cell
Lymph node
Activated CD4+ or CD8+ T cell
Tumor antigen
IL-2
IFNγ
TNF
Tumor cell
Tumor site
1
A tumor antigen is injected into a patient, where it uptaken by an APC
2
The APC presents the tumor antigen to a T cell during binding, initiating T cell activation
3
The activated T cell utilizes the bound tumor antigen to recognize and eliminate the designated tumor cell

Cytokine	Metabolic impact	Role in immunometabolism
IL-2	• Promotes glycolysis and OXPHOS in T cells, providing energy for proliferation and effector functions • Enhances glucose uptake and supports metabolic pathways critical for T cell activation	• Regulates nutrient utilization, impacting T cell bioenergetics and differentiation • Essential for promoting effector T cell metabolism during immune responses
IL-12	• Induces IFN-γ production in T cells, enhancing glycolysis and supporting cytotoxic activity against infected or transformed cells • Contributes to metabolic reprogramming in immune cells to promote antitumor and antimicrobial responses	• Stimulates metabolic pathways that support effector functions in T cells and NK cells, promoting their activation and survival during immune responses
IL-15	• Supports memory T cell survival and homeostasis by promoting OXPHOS and fatty acid metabolism • Enhances mitochondrial function in memory CD8+ T cells and NK cells	• Regulates energy metabolism in memory T cells, promoting their long-term persistence and enhancing antitumor immunity
IL-17	• Influence metabolic pathways in various cell types, promoting glycolysis and inflammatory responses • Contributes to metabolic changes in neutrophils and epithelial cells during inflammation	• IL-17-mediated inflammation alters metabolic pathways in immune cells, impacting their effector functions and immune responses
IL-21	• Modulates metabolic pathways in T and B cells, promoting glycolysis and enhancing effector functions • Supports the expansion and differentiation of antigen-specific T cells	• Influences T cell metabolism, regulating immune cell activation and differentiation during immune responses
IL-27	• Suppresses glycolysis in effector T cells, promoting a quiescent metabolic state and regulatory T cell development • Modulates macrophage polarization towards an anti-inflammatory phenotype	• Regulates immune cell metabolism to promote immune tolerance and suppress inflammatory responses
TGF-β	• Inhibits glycolysis and supports FAO in T cells, contributing to immune tolerance and Treg stability. • Suppresses effector T cell activation and promotes a regulatory phenotype	• Modulates immune cell metabolism to promote tissue repair and limit inflammation • Influences T cell differentiation and function through metabolic regulation
TNF-α	• Induces metabolic changes in immune cells, promoting glycolysis and altering mitochondrial function • Enhances inflammatory responses and regulates immune cell survival	• Influences immune cell metabolism to support inflammation and antimicrobial responses • Contributes to tissue remodeling and immune cell effector functions

Acknowledgments

This work was supported by National Institutes of Health Grant R01CA221867, R01CA273002, and R21AI167793 to J. S., and Department of Defense Grant LC210150 to J. S.

References

[1] Jones DM, Read KA, Oestreich KJ. Dynamic roles for IL-2-STAT5 signaling in effector and regulatory CD4+ T cell populations. J Immunol 2020;205(7):1721–30.

[2] Coppola C, Hopkins B, Huhn S, Du Z, Huang Z, Kelly WJ. Investigation of the impact from IL-2, IL-7, and IL-15 on the growth and signaling of activated CD4+ T cells. Int J Mol Sci 2020;21(21):7814.

[3] Widowati W, Jasaputra DK, Sumitro SB, Widodo MA, Mozef T, Rizal R, et al. Effect of interleukins (IL-2, IL-15, IL-18) on receptors activation and cytotoxic activity of natural killer cells in breast cancer cell. Afr Health Sci 2020;20(2):822–32.

[4] Sharapova TN, Romanova EA, Ivanova OK, Sashchenko LP, Yashin DV. Cytokines TNFα, IFNγ and IL-2 are responsible for signal transmission from the innate immunity protein Tag7 (PGLYRP1) to cytotoxic effector lymphocytes. Cells 2020;9(12):2602.

[5] Yagi Y, Kuwahara M, Suzuki J, Imai Y, Yamashita M. Glycolysis and subsequent mevalonate biosynthesis play an important role in Th2 cell differentiation. Biochem Biophys Res Commun 2020;530(2):355–61.

[6] Khan AUH, Ali AK, Marr B, Jo D, Ahmadvand S, Fong-McMaster C, et al. The TNFα/TNFR2 axis mediates natural killer cell proliferation by promoting aerobic glycolysis. Cell Mol Immunol 2023;20(10):1140–55.

[7] Lv C, Zhang H, Li Z, Zhang H, Zhang Y, Li R, et al. Low-dose IL-2 therapy compensates for metabolic shifts and reverses anxiety-like behavior in PD-1 deficiency-induced autoimmunity. Cell Mol Immunol 2021;18(5):1336–8.

[8] Zhao M, Xing J, Tang X, Sheng X, Chi H, Zhan W. Expression of Interleukin-2 receptor subunit gamma (IL-2Rγ) and its binding with IL-2 induced activation of CD4 T lymphocytes in flounder (*Paralichthys olivaceus*). Fish Shellfish Immunol 2022;122:426–36.

[9] Wenzel TJ, Haskey N, Kwong E, Greuel BK, Gates EJ, Gibson DL, et al. Dietary fats modulate neuroinflammation in mucin 2 knock out mice model of spontaneous colitis. Biochim Biophys Acta (BBA) Mol Basis Dis 2022;1868(3)166336.

[10] Moro A, Gao Z, Wang L, Yu A, Hsuing S, Yunguang B, et al. Dynamic transcriptional activity and chromatin remodeling of regulatory T cells after varied duration of interleukin-2 receptor signaling | nature immunology. Nat Immunol 2022;23:802–13.

[11] Shi L, Sheng J, Chen G, Zhu P, Shi C, Li B, et al. Combining IL-2-based immunotherapy with commensal probiotics produces enhanced antitumor immune response and tumor clearance. J Immunother Cancer 2020;8(2)e000973.

[12] Wen L, Shi L, Kong XL, Li KY, Li H, Jiang DX, et al. Gut microbiota protected against *Pseudomonas aeruginosa* pneumonia via restoring Treg/Th17 balance and metabolism, Front Cell Infect Microbiol 2022;12: [cited 2024 Apr 16]. Available from: https://www.frontiersin.org/articles/10.3389/fcimb.2022.856633.

[13] Dong S, Hiam-Galvez KJ, Mowery CT, Herold KC, Gitelman SE, Esensten JH, et al. The effect of low-dose IL-2 and Treg adoptive cell therapy in patients with type 1 diabetes. JCI Insight 2021;6(18)e147474.

[14] Aram J, Frakich N, Morandi E, Alrouji M, Samaraweera A, Onion D, et al. Increased IL-2 and reduced TGF-β upon T-cell stimulation are associated with GM-CSF upregulation in multiple immune cell types in multiple sclerosis. Biomedicine 2020;8(7):226.

[15] Hashimoto M, Araki K, Cardenas MA, Li P, Jadhav RR, Kissick HT, et al. PD-1 combination therapy with IL-2 modifies CD8+ T cell exhaustion program. Nature 2022;610(7930):173–81.

[16] Lopes JE, Fisher JL, Flick HL, Wang C, Sun L, Ernstoff MS, et al. ALKS 4230: a novel engineered IL-2 fusion protein with an improved cellular selectivity profile for cancer immunotherapy. J Immunother Cancer 2020;8(1)e000673.

[17] Hernandez R, Põder J, LaPorte KM, Malek TR. Engineering IL-2 for immunotherapy of autoimmunity and cancer. Nat Rev Immunol 2022;22(10):614–28.

[18] Yang Y, Lundqvist A. Immunomodulatory effects of IL-2 and IL-15; implications for cancer immunotherapy. Cancer 2020;12(12):3586.

[19] Kennedy LB, Salama AKS. A review of cancer immunotherapy toxicity. CA Cancer J Clin 2020;70(2):86–104.

[20] Mirlekar B, Pylayeva-Gupta Y. IL-12 family cytokines in cancer and immunotherapy. Cancer 2021;13(2):167.

[21] Ullrich KAM, Schulze LL, Paap EM, Müller TM, Neurath MF, Zundler S. Immunology of IL-12: an update on functional activities and implications for disease. EXCLI J 2020;19:1563–89.

[22] Terrén I, Orrantia A, Mosteiro A, Vitallé J, Zenarruzabeitia O, Borrego F. Metabolic changes of Interleukin-12/15/18-stimulated human NK cells. Sci Rep 2021;11(1):6472.

[23] Kang S, Kumanogoh A. The spectrum of macrophage activation by immunometabolism. Int Immunol 2020;32 (7):467–73.

[24] Agliardi G, Liuzzi AR, Hotblack A, De Feo D, Núñez N, Stowe CL, et al. Intratumoral IL-12 delivery empowers CAR-T cell immunotherapy in a pre-clinical model of glioblastoma. Nat Commun 2021;12(1):444.

[25] Zarrin AA, Bao K, Lupardus P, Vucic D. Kinase inhibition in autoimmunity and inflammation. Nat Rev Drug Discov 2021;20(1):39–63.

[26] Kondo N, Kuroda T, Kobayashi D. Cytokine networks in the pathogenesis of rheumatoid arthritis. Int J Mol Sci 2021;22(20):10922.

[27] Jahan T, Saleh AA, Anwar S. Association of cytokine IL-17, IL-4, IL-6, and IL-12 gene polymorphisms in rheumatoid arthritis patients in a tertiary care hospital in Bangladesh. Int J Rheumatol 2024;2024:e3728179.

[28] Jana M, Mondal S, Jana A, Pahan K. Induction of IL-2 by interleukin-12 p40 homodimer and IL-12, but not IL-23, in microglia and macrophages: implications for multiple sclerosis. Cytokine 2024;174:156457.

[29] Chang JT. Pathophysiology of inflammatory bowel diseases. N Engl J Med 2020;383(27):2652–64.

[30] Mortezaee K, Majidpoor J. Checkpoint inhibitor/interleukin-based combination therapy of cancer. Cancer Med 2022;11(15):2934–43.

[31] Zhang L, Davies JS, Serna C, Yu Z, Restifo NP, Rosenberg SA, et al. Enhanced efficacy and limited systemic cytokine exposure with membrane-anchored interleukin-12 T-cell therapy in murine tumor models. J Immunother Cancer 2020;8(1)e000210.

[32] Nolz JC, Richer MJ. Control of memory CD8+ T cell longevity and effector functions by IL-15. Mol Immunol 2020;117:180–8.

[33] Xu Y, Carrascosa LC, Yeung YA, Chu MLH, Yang W, Djuretic I, et al. An engineered IL15 cytokine mutein fused to an anti-PD1 improves intratumoral T-cell function and antitumor immunity. Cancer Immunol Res 2021;9 (10):1141–57.

[34] Fang W, Li L, Lin Z, Zhang Y, Jing Z, Li Y, et al. Engineered IL-15/IL-15Rα-expressing cellular vesicles promote T cell anti-tumor immunity. Extracell Vesicle 2023;2:100021.

[35] Allard-Chamard H, Mishra HK, Nandi M, Mayhue M, Menendez A, Ilangumaran S, et al. Interleukin-15 in autoimmunity. Cytokine 2020;136:155258.

[36] Waldmann TA, Miljkovic MD, Conlon KC. Interleukin-15 (dys)regulation of lymphoid homeostasis: implications for therapy of autoimmunity and cancer. J Exp Med 2019;217(1)e20191062.

[37] Gunst JD, Goonetilleke N, Rasmussen TA, Søgaard OS. Immunomodulation with IL-7 and IL-15 in HIV-1 infection. J Virus Erad 2023;9(3)100347.

[38] Ma S, Caligiuri MA, Yu J. Harnessing IL-15 signaling to potentiate NK cell-mediated cancer immunotherapy. Trends Immunol 2022;43(10):833–47.

[39] Jung SM, Kim WU. Targeted immunotherapy for autoimmune disease. Immune Netw 2022;22(1)e9.

[40] Bergamaschi C, Terpos E, Rosati M, Angel M, Bear J, Stellas D, et al. Systemic IL-15, IFN-γ, and IP-10/CXCL10 signature associated with effective immune response to SARS-CoV-2 in BNT162b2 mRNA vaccine recipients. Cell Rep 2021;36(6)109504.

[41] Berry SPDG, Dossou C, Kashif A, Sharifinejad N, Azizi G, Hamedifar H, et al. The role of IL-17 and anti-IL-17 agents in the immunopathogenesis and management of autoimmune and inflammatory diseases. Int Immunopharmacol 2022;102:108402.

[42] Mills KHG. IL-17 and IL-17-producing cells in protection versus pathology. Nat Rev Immunol 2023;23(1):38–54.

[43] Bechara R, McGeachy MJ, Gaffen SL. The metabolism-modulating activity of IL-17 signaling in health and disease. J Exp Med 2021;218(5)e20202191.

[44] Kono M. New insights into the metabolism of Th17 cells. Immunol Med 2023;46(1):15–24.

[45] Crawford MP, Sinha S, Renavikar PS, Borcherding N, Karandikar NJ. CD4 T cell-intrinsic role for the T helper 17 signature cytokine IL-17: effector resistance to immune suppression. Proc Natl Acad Sci 2020;117(32):19408–14.

[46] Taams LS. Interleukin-17 in rheumatoid arthritis: trials and tribulations. J Exp Med 2020;217(3)e20192048.

[47] Mosca M, Hong J, Hadeler E, Hakimi M, Liao W, Bhutani T. The role of IL-17 cytokines in psoriasis. ImmunoTargets Ther 2021;10:409–18.

[48] Schmitt H, Neurath MF, Atreya R. Role of the IL23/IL17 pathway in Crohn's disease, Front Immunol 2021;12: [cited 2024 Apr 16]. Available from: https://www.frontiersin.org/journals/immunology/articles/10.3389/fimmu.2021.622934/full.

[49] Furue M, Furue K, Tsuji G, Nakahara T. Interleukin-17A and keratinocytes in psoriasis. Int J Mol Sci 2020; 21(4):1275.

[50] Chimenz R, Tropeano A, Chirico V, Ceravolo G, Salpietro C, Cuppari C. IL-17 serum level in patients with chronic mucocutaneous candidiasis disease. Pediatr Allergy Immunol 2022;33(S27):77–9.

[51] Davidson L, van den Reek JMPA, Bruno M, van Hunsel F, Herings RMC, Matzaraki V, et al. Risk of candidiasis associated with interleukin-17 inhibitors: a real-world observational study of multiple independent sources. Lancet Reg Health Eur 2022;13:100266.

[52] di Carlo E, de Totero D, Piazza T, Fabbi M, Ferrini S. Role of IL-21 in immune-regulation and tumor immunotherapy. Cancer Immunol Immunother 2007;56(9):1323–34.

[53] Topchyan P, Xin G, Chen Y, Zheng S, Burns R, Shen J, et al. Harnessing the IL-21-BATF pathway in the CD8+ T cell anti-tumor response. Cancer 2021;13(6):1263.

[54] Shen S, Sckisel G, Sahoo A, Lalani A, Otter DD, Pearson J, et al. Engineered IL-21 cytokine muteins fused to anti-PD-1 antibodies can improve CD8+ T cell function and anti-tumor immunity, Front Immunol 2020;11: [cited 2024 Apr 18]. Available from: https://www.frontiersin.org/journals/immunology/articles/10.3389/fimmu.2020.00832/full.

[55] Lechner MG, Zhou Z, Hoang AT, Huang N, Ortega J, Scott LN, et al. Clonally expanded, thyrotoxic effector CD8 + T cells driven by IL-21 contribute to checkpoint inhibitor thyroiditis. Sci Transl Med 2023;15(696)eadg0675.

[56] Sun Y, Zhang Z, Zhang C, Zhang N, Wang P, Chu Y, et al. An effective therapeutic regime for treatment of glioma using oncolytic vaccinia virus expressing IL-21 in combination with immune checkpoint inhibition. Mol Ther Oncolytics 2022;26:105–19.

[57] Wilz SW. A clinical trial of IL-15 and IL-21 combination therapy for COVID-19 is warranted. Cytokine Growth Factor Rev 2021;58:49–50.

[58] Chen T, Ding X, Liao Q, Gao N, Chen Y, Zhao C, et al. IL-21 arming potentiates the anti-tumor activity of an oncolytic vaccinia virus in monotherapy and combination therapy. J Immunother Cancer 2021;9(1)e001647.

[59] Toskas A, Milias S, Papamitsou T, Meditskou S, Kamperidis N, Sioga A. The role of IL-19, IL-24, IL-21 and IL-33 in intestinal mucosa of inflammatory bowel disease: a narrative review, Arab J Gastroenterol 2024; [cited 2024 Apr 18]. Available from: https://www.sciencedirect.com/science/article/pii/S1687197924000029.

[60] Wang M, Guo J, Zhao YQ, Wang JP. IL-21 mediates microRNA-423-5p/claudin-5 signal pathway and intestinal barrier function in inflammatory bowel disease. Aging (Albany NY) 2020;12(16):16099–110.

[61] Ren HM, Lukacher AE, Rahman ZSM, Olsen NJ. New developments implicating IL-21 in autoimmune disease. J Autoimmun 2021;122:102689.

[62] Sengupta S, Bhattacharya G, Mohanty S, Shaw SK, Jogdand GM, Jha R, et al. IL-21, inflammatory cytokines and hyperpolarized CD8+ T cells are central players in lupus immune pathology. Antioxidants 2023;12(1):181.

[63] Khan HA, Acar U, Stephens AV, Thauland TJ, Butte MJ. Complicated common variable immunodeficiency is driven by Aberrant IL-10/IL-21 signaling and predisposed polygenic risk, medRxiv 2022; [cited 2024 Apr 18]. p. 2022.06.20.22276681. Available from: https://www.medrxiv.org/content/10.1101/2022.06.20.22276681v1.

[64] Cagdas D, Mayr D, Baris S, Worley L, Langley DB, Metin A, et al. Genomic spectrum and phenotypic heterogeneity of human IL-21 receptor deficiency. J Clin Immunol 2021;41(6):1272–90.

[65] Liu M, Wang Z, Liu X, Xiao H, Liu Y, Wang J, et al. Therapeutic effect of Yiyi Fuzi Baijiang formula on TNBS-induced ulcerative colitis via metabolism and Th17/Treg cell balance. J Ethnopharmacol 2023;309:116301.

[66] Bamidele A, Sagstetter M, Hirsova P, Ramos GP, Klatt E, Westphal M, et al. Interleukin 21-induced metabolic reprogramming drives CD4 t cell inflammatory response during intestinal inflammation. Gastroenterology 2022;162(3):S58.

[67] Povroznik JM, Robinson CM. IL-27 regulation of innate immunity and control of microbial growth. Future Sci OA 2020;6(7)FSO588.

[68] Jin Y, Fyfe PK, Gardner S, Wilmes S, Bubeck D, Moraga I. Structural insights into the assembly and activation of the IL-27 signaling complex. EMBO Rep 2022;23(10)e55450.

[69] Mei Y, Lv Z, Xiong L, Zhang H, Yin N, Qi H. The dual role of IL-27 in CD4+T cells. Mol Immunol 2021;138:172–80.

[70] Lin CH, Wu CJ, Cho S, Patkar R, Huth WJ, Lin LL, et al. Selective IL-27 production by intestinal regulatory T cells permits gut-specific regulation of TH17 cell immunity. Nat Immunol 2023;24(12):2108–20.

[71] de Oca MM, Rivera FdL, Winterford C, TCM F, Ng SS, Amante FH, et al. IL-27 signalling regulates glycolysis in Th1 cells to limit immunopathology during infection. PLoS Pathog 2020;16(10)e1008994.
[72] Clénet ML, Laurent C, Lemaitre F, Farzam-kia N, Tastet O, Devergne O, et al. The IL-27/IL-27R axis is altered in CD4+ and CD8+ T lymphocytes from multiple sclerosis patients. Clin Transl Immunol 2021;10(3)e1262.
[73] Zha X, Yang S, Niu W, Tan L, Xu Y, Zeng J, et al. IL-27/IL-27R mediates protective immunity against chlamydial infection by suppressing excessive Th17 responses and reducing neutrophil inflammation. J Immunol 2021;206(9):2160–9.
[74] Ran Y, Huang D, Mei Y, Liu Z, Zhou Y, He J, et al. Identification of the correlations between interleukin-27 (IL-27) and immune-inflammatory imbalance in preterm birth. Bioengineered 2021;12(1):3201–18.
[75] Aioi A, Imamichi T. IL-27 regulates cytokine production as a double-edged sword in keratinocytes. Trends Immunother 2022;6(1):23–31.
[76] Coppock GM, Aronson LR, Park J, Qiu C, Park J, DeLong JH, et al. Loss of IL-27Rα results in enhanced tubulointerstitial fibrosis associated with elevated Th17 responses. J Immunol 2020;205(2):377–86.
[77] Okuzumi S, Miyata J, Kabata H, Mochimaru T, Kagawa S, Masaki K, et al. TLR7 agonist suppresses group 2 innate lymphoid cell-mediated inflammation via IL-27-producing interstitial macrophages. Am J Respir Cell Mol Biol 2021;65(3):309–18.
[78] Kumar S, Mulia GE, Figueiredo ML. Cabozantinib and IL-27 combinatorial therapy for bone-metastatic prostate cancer, Front Mol Biosci 2023;10: [cited 2024 Apr 18]. Available from: https://www.frontiersin.org/articles/10.3389/fmolb.2023.1259336.
[79] Baba AB, Rah B, Bhat GR, Mushtaq I, Parveen S, Hassan R, et al. Transforming growth factor-beta (TGF-β) signaling in cancer—a betrayal within, Front Pharmacol 2022;13: [cited 2024 Apr 18]. Available from: https://www.frontiersin.org/journals/pharmacology/articles/10.3389/fphar.2022.791272/full.
[80] Gneo L, Rizkalla N, Hejmadi R, Mussai F, de Santo C, Middleton G. TGF-β orchestrates the phenotype and function of monocytic myeloid-derived suppressor cells in colorectal cancer. Cancer Immunol Immunother 2022;71(7):1583–96.
[81] Lainé A, Labiad O, Hernandez-Vargas H, This S, Sanlaville A, Léon S, et al. Regulatory T cells promote cancer immune-escape through integrin αvβ8-mediated TGF-β activation. Nat Commun 2021;12(1):6228.
[82] Frangogiannis NG. Transforming growth factor-β in tissue fibrosis. J Exp Med 2020;217(3)e20190103.
[83] Stolfi C, Troncone E, Marafini I, Monteleone G. Role of TGF-beta and Smad7 in gut inflammation, fibrosis and cancer. Biomolecules 2021;11(1):17.
[84] Gómez-Bernal F, Quevedo-Abeledo JC, García-González M, Fernández-Cladera Y, González-Rivero AF, de Vera-González A, et al. Serum levels of transforming growth factor Beta 1 in systemic lupus erythematosus patients. Biomol Ther 2023;13(1):73.
[85] Peng D, Fu M, Wang M, Wei Y, Wei X. Targeting TGF-β signal transduction for fibrosis and cancer therapy. Mol Cancer 2022;21(1):104.
[86] Zhang M, Zhang YY, Chen Y, Wang J, Wang Q, Lu H. TGF-β signaling and resistance to cancer therapy, Front Cell Dev Biol 2021;(9) [cited 2024 Apr 18]. Available from: https://www.frontiersin.org/articles/10.3389/fcell.2021.786728.
[87] Ciardiello D, Elez E, Tabernero J, Seoane J. Clinical development of therapies targeting TGFβ: current knowledge and future perspectives. Ann Oncol 2020;31(10):1336–49.
[88] Sethi JK, Hotamisligil GS. Metabolic messengers: tumour necrosis factor. Nat Metab 2021;3(10):1302–12.
[89] van Loo G, Bertrand MJM. Death by TNF: a road to inflammation. Nat Rev Immunol 2023;23(5):289–303.
[90] Mahdavi Sharif P, Jabbari P, Razi S, Keshavarz-Fathi M, Rezaei N. Importance of TNF-alpha and its alterations in the development of cancers. Cytokine 2020;130:155066.
[91] Jasinski-Bergner S, Mandelboim O, Seliger B. Molecular mechanisms of human herpes viruses inferring with host immune surveillance. J Immunother Cancer 2020;8(2)e000841.
[92] Ge Z, Wu S, Zhang Z, Ding S. Mechanism of tumor cells escaping from immune surveillance of NK cells. Immunopharmacol Immunotoxicol 2020;42(3):187–98.
[93] Jang D, Lee AH, Shin HY, Song HR, Park JH, Kang TB, et al. The role of tumor necrosis factor alpha (TNF-α) in autoimmune disease and current TNF-α inhibitors in therapeutics. Int J Mol Sci 2021;22(5):2719.
[94] Gareb B, Otten AT, Frijlink HW, Dijkstra G, Kosterink JGW. Review: local tumor necrosis factor-α inhibition in inflammatory bowel disease. Pharmaceutics 2020;12(6):539.
[95] Yamanaka K, Yamamoto O, Honda T. Pathophysiology of psoriasis: a review. J Dermatol 2021;48(6):722–31.

[96] Leone GM, Mangano K, Petralia MC, Nicoletti F, Fagone P. Past, present and (foreseeable) future of biological anti-TNF alpha therapy. J Clin Med 2023;12(4):1630.

[97] Zhou Y, Fan R, Botchway BOA, Zhang Y, Liu X. Infliximab can improve traumatic brain injury by suppressing the tumor necrosis factor alpha pathway. Mol Neurobiol 2021;58(6):2803–11.

[98] Kciuk M, Yahya EB, Mohamed Ibrahim Mohamed M, Rashid S, Iqbal MO, Kontek R, et al. Recent advances in molecular mechanisms of cancer immunotherapy. Cancer 2023;15(10):2721:.

[99] Bergman PJ. Cancer immunotherapy. Vet Clin Small Anim Pract 2024;54(3):441–68.

[100] Ren A, Tong X, Xu N, Zhang T, Zhou F, Zhu H. CAR T-cell immunotherapy treating T-ALL: challenges and opportunities. Vaccine 2023;11(1):165.

[101] Marei HE, Hasan A, Pozzoli G, Cenciarelli C. Cancer immunotherapy with immune checkpoint inhibitors (ICIs): potential, mechanisms of resistance, and strategies for reinvigorating T cell responsiveness when resistance is acquired. Cancer Cell Int 2023;23(1):64.

[102] Wu Y, Zhang Z, Wei Y, Qian Z, Wei X. Nanovaccines for cancer immunotherapy: current knowledge and future perspectives. Chin Chem Lett 2023;34(8)108098.

[103] Ng JJQ, Goh D, Joseph CR, Lim JCT, Lau MC, Yeong JPS. The hurdle of precision medicine in Cancer immunotherapy: personalization now or then?, In: Rezaei N, editor. Handbook of cancer and immunology. Cham: Springer International Publishing; 2022. p. 1–32. [cited 2024 Apr 18], https://doi.org/10.1007/978-3-030-80962-1_289-1.

[104] Saijo H, Hirohashi Y, Honjo O, Saikai T, Shijubo N, Takabatake H, et al. Anti-CTLA-4 antibody might be effective against non-small cell lung cancer with large size tumor. Anticancer Res 2023;43(9):4155–60.

[105] Fazel M, Dufresne A, Vanacker H, Waissi W, Blay JY, Brahmi M. Immunotherapy for soft tissue sarcomas: anti-PD1/PDL1 and beyond. Cancer 2023;15(6):1643.

[106] García-Corbacho J, Indacochea A, González Navarro AE, Victoria I, Moreno D, Pesántez D, et al. Determinants of activity and efficacy of anti-PD1/PD-L1 therapy in patients with advanced solid tumors recruited in a clinical trials unit: a longitudinal prospective biomarker-based study. Cancer Immunol Immunother 2023;72(6):1709–23.

[107] Chen AP, Elad S, Geraldine O'S-C, Nancy M, Foster JC, Hu JS, et al. Atezolizumab for advanced alveolar soft part sarcoma. N Engl J Med 2023;389(10):911–21.

[108] Heather W, Moishe L, Terufumi K, Masahiro T, Se-Hoon L, Shugeng G, et al. Perioperative pembrolizumab for early-stage non-small-cell lung cancer. N Engl J Med 2023;389(6):491–503.

[109] Albrecht LJ, Livingstone E, Zimmer L, Schadendorf D. The latest option: nivolumab and relatlimab in advanced melanoma. Curr Oncol Rep 2023;25(6):647–57.

[110] Li B, Jin J, Guo D, Tao Z, Hu X. Immune checkpoint inhibitors combined with targeted therapy: the recent Advances and future potentials. Cancer 2023;15(10):2858.

[111] Borgeaud M, Sandoval J, Obeid M, Banna G, Michielin O, Addeo A, et al. Novel targets for immune-checkpoint inhibition in cancer. Cancer Treat Rev 2023;120:102614.

[112] Dutta S, Ganguly A, Chatterjee K, Spada S, Mukherjee S. Targets of immune escape mechanisms in cancer: basis for development and evolution of cancer immune checkpoint inhibitors. Biology 2023;12(2):218.

[113] Alturki NA. Review of the immune checkpoint inhibitors in the context of cancer treatment. J Clin Med 2023;12(13):4301.

[114] Yang F, Shay C, Abousaud M, Tang C, Li Y, Qin Z, et al. Patterns of toxicity burden for FDA-approved immune checkpoint inhibitors in the United States. J Exp Clin Cancer Res 2023;42(1):4.

[115] Abbasi S, Totmaj MA, Abbasi M, Hajazimian S, Goleij P, Behroozi J, et al. Chimeric antigen receptor T (CAR-T) cells: novel cell therapy for hematological malignancies. Cancer Med 2023;12(7):7844–58.

[116] Awasthi R, Maier HJ, Zhang J, Lim S. Kymriah® (tisagenlecleucel)—an overview of the clinical development journey of the first approved CAR-T therapy. Hum Vaccin Immunother 2023;19(1):2210046.

[117] Cappell KM, Kochenderfer JN. Long-term outcomes following CAR T cell therapy: what we know so far. Nat Rev Clin Oncol 2023;20(6):359–71.

[118] Flugel CL, Majzner RG, Krenciute G, Dotti G, Riddell SR, Wagner DL, et al. Overcoming on-target, off-tumour toxicity of CAR T cell therapy for solid tumours. Nat Rev Clin Oncol 2023;20(1):49–62.

[119] Seclì L, Leoni G, Ruzza V, Siani L, Cotugno G, Scarselli E, et al. Personalized cancer vaccines go viral: viral vectors in the era of personalized immunotherapy of cancer. Int J Mol Sci 2023;24(23):16591.

[120] Kaczmarek M, Poznańska J, Fechner F, Michalska N, Paszkowska S, Napierała A, et al. Cancer vaccine therapeutics: limitations and effectiveness—a literature review. Cells 2023;12(17):2159.
[121] Katsikis PD, Ishii KJ, Schliehe C. Challenges in developing personalized neoantigen cancer vaccines. Nat Rev Immunol 2024;24(3):213–27.
[122] Baulu E, Gardet C, Chuvin N, Depil S. TCR-engineered T cell therapy in solid tumors: state of the art and perspectives. Sci Adv 2023;9(7)eadf3700.
[123] Einsele H, Borghaei H, Orlowski RZ, Subklewe M, Roboz GJ, Zugmaier G, et al. The BiTE (bispecific T-cell engager) platform: development and future potential of a targeted immuno-oncology therapy across tumor types. Cancer 2020;126(14):3192–201.
[124] Yang X, Lin X, Shi W, Xie S, Huang X, Yin S, et al. Nanobody-based bispecific T-cell engager (Nb-BiTE): a new platform for enhanced T-cell immunotherapy. Sig Transduct Target Ther 2023;8(1):1–4.
[125] Arvedson T, Bailis JM, Britten CD, Klinger M, Nagorsen D, Coxon A, et al. Targeting solid tumors with bispecific T cell engager immune therapy. Annu Rev Cancer Biol 2022;6:17–34.
[126] Wei F, Cheng XX, Xue JZ, Xue SA. Emerging strategies in TCR-engineered T cells, Front Immunol 2022;13: [cited 2024 Apr 18]. Available from: https://www.frontiersin.org/journals/immunology/articles/10.3389/fimmu.2022.850358/full.
[127] van de Donk NWCJ, Zweegman S. T-cell-engaging bispecific antibodies in cancer. Lancet 2023;402(10396):142–58.
[128] Hiltensperger M, Krackhardt AM. Current and future concepts for the generation and application of genetically engineered CAR-T and TCR-T cells, Front Immunol 2023;14: [cited 2024 Apr 18]. Available from: https://www.frontiersin.org/journals/immunology/articles/10.3389/fimmu.2023.1121030/full.
[129] Oliveira G, Wu CJ. Dynamics and specificities of T cells in cancer immunotherapy. Nat Rev Cancer 2023;23(5):295–316.
[130] El-Sayes N, Vito A, Mossman K. Tumor heterogeneity: a great barrier in the age of cancer immunotherapy. Cancer 2021;13(4):806.
[131] Vitale I, Shema E, Loi S, Galluzzi L. Intratumoral heterogeneity in cancer progression and response to immunotherapy. Nat Med 2021;27(2):212–24.
[132] Megyesfalvi Z, Gay CM, Popper H, Pirker R, Ostoros G, Heeke S, et al. Clinical insights into small cell lung cancer: tumor heterogeneity, diagnosis, therapy, and future directions. CA Cancer J Clin 2023;73(6):620–52.
[133] Li J, Li L, You P, Wei Y, Xu B. Towards artificial intelligence to multi-omics characterization of tumor heterogeneity in esophageal cancer. Semin Cancer Biol 2023;91:35–49.
[134] Song X, Xiong A, Wu F, Li X, Wang J, Jiang T, et al. Spatial multi-omics revealed the impact of tumor ecosystem heterogeneity on immunotherapy efficacy in patients with advanced non-small cell lung cancer treated with bispecific antibody. J Immunother Cancer 2023;11(2)e006234.
[135] Chen L, Wang Y, Hu Q, Liu Y, Qi X, Tang Z, et al. Unveiling tumor immune evasion mechanisms: abnormal expression of transporters on immune cells in the tumor microenvironment, Front Immunol 2023;14: [cited 2024 Apr 18]. Available from: https://www.frontiersin.org/journals/immunology/articles/10.3389/fimmu.2023.1225948/full.
[136] Kim SK, Cho SW. The evasion mechanisms of cancer immunity and drug intervention in the tumor microenvironment, Front Pharmacol 2022;13: [cited 2024 Apr 18]. Available from: https://www.frontiersin.org/journals/pharmacology/articles/10.3389/fphar.2022.868695/full.
[137] Henke E, Nandigama R, Ergün S. Extracellular matrix in the tumor microenvironment and its impact on cancer therapy, Front Mol Biosci 2020;(6) [cited 2024 Apr 18]. Available from: https://www.frontiersin.org/articles/10.3389/fmolb.2019.00160.
[138] Xiao Y, Yu D. Tumor microenvironment as a therapeutic target in cancer. Pharmacol Ther 2021;221:107753.
[139] Cindy Yang SY, Lien SC, Wang BX, Clouthier DL, Hanna Y, Cirlan I, et al. Pan-cancer analysis of longitudinal metastatic tumors reveals genomic alterations and immune landscape dynamics associated with pembrolizumab sensitivity. Nat Commun 2021;12(1):5137.
[140] Bailey C, Black JRM, Reading JL, Litchfield K, Turajlic S, McGranahan N, et al. Tracking cancer evolution through the disease course. Cancer Discov 2021;11(4):916–32.
[141] Bejarano L, Jordão MJC, Joyce JA. Therapeutic targeting of the tumor microenvironment. Cancer Discov 2021;11(4):933–59.
[142] Galli F, Aguilera JV, Palermo B, Markovic SN, Nisticò P, Signore A. Relevance of immune cell and tumor microenvironment imaging in the new era of immunotherapy. J Exp Clin Cancer Res 2020;39(1):89.

[143] Ramos-Casals M, Brahmer JR, Callahan MK, Flores-Chávez A, Keegan N, Khamashta MA, et al. Immune-related adverse events of checkpoint inhibitors. Nat Rev Dis Primers 2020;6(1):1–21.

[144] Khan S, Gerber DE. Autoimmunity, checkpoint inhibitor therapy and immune-related adverse events: a review. Semin Cancer Biol 2020;64:93–101.

[145] Darnell EP, Mooradian MJ, Baruch EN, Yilmaz M, Reynolds KL. Immune-related adverse events (irAEs): diagnosis, management, and clinical pearls. Curr Oncol Rep 2020;22(4):39.

[146] Esfahani K, Elkrief A, Calabrese C, Lapointe R, Hudson M, Routy B, et al. Moving towards personalized treatments of immune-related adverse events. Nat Rev Clin Oncol 2020;17(8):504–15.

[147] Choi J, Lee SY. Clinical characteristics and treatment of immune-related adverse events of immune checkpoint inhibitors. Immune Netw 2020;20(1)e9.

[148] Schneider BJ, Naidoo J, Santomasso BD, Lacchetti C, Adkins S, Anadkat M, et al. Management of immune-related adverse events in patients treated with immune checkpoint inhibitor therapy: ASCO guideline update. J Clin Oncol 2021;39(36):4073–126.

[149] Luo J, Beattie JA, Fuentes P, Rizvi H, Egger JV, Kern JA, et al. Beyond steroids: immunosuppressants in steroid-refractory or resistant immune-related adverse events. J Thorac Oncol 2021;16(10):1759–64.

[150] Geethakumari PR, Ramasamy DP, Dholaria B, Berdeja J, Kansagra A. Balancing quality, cost, and access during delivery of newer cellular and immunotherapy treatments. Curr Hematol Malig Rep 2021;16(4):345–56.

[151] Schaft N, Dörrie J, Schuler G, Schuler-Thurner B, Sallam H, Klein S, et al. The future of affordable cancer immunotherapy, Front Immunol 2023;14: [cited 2024 Apr 18]. Available from: https://www.frontiersin.org/journals/immunology/articles/10.3389/fimmu.2023.1248867/full.

[152] Nesline MK, Knight T, Colman S, Patel K. Economic burden of checkpoint inhibitor immunotherapy for the treatment of non-small cell lung cancer in US Clinical Practice. Clin Ther 2020;42(9):1682–1698.e7.

[153] Mirsky MM, Mitchell C, Hong A, Cao S, Fu P, Margevicius S, et al. Outcomes of antineoplastic immunotherapy at a large healthcare organization: impact of provider, race and socioeconomic status. Cancer Manag Res 2023;15:913–27.

[154] Gupta A, Omeogu C, Islam JY, Joshi A, Zhang D, Braithwaite D, et al. Socioeconomic disparities in immunotherapy use among advanced-stage non-small cell lung cancer patients: analysis of the National Cancer Database. Sci Rep 2023;13(1):8190.

[155] Sahara K, Farooq SA, Tsilimigras DI, Merath K, Paredes AZ, Wu L, et al. Immunotherapy utilization for hepatobiliary cancer in the United States: disparities among patients with different socioeconomic status. Hepatobiliary Surg Nutr 2020;9(1):13–24.

[156] Patel A, Goldstein DA, Tannock IF. Improving access to immunotherapy in low- and middle-income countries. Ann Oncol 2022;33(4):360–1.

[157] Lane AN, Higashi RM, Fan TWM. Metabolic reprogramming in tumors: contributions of the tumor microenvironment. Genes Dis 2020;7(2):185–98.

[158] Missiaen R, Lesner NP, Simon MC. HIF: a master regulator of nutrient availability and metabolic cross-talk in the tumor microenvironment. EMBO J 2023;42(6)e112067.

[159] Zhou D, Duan Z, Li Z, Ge F, Wei R, Kong L. The significance of glycolysis in tumor progression and its relationship with the tumor microenvironment, Front Pharmacol 2022;13: [cited 2024 Apr 19]. Available from: https://www.frontiersin.org/journals/pharmacology/articles/10.3389/fphar.2022.1091779/full.

[160] Angiari S, Runtsch MC, Sutton CE, Palsson-McDermott EM, Kelly B, Rana N, et al. Pharmacological activation of pyruvate kinase M2 inhibits CD4+ T cell pathogenicity and suppresses autoimmunity. Cell Metab 2020;31(2) 391–405.e8.

[161] St Paul M, Saibil SD, Han S, Israni-Winger K, Lien SC, Laister RC, et al. Coenzyme a fuels T cell anti-tumor immunity. Cell Metab 2021;33(12):2415–2427.e6.

[162] Shyer JA, Flavell RA, Bailis W. Metabolic signaling in T cells. Cell Res 2020;30(8):649–59.

[163] Geiger R, Rieckmann JC, Wolf T, Basso C, Feng Y, Fuhrer T, et al. L-arginine modulates T cell metabolism and enhances survival and anti-tumor activity. Cell 2016;167(3):829–842.e13.

[164] Ren W, Liu G, Yin J, Tan B, Wu G, Bazer FW, et al. Amino-acid transporters in T-cell activation and differentiation. Cell Death Dis 2017;8(3):e2655.

[165] Mellor AL, Munn D, Chandler P, Keskin D, Johnson T, Marshall B, et al. Tryptophan catabolism and T cell responses, In: Allegri G, Costa CVL, Ragazzi E, Steinhart H, Varesio L, editors. Developments in tryptophan and serotonin metabolism. Boston, MA: Springer US; 2003. p. 27–35. [cited 2024 Apr 19] https://doi.org/10.1007/978-1-4615-0135-0_3.

[166] Sikalidis AK. Amino acids and immune response: a role for cysteine, glutamine, phenylalanine, tryptophan and arginine in T-cell function and Cancer? Pathol Oncol Res 2015;21(1):9–17.
[167] Han C, Ge M, Ho PC, Zhang L. Fueling T-cell antitumor immunity: amino acid metabolism revisited. Cancer Immunol Res 2021;9(12):1373–82.
[168] Sukumar M, Kishton RJ, Restifo NP. Metabolic reprograming of anti-tumor immunity. Curr Opin Immunol 2017;46:14–22.
[169] Guo Y, Xie YQ, Gao M, Zhao Y, Franco F, Wenes M, et al. Metabolic reprogramming of terminally exhausted CD8+ T cells by IL-10 enhances anti-tumor immunity. Nat Immunol 2021;22(6):746–56.
[170] Van Wilpe S, Koornstra R, Den Brok M, De Groot JW, Blank C, De Vries J, et al. Lactate dehydrogenase: a marker of diminished antitumor immunity. Onco Targets Ther 2020;9(1):1731942.
[171] Feng Q, Liu Z, Yu X, Huang T, Chen J, Wang J, et al. Lactate increases stemness of CD8 + T cells to augment anti-tumor immunity. Nat Commun 2022;13(1):4981.
[172] Finetti F, Travelli C, Ercoli J, Colombo G, Buoso E, Trabalzini L. Prostaglandin E2 and cancer: insight into tumor progression and immunity. Biology 2020;9(12):434.
[173] Allard B, Beavis PA, Darcy PK, Stagg J. Immunosuppressive activities of adenosine in cancer. Curr Opin Pharmacol 2016;29:7–16.
[174] Ray A, Song Y, Du T, Tai YT, Chauhan D, Anderson KC. Targeting tryptophan catabolic kynurenine pathway enhances antitumor immunity and cytotoxicity in multiple myeloma. Leukemia 2020;34(2):567–77.
[175] Kabouridis PS, Janzen J, Magee AL, Ley SC. Cholesterol depletion disrupts lipid rafts and modulates the activity of multiple signaling pathways in T lymphocytes. Eur J Immunol 2000;30(3):954–63.
[176] Ma X, Bi E, Lu Y, Su P, Huang C, Liu L, et al. Cholesterol induces CD8+ T cell exhaustion in the tumor microenvironment. Cell Metab 2019;30(1) 143–156.e5.
[177] Sestili P, Fimognari C. Cytotoxic and antitumor activity of sulforaphane: the role of reactive oxygen species. Biomed Res Int 2015;2015:e402386.
[178] Shen C, Zhang Z, Tian Y, Li F, Zhou L, Jiang W, et al. Sulforaphane enhances the antitumor response of chimeric antigen receptor T cells by regulating PD-1/PD-L1 pathway. BMC Med 2021;19(1):283.

CHAPTER

6

TCR-T cells for cancer immunotherapy

Mohamed Shawky Hussein[a], *Rui Mao*[a], *and Yukai He*[a,b]

[a]Georgia Cancer Center, Medical College of Georgia, Augusta, GA, United States [b]Department of Medicine, Medical College of Georgia, Augusta University, Augusta, GA, United States

Abstract

In this chapter, we are discussing the advances in cancer immunotherapy with a particular focus on adoptive cell therapy (ACT) using T cell receptor-engineered T (TCR-T) cells. Despite the remarkable success of chimeric antigen receptor T (CAR-T) cells in the treatment of hematological malignancies, they demonstrated modest results when applied to solid tumors. Unlike CAR-T cells, TCR-T cells recognize both surface and intracellular proteins, making them a great alternative as they can detect a broad range of potential targetable antigens. Additionally, TCRs respond to much lower epitope density than CARs and have signaling pathways similar to the endogenous natural receptor. Herein, we summarize the current status of TCR-T cell therapy and critical factors required to develop effective TCR-T immunotherapy, including antigen selection, TCR identification and optimization, and toxicities associated with TCR-T engineering. Also, we discuss potential challenges and immune resistance facing TCR-T therapy.

Abbreviations

ACT	adoptive cell therapy
AFP	alfa-fetoprotein
AP-1	activator protein-1
CAR	chimeric antigen receptor
CAR-T	chimeric antigen receptor T cell
CD	cluster of differentiation
CEA	carcinoembryonic antigen
CIK	cytokine-induced killer cells
CRS	cytokine release syndrome
CTLA-4	cytotoxic T lymphocyte-associated antigen 4
DAG	diacylglycerol
EBV	Epstein-Barr virus
ER	endoplasmic reticulum
FDA	Food and Drug Administration
gp	glycoprotein

T Cell Metabolism and Cancer Immunotherapy
https://doi.org/10.1016/B978-0-443-13827-0.00013-5

HLA	human leukocyte antigen
HPV	human papillomaviruses
HTLV-1	human T cell lymphotropic virus-1
ICIs	immune checkpoint inhibitors
ICOS	inducible costimulator
IFN- γ	interferon-γ
IL-12	interleukin-12
IL-15	interleukin-15
IL-2	interleukin-2
IL-7	interleukin-7
IL-8	interleukin-8
IP3	inositol trisphosphate
ITAMs	immunoreceptor tyrosine-based activation motifs
LAG-3	lymphocyte-activation gene 3
LAGE-1	L antigen family member-1
LAK	lymphokine-activated killer cells
LAT	linker for activation of T cell family member 1
LFA-1	lymphocyte function-associated antigen 1
LMP	latent membrane protein
MAGE-A1	melanoma-associated gene-A1
MAPK	mitogen-activated protein kinase
MART-1	melanoma antigen recognized by T cells 1
MHC	major histocompatibility complex
NFAT	nuclear factor of activated T cells
NF-κB	nuclear factor-κB
NY-ESO-1	New York-esophageal squamous cell carcinoma-1
PD-1	programmed cell death 1
PD-L1	programmed death ligand 1
PIP_3	phosphatidylinositol triphosphate
PKCθ	protein kinase C
PLCγ	phospholipase Cγ
RasGRP	RAS guanyl nucleotide-releasing protein
scFv	single-chain variable fragment
Syk	spleen tyrosine kinase
TCR	T cell receptor
TCR-T	T cell receptor-engineered T cell
TGF-β	transforming growth factor-β
TIGIT	T cell immunoreceptor with immunoglobulin and ITIM domains
TILs	tumor-infiltrating lymphocytes
TIM-3	T cell immunoglobulin and mucin domain-containing-3
TM	transmembrane domain
VH	variable heavy
VL	variable light
ZAP70	ζ-associated protein of 70 kDa

No conflict statement

Augusta University patented the AFP-specific TCRs that were licensed by Cellular Biomedicine Group, Inc., which is sponsoring a clinical trial for treating HCC NCT03971747. YH is one of the inventors of the patent. M.S.H. and R.M. declare no potential conflicts of interest.

Introduction

Immunotherapy with checkpoint inhibitors (ICIs) and adoptive cell therapy (ACT) have become a frontline cancer treatment in the past decade. ICIs have been used as first-line treatment for several forms of cancer and viral infections [1–3]. ICI's efficacy relies on the presence of effective tumor-infiltrating T cells, which is usually limited in most solid tumors. This represents an obstacle to several of the recent Food and Drug Administration (FDA)-approved ICIs in which only a small fraction of patients benefit from them [4–6].

On the other hand, ACT is a therapeutic approach that involves the administration of ex vivo sensitized and/or expanded immune cells then re-infused into patients [7]. Several ACT approaches have been described, including the transfer of autologous $CD8^+$ T cells, lymphokine-activated killer cells (LAK) [8,9], cytokine-induced killer cells (CIK) [10,11], natural killer (NK) cells [12,13], γδ T cells [14,15], tumor-infiltrating lymphocytes (TILs) [16,17], and engineering peripheral blood T cells with specific receptors such as tumor antigen-specific T cell receptors (TCR-T cells) [18–20] and chimeric antigen receptors (CAR-T cells) [21–23]. TCR-Ts and TILs are major histocompatibility complex (MHC)-dependent, thus limiting the number of patients who can benefit from them, while CAR-Ts, LAK, and CIK are MHC-independent [24–26].

ACT using TCR-T and CAR-T cells is a rapidly growing research field and represents a promising approach to treat human cancers [7,27,28]. Although CAR-T cell therapies are FDA-approved for hematological malignancies targeting CD19 or B cell maturation antigen, they are still in development to effectively treat solid tumors [21,29–31]. This could be attributed to the limited availability of targetable antigens, tumor heterogeneity, and/or the immunosuppressive tumor microenvironment [32]. In contrast, TCR-T cells such as the New York-esophageal squamous cell carcinoma-1 (NY-ESO-1)-specific TCR-Ts have demonstrated great potential and significant clinical outcomes in treating patients with advanced multiple myeloma and metastatic synovial sarcoma [33,34]. Recently, the FDA-approved tebentafusp, a bispecific glycoprotein100 (gp100) and CD3 TCR-T cell therapy for adult patients with HLA-A*02:01-positive unresectable or metastatic uveal melanoma (NCT03070392) [35]. In this chapter, we will focus on reviewing the recent development of TCR-T cell therapy, and in some cases, comparisons to CAR-T cells will be made.

TCR-T and CAR-T cell therapy

TCR-T therapy might have broader applicability and several advantages over CAR-T therapy. More than 90% of tumor antigens are intracellular proteins [36], which are targetable by TCR-T cells but not by CAR-T cells. Additionally, CAR-T cells require more antigen density (1-5 $\times 10^3$ target antigen/cell) than TCR-T cells (1–50 target antigen/cell) to induce activation and antigen-specific cytokine release [18,19].

TCR structure and signaling

In general, genetically engineered TCRs are structurally similar to the endogenous TCR, consisting of TCR_α and TCR_β chains associated with CD3 γ, δ, and ε chains, and signaling ζ chains (Fig.1). TCRs recognize intracellular antigens processed and presented by MHC to provide signal 1 for T cell activation. In addition to antigen recognition, T cells require a costimulatory signal (signal 2) from costimulatory domains such as CD28, 4-1BB, or OX-40 to become fully activated and initiate T cell response [28].

TCR signaling depends on the recruitment of downstream signaling molecules such as linker for activation of T cell family member 1 (LAT), spleen tyrosine kinase (Syk), and ζ-associated protein of 70 kDa (ZAP70) to transduce the activation signal to the nucleus [37]. Briefly, CD4 and CD8 coreceptors facilitate signaling through leukocyte-specific tyrosine kinase (Lck), which phosphorylates tyrosine residues within the immunoreceptor tyrosine-based activation motif (ITAMs) of CD3 γ, δ, and ε chains and signaling ζ chains. The ζ chains act as docking sites for ZAP70, which is phosphorylated and activated by Lck. ZAP70 phosphorylates several adaptor proteins and enzymes of major signaling pathways. ZAP70 phosphorylates and activates phospholipase Cγ (PLCγ), which in turn cleaves phosphatidylinositol triphosphate (PIP_3) to generate inositol trisphosphate (IP3) and diacylglycerol (DAG). IP3 stimulates Ca^{2+} efflux from the endoplasmic reticulum (ER) to increase cytoplasmic Ca^{2+} concentration to activate calcineurin, which dephosphorylates nuclear factor of activated T cells (NFAT), enabling its translocation to the nucleus. On the other side, DAG activates protein kinase C (PKCθ), participating in the activation of the transcription factor nuclear factor-κB (NF-κB). DAG also activates RAS guanyl nucleotide-releasing protein (RasGRP)-MAPK pathways, leading to activation of activating protein 1 (AP-1). NFAT, NFκB, and AP-1 bind to the promoter and induce the transcription of several transcription factors required for T cell activation and proliferation, such as interleukin 2 (IL-2) [21,38].

CAR structure and signaling

CARs have four main parts: (a) an extracellular antigen-binding domain usually consisting of a single-chain variable fragment (scFv) derived from a monoclonal antibody specific for a cell surface tumor antigen; (b) a flexible hinge; (c) a transmembrane domain (TM); and (d) an intracellular signaling domain with one or more costimulatory signaling domains through which the cell receives both signals 1 and 2 [26,28] (Fig.1). The antigen-binding domain grants specificity to CAR-Ts. The scFv consists of variable heavy (VH) and light (VL) chains connected by a flexible linker and can allow the recognition of both cell surface and soluble antigens. The hinge provides flexibility for the antigen-binding domain and facilitates synapse formation between CAR-Ts and target cells. The transmembrane domain anchors the CAR to the cell membrane and plays an important role in CAR expression, stability, and overall functionality. The intracellular signaling domain triggers T cell activation [19,21]. The first generation CARs have only a single CD3ζ domain, and despite their ability to elicit an immune response in vitro, the in vivo antitumor effects were minimal [39,40]. The second generation CARs incorporate one costimulatory signaling domain proximal to the cell membrane followed by CD3ζ. CD28 and 4-1BB are the most commonly used costimulatory signaling domains for 2nd-generation CARs. This configuration enhanced T cell functionality in vitro and the in vivo antitumor response [41,42]. Other costimulatory signaling domains include OX40, ICOS, CD27, 2B4, and CD3ε [21,43]. The third generation CARs incorporate two or more costimulatory signaling domains with CD3ζ. The third generation CAR-Ts using both 4-1BB and ICOS showed significant antitumor effects with increased T cell persistence in vivo compared to the second generation with either 41BB or ICOS alone [44].

CAR signaling works differently from TCR signaling. Although CAR combines both signal 1 (CD3ζ domain) and signal 2 (one or more costimulatory signaling domains), it requires

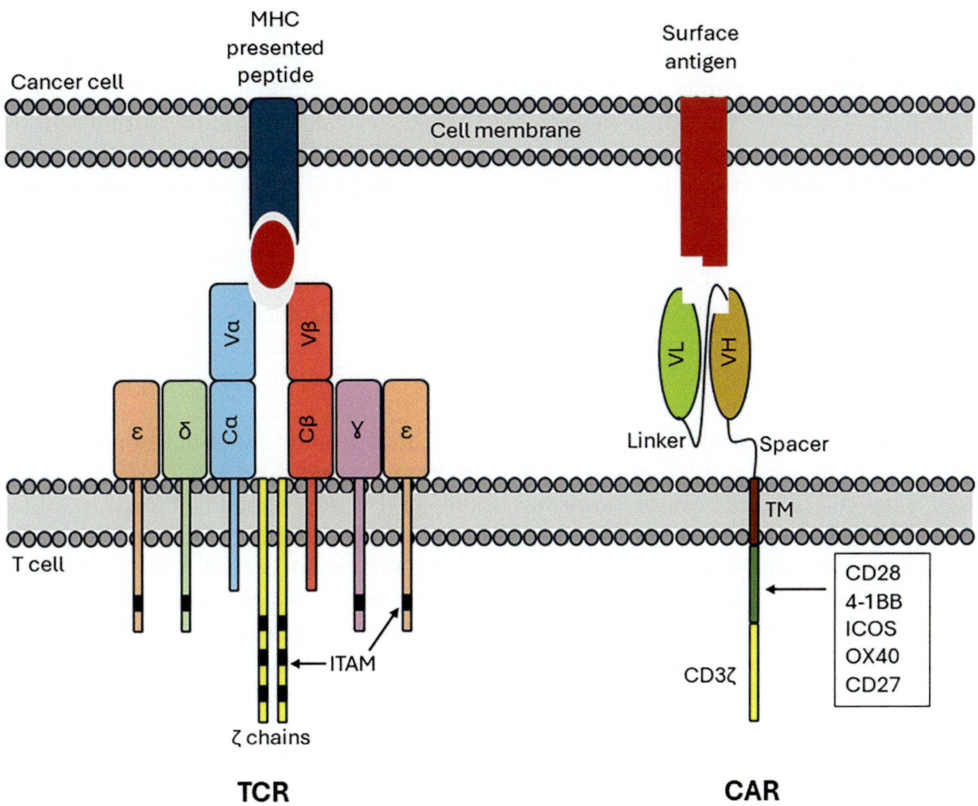

FIG. 1 Schematic diagram of basic structure of engineered TCR and CAR.

oligomerization to initiate signaling [45]. The immunological synapse formed between a CAR and a cell expressing target antigen is different from that of a TCR and less defined [46]. CAR signals less efficiently than a TCR and is characterized by rapid downstream signaling less dependent on lymphocyte function-associated antigen 1 (LFA-1) [19]. CAR requires greater antigen levels, although it has a higher affinity [47].

TCR-T engineering

Antigen selection

The first and most critical step in generating effective and safe TCR-T therapy is antigen selection. An ideal antigen should fulfill the following criteria: (a) tumor specific: to be expressed mainly on tumor cells but not on normal tissue; (b) tumor addiction: to be related to the ongoing process of oncogenesis; (c) immunogenic: to be capable of activating T cells and evoke an adequate immune response [27]. Antigens with selective tumor expression include neo-antigens derived from somatic mutations, cancer germline antigens, and viral antigens.

Neo-antigens are highly immunogenic tumor antigens that result from point mutations or gene rearrangements that occur during the process of oncogenesis, and their frequency is directly correlated to the mutational load. The higher the mutational load, the higher the presence of neo-antigens within tumor tissues. Neo-antigens related to cellular proliferation and tumor growth are promising therapeutic targets for TCR-T therapy, and studies targeting such antigens resulted in significant antitumor responses [48]. In colorectal cancers, the density of tumor-infiltrating CD8 T cells was found to correlate with the number of frameshift mutations [49]. Approaches such as whole genome or exome sequencing and immunopeptidomics-based screening help in identifying optimal neo-antigens and tumor-specific mutations [50,51]. Neo-antigens arising from the mutation of driver genes such as TP53 and KRAS that are shared between different cancer patients are considered public neo-antigens, and creating TCRs against these antigens can help in increasing the availability of TCR-T therapy to more cancer patients [51].

Oncoviral antigens have high immunogenicity and provide potential nonself-antigens with no toxicity toward normal tissues [18]. Human endogenous retroviruses are expressed in most renal cell cancers but not in normal tissues [52]. Viruses such as human papillomaviruses (HPV), Epstein-Barr virus (EBV), and human T cell lymphotropic virus-1 (HTLV-1) account for more than 10% of human cancers, and their products contribute to the process of oncogenesis [53]. TCR-Ts against E6 and E7 viral antigens expressed in HPV-associated cancers have been developed and showed effective clinical responses with no apparent toxicities [54,55]. EBV-specific TCR-T against latent membrane proteins (LMP1 and LMP2) in EBV-associated nasopharyngeal carcinoma is now tested in the clinic (NCT04509726, NCT03648697, and NCT03925896).

Cancer germline antigens are also called cancer testis antigens, as they are encoded by genes that are normally only expressed in male testis germ cells. Melanoma-associated antigens (MAGE) and New York-esophageal squamous cell carcinoma-1 (NY-ESO-1) are cancer germline antigens that are de-repressed in some types of cancers while being repressed in normal cells, making them good candidates to avoid on-target toxicity. Cancer germline antigens have been associated with the process of oncogenesis and metastasis by inhibiting p53-dependent apoptosis and inducing epithelial-to-mesenchymal transition [56,57]. Melanoma patients who received the MAGE tumor antigen vaccine showed favorable clinical outcomes without major toxicity [58]. Additionally, cancer germline antigen immunogenicity can be enhanced with epigenetic drugs that control DNA demethylation and histone acetylation, sensitizing tumor cells to CD8 T cells [59,60]. Clinical trials of TCR-T therapy against NY-ESO-1 showed promising outcomes in patients with melanoma and synovial sarcoma [33,34,61].

Tissue differentiation antigens such as carcinoembryonic antigen (CEA), melanoma antigen recognized by T cells 1 (MART-1), and glycoprotein 100 (gp100) are overexpressed in cancers and often shared between patients, making them attractive targets for TCR-T therapy with broader application. Since they are tumor-associated and not tumor-specific, they are also expressed in normal tissues. Targeting these antigens could be associated with on-target/off-tumor toxicity. Despite observing some clinical responses from TCR-T cells targeting differentiation antigens, several toxicities of varying intensities have been described [62–64].

TCRs for TCR-T therapy

Generating an epitope-specific TCR to obtain antigen-specific T cells is the second important step in TCR-T therapy. In recent advances in generating mouse models with humanized T cell repertoire, different in vivo approaches have been used to identify epitope-specific TCR, such as (a) immunizing human HLA transgenic mice with human antigens to generate TCRs against human antigens, as with Alfa-fetoprotein $(AFP)_{158}$-specific TCRs and gp100-specific TCRs [65,66], and (b) using transgenic mice for human TCR and HLA-A2 genes to obtain humanized TCRs, as with MAGE-A1- and NY-ESO-1-specific TCRs that resulted in improved functionality [67]. After getting antigen-specific T cells by any of these approaches, T cell hybridomas are established, and molecular techniques such as 5′ RACE are used to determine the paired TCRα and TCRβ chain genes from T cell hybridomas [68]. Utilizing this approach, we identified a large number of HLA-A2/AFP158-specific TCRs from HLA-A2 AAD transgenic mice by priming with recombinant lentiviral vector followed by peptide boosting [65]. We tested 3 TCRs after transduction into human T cells. Both in vitro and in vivo tests revealed strong antitumor efficacy. We further examined the cross-reactivity of AFP-specific TCR-T cells by comprehensive X-scans, and we showed that there is little to no cross-reactivity between these TCR-T cells and human cells [69]. One TCR was then selected [70] and moved into clinical trial for the treatment of HCC patients (NCT03971747).

Other approaches include T cell priming by coculturing T cells from cancer patients or healthy donors with target antigen or artificial antigen-presenting cells genetically engineered to express human leucocyte antigen (HLA) of choice and costimulatory ligands CD80 and CD83 loaded with the peptide of choice [18,71], followed by sorting and sequencing to identify highly functional epitope-specific TCRs from primed T cells.

In addition, a latest study reported the identification of TCR that could recognize and bind the HLA-DRA/DRB1*09:01 presented HPV18 $E7_{84-98}$ peptide, demonstrating that long-time cancer survivors successfully treated by immunotherapy may harbor a vast source of effective TCRs that remain to be identified to generate TCR-T cells [72].

The identified new tumor-specific TCRs have to be tested for their sensitivity, affinity, and safety after introduction into T cells before moving into clinical application. First, TCR assembly and conformation are validated in vitro using a tetramer binding assay. Next, TCR-T cells' specificity, efficacy, and avidity are tested by coculture in vitro against their cognate peptide to measure cytokine release and determine the effective concentration to achieve 50% efficacy as a measure for T cell avidity. Additionally, they are tested against autologous tumor cells expressing the HLA and antigens specific to TCRs. Furthermore, in vivo administration of TCR-T cells in engineered mouse models is used to study T cell tumor infiltration and persistence and the overall antitumor immune response. The most widely used mouse models are the immunocompromised NSG mice transplanted with human tumor expressing the HLA and antigen of interest and human HLA-A2 transgenic mice transplanted with a syngeneic tumor expressing the human antigen of interest [27,73].

Several efforts have been carried out to enhance TCR-T therapy efficacy and functional outcomes. The interactions between variable and constant regions of TCR are important for TCR conformation, and strategies to improve the pairing of TCRα and TCRβ chains resulted in improved surface expression of the TCRs [74,75]. TCR and peptide-MHC (p-MHC) affinity is thought to play a critical role in antigen recognition and T cell cytotoxicity [76,77]. TCR affinities can be enhanced through mutations in the complementarity-determining region, phage display, or immunizing humanized mice with the epitope of choice [18,78], but high-affinity TCRs are a double-edged sword and may cause severe on-target and off-target toxicities [73,79–81]. Also, TCR-Ts with high-affinity TCRs are susceptible to activation-induced cell death following antigen recognition and stimulation [82]. Approaches to enhance TCR signaling after an antigen encounter through inserting a signaling cassette with costimulatory molecules, such as CD28 or 4-1BB, similar to the approach used in CAR-T cells to improve TCR-T cell function, have been done but with varying outcomes [83–85].

Challenges and side effects of TCR-T therapy

TCR-T toxicity

Toxicities associated with TCR-T cells represent critical barriers to the safety and efficacy of TCR-T therapy. Toxicity caused by recognizing the target antigen expressed on normal healthy tissues is known as on-target/off-tumor toxicity. Examples include uveitis, hearing loss, and skin inflammation with high avidity TCR-T cells recognizing melanoma antigens HLA-A2+/MART-1+ and HLA-A2+/gp 100+ [63,64]. Severe acute inflammatory colitis was reported with the use of high-affinity HLA-A2+/ CEA+ TCR-Ts targeting CEA+ colon cancer due to its expression on normal intestinal tissues [62]. On the other hand, toxicity caused by recognizing a highly similar or a different antigen expressed by normal healthy cells due to TCR cross-reactivity is known as off-target/off-tumor toxicity [86]. Examples include neurological toxicity reported with high avidity HLA-A2+/MAGEA3+ TCR-Ts from cross-reactivity against MAGE-A12 protein [80], and cardiac tissue destruction and cardiogenic shock associated with high-affinity HLA-A1+/MAGE-A3+ TCR-Ts from cross-reactivity against Titin protein expressed in cardiomyocytes leading to the death of four patients [81,87]. Furthermore, cytokine release syndrome (CRS) has been described with the use of TCR-T cells against MART-I, NY-ESO-1, and L antigen family member 1 (LAGE-1) [88]. Thus, it is essential to optimize TCR affinity within a certain range to avoid unpredicted cross-reactivities and serious adverse reactions.

Several methods can be applied to predict TCR-T cell toxicity. These include the coculture of TCR-T cells with primary normal cells in vitro to assess cytotoxicity and cytokines production [89]; the use of TCR-like antibodies to identify p-MHC targets in normal tissues [90]; proteomic analysis and immunopeptidomics to detect TCR-T cells recognizing normal tissues [18]; and TCR clustering using complementarity-determining region similarities to elucidate similarities between TCRs against the same epitope and can help to predict TCR-T cells cross-reactivity [91,92].

TCR mispairing

Transgenic TCR α or β chains mispairing with natural/endogenous TCR α or β chains can lead to the formation of a new receptor with altered functionality due to competition for signaling molecules or even a receptor with new specificity. To minimize mispairing and improve the expression of the transgenic TCR, several approaches have been described. These include: introducing an interchain disulfide bond between the cysteine residues of the TCR constant regions to facilitate the assembly of the transgenic TCR α and β chains [93]; substituting all or some murine residues in place of the human ones in the TCR constant regions [94,95]; knocking out endogenous TCR using CRISPR technique or by cotransferring endogenous TCR siRNA or TCR chain-specific zinc finger nucleases [96,97]; inverting the human sequences of the TCR α and β constant regions [98]; generating mutations in the transgenic TCR α and β chain C-terminal domains to promote correct binding and increase TCR transduction levels [99]; modifying the TCRα transmembrane region hydrophobicity [100]; and using single-stranded TCR by linking the variable domains of the TCR chains or TCR/CD3 fusion products to compete with endogenous TCR [101,102].

Immune evasion and therapeutic resistance

TCR-T cells are MHC dependent and require antigen presentation by MHC class I to recognize their target epitope. Some tumors lose the expression of MHC class I molecules and are no longer recognized by TCR-T cells. This could result from mutations of β2-microglobulin or HLA genes. Loss of heterozygosity and epigenetic silencing of HLA have been reported in recent studies, and the use of hypomethylating agents can help to restore the expression of epigenetic silenced HLA [18,103,104].

Downregulation or loss of target tumor antigens after TCR-T cell therapy can result in immune evasion and tumor recurrence. Strategies to overcome low expression of target antigen include using T cells targeting more than one tumor antigen; infusing various T cell clones with TCRs specific for different tumor antigens; or generating TCRs to target tumor antigens required for the ongoing process of oncogenesis and tumor survival [105].

T cell exhaustion

T cell exhaustion represents another major obstacle for effective TCR-T therapy, particularly for solid tumors. It is a dysfunctional state of T cells that occurs during cancer from continuous antigen exposure and inflammation [106]. The loss of T cell effector functions occurs in a hierarchical manner with early stages of exhaustion characterized by loss of IL-2 production, proliferative capacity, cytokine production, and less persistence, followed by defects in tumor necrosis factor (TNF), interferon-γ (IFN-γ), and chemokines production [107,108]. Exhausted T cells are accompanied by progressive upregulation of inhibitory markers, such as programmed cell death 1 (PD-1), cytotoxic T lymphocyte-associated antigen 4 (CTLA-4), lymphocyte-activation gene 3 (LAG-3), T cell immunoglobulin and mucin domain containing-3 (TIM-3), and T cell immunoreceptor with immunoglobulin and ITIM domains (TIGIT) [109]. Advances in single-cell transcriptomics and genome-wide epigenetic profiling showed that exhausted CD8 T cells isolated from tumor possess widespread epigenetic changes and significant enrichment of recent TCR signaling genes such as *Nr4a*, *Nfatc*, *Batf*, and *Irf4*, suggesting that prolonged antigen persistence and stimulation is a dominant driver of T cell exhaustion [110–112].

Several approaches have been utilized to overcome exhaustion and improve the outcome of TCR-T therapy. TCR-T cells armored with pro-inflammatory cytokines such as IL-12 or IL-

18 to enhance T cell activation and effector function have been developed [113,114]. IL-18-secreting T cells were found to resist exhaustion and persist in vivo to enhance survival of the melanoma xenograft mouse model [114]. TCR-T cells engineered to overexpress c-Jun demonstrated enhanced expansion potential and cytokine production together with improved tumor infiltration and overall survival of the hepatocellular carcinoma xenograft mouse model [20]. Additionally, combining ICIs such as anti-PD1/PD-L1 or CTLA4 blockade with ACT can help to resist T cell exhaustion [115,116]. Furthermore, generating TCR-T cells from the minimally differentiated T stem cell memory (T_{SCM}) and T central memory (T_{CM}) can improve the persistence and the antitumor effects of the infused T cells [117–119].

Summary and conclusion

TCR-T immunotherapy has great potential and has already shown promising responses in patients with various solid tumors. Tebentafusp is the first FDA-approved TCR-T therapy for patients with unresectable or metastatic uveal melanoma. It consists of a TCR molecule fused to an antibody fragment. The TCR arm binds to gp100 presented by HLA-A*02:01, and the antibody arm binds to CD3 on T cells. The median 3-year overall survival was 21.6 months in the treatment group compared to 16.19 months in the control group [35]. More recently, the FDA approved a priority review for Afami-cel, a TCR-T therapy against MAGE-A4 for HLA-A*02 positive patients with advanced synovial sarcoma based on encouraging results from the pivotal trial SPEARHEAD-1 (NCT04044768) with an anticipated decision by August 2024 [120] (Note: at the time of publication, FDA has just approved the TCR-T cells for unresectable or metastatic synovial sarcoma).

Several challenges associated with TCR-T therapy need to be addressed to resolve its complexity and enhance its function and safety. Special considerations should be implemented regarding the selection of TCR-T target antigen, optimization of TCRs to enhance the antitumor immune response, and toxicity problems associated with TCR-T therapy. Combining TCR-T therapy with other immunotherapies such as ICIs, immune-stimulatory cytokines, or genetically engineering TCR-T cells to express a PD-1–CD28 chimeric construct to transform inhibitory signal into stimulatory signal or to express a dominant negative TGF-β receptor can help to overcome the immunosuppressive tumor microenvironment [115,116,121]. The addition of cytokines such as IL-7 and IL-15 to the T cell culture medium enables the expression of T_{SCM} with enhanced proliferation and survival capacity [122]. We believe with the advances in genetic engineering, genomic and proteomic analysis, and cell manufacturing techniques, other TCR-T cell therapies will be approved for other solid tumors.

Acknowledgments

The authors would like to thank current and previous members of Dr. He's laboratory. In addition, over the years, research in He's lab has been funded by NIH/NCI grants (CA168912 and CA235159) and the Augusta University startup package.

References

[1] Pan Y, Tan J, Li J, Li T, Li J, Cao Y, et al. Immune checkpoint inhibitors in cancer patients with COVID-19. Open Life Sci 2023;18(1):20220641.

[2] Dyck L, Mills KHG. Immune checkpoints and their inhibition in cancer and infectious diseases. Eur J Immunol 2017;47(5):765–79.

[3] Wykes MN, Lewin SR. Immune checkpoint blockade in infectious diseases. Nat Rev Immunol 2018;18(2):91–104.

[4] Greten TF, Lai CW, Li G, Staveley-O'Carroll KF. Targeted and immune-based therapies for hepatocellular carcinoma. Gastroenterology 2019;156(2):510–24.

[5] Pico de Coaña Y, Choudhury A, Kiessling R. Checkpoint blockade for cancer therapy: revitalizing a suppressed immune system. Trends Mol Med 2015;21(8):482–91.

[6] Beaver JA, Pazdur R. The wild West of checkpoint inhibitor development. New England Journal of Medicine 2021;386(14):1297–301.

[7] Rosenberg SA, Restifo NP, Yang JC, Morgan RA, Dudley ME. Adoptive cell transfer: a clinical path to effective cancer immunotherapy. Nat Rev Cancer 2008;8(4):299–308.

[8] Saito H, Ando S, Morishita N, Lee K-M, Dator D, Dy D, et al. A combined lymphokine-activated killer (LAK) cell Immunotherapy and Adenovirus-<em>p53</em> gene therapy for head and neck squamous cell carcinoma. Anticancer Res 2014;34(7):3365–70.

[9] West EJ, Scott KJ, Jennings VA, Melcher AA. Immune activation by combination human lymphokine-activated killer and dendritic cell therapy. Br J Cancer 2011;105(6):787–95.

[10] Gao X, Mi Y, Guo N, Xu H, Xu L, Gou X, Jin W. Cytokine-induced killer cells as pharmacological tools for Cancer immunotherapy. Front Immunol 2017;8.

[11] Garofano F, Gonzalez-Carmona MA, Skowasch D, Schmidt-Wolf R, Abramian A, Hauser S, et al. Clinical trials with combination of cytokine-induced killer cells and dendritic cells for Cancer therapy. Int J Mol Sci 2019;20(17).

[12] Liu S, Galat V, Galat Y, Lee YKA, Wainwright D, Wu J. NK cell-based cancer immunotherapy: from basic biology to clinical development. J Hematol Oncol 2021;14(1):7.

[13] Chu J, Gao F, Yan M, Zhao S, Yan Z, Shi B, Liu Y. Natural killer cells: a promising immunotherapy for cancer. J Transl Med 2022;20(1):240.

[14] Saura-Esteller J, de Jong M, King LA, Ensing E, Winograd B, de Gruijl TD, et al. Gamma Delta T-cell based Cancer immunotherapy: past-present-future. Front Immunol 2022;13.

[15] Mensurado S, Blanco-Domínguez R, Silva-Santos B. The emerging roles of γδ T cells in cancer immunotherapy. Nat Rev Clin Oncol 2023;20(3):178–91.

[16] Zhao Y, Deng J, Rao S, Guo S, Shen J, Du F, et al. Tumor infiltrating lymphocyte (TIL) therapy for solid tumor treatment: progressions and challenges. Cancers (Basel) 2022;14(17).

[17] Reardon S. First cell therapy for solid tumours heads to the clinic: what it means for cancer treatment. Nature 2024.

[18] Baulu E, Gardet C, Chuvin N, Depil S. TCR-engineered T cell therapy in solid tumors: state of the art and perspectives. Sci Adv 2023;9(7):eadf3700.

[19] Chandran SS, Klebanoff CA. T cell receptor-based cancer immunotherapy: emerging efficacy and pathways of resistance. Immunol Rev 2019;290(1):127–47.

[20] Hussein MS, Li Q, Mao R, Peng Y, He Y. TCR T cells overexpressing c-Jun have better functionality with improved tumor infiltration and persistence in hepatocellular carcinoma. Front Immunol 2023;14:1114770.

[21] Mao R, Hussein MS, He Y. Chimeric antigen receptor engineered T cells and their application in the immunotherapy of solid tumours. Expert Rev Mol Med 2022;24, e7.

[22] Jogalekar MP, Rajendran RL, Khan F, Dmello C, Gangadaran P, Ahn BC. CAR T-cell-based gene therapy for cancers: new perspectives, challenges, and clinical developments. Front Immunol 2022;13, 925985.

[23] Caraballo Galva LD, Jiang X, Hussein MS, Zhang H, Mao R, Brody P, et al. Novel low-avidity glypican-3 specific CARTs resist exhaustion and mediate durable antitumor effects against HCC. Hepatology 2022;76(2):330–44.

[24] Docta RY, Ferronha T, Sanderson JP, Weissensteiner T, Pope GR, Bennett AD, et al. Tuning T-cell receptor affinity to optimize clinical risk-benefit when targeting alpha-fetoprotein-positive liver Cancer. Hepatology 2019;69(5):2061–75.

[25] Giraud J, Chalopin D, Blanc J-F, Saleh M. Hepatocellular carcinoma immune landscape and the potential of immunotherapies. Front Immunol 2021;12.

[26] June CH, Sadelain M. Chimeric antigen receptor therapy. New England Journal of Medicine 2018;379(1):64–73.

[27] Debets R, Donnadieu E, Chouaib S, Coukos G. TCR-engineered T cells to treat tumors: seeing but not touching? Semin Immunol 2016;28(1):10–21.

[28] Fesnak AD, June CH, Levine BL. Engineered T cells: the promise and challenges of cancer immunotherapy. Nat Rev Cancer 2016;16(9):566–81.

[29] Maude SL, Frey N, Shaw PA, Aplenc R, Barrett DM, Bunin NJ, et al. Chimeric antigen receptor T cells for sustained remissions in leukemia. New England Journal of Medicine 2014;371(16):1507–17.

[30] Grupp SA, Kalos M, Barrett D, Aplenc R, Porter DL, Rheingold SR, et al. Chimeric Antigen Receptor–Modified T Cells for Acute Lymphoid Leukemia. New England Journal of Medicine 2013;368(16):1509–18.
[31] Holstein SA, Lunning MA. CAR T-cell therapy in hematologic malignancies: a voyage in Progress. Clin Pharmacol Ther 2020;107(1):112–22.
[32] Zhao L, Cao YJ. Engineered T cell therapy for Cancer in the clinic. Front Immunol 2019;10:2250.
[33] Robbins PF, Morgan RA, Feldman SA, Yang JC, Sherry RM, Dudley ME, et al. Tumor regression in patients with metastatic synovial cell sarcoma and melanoma using genetically engineered lymphocytes reactive with NY-ESO-1. J Clin Oncol 2011;29(7):917–24.
[34] Rapoport AP, Stadtmauer EA, Binder-Scholl GK, Goloubeva O, Vogl DT, Lacey SF, et al. NY-ESO-1–specific TCR–engineered T cells mediate sustained antigen-specific antitumor effects in myeloma. Nat Med 2015;21(8):914–21.
[35] Hassel JC, Piperno-Neumann S, Rutkowski P, Baurain JF, Schlaak M, Butler MO, et al. Three-year overall survival with Tebentafusp in metastatic uveal melanoma. N Engl J Med 2023;389(24):2256–66.
[36] He Q, Liu Z, Liu Z, Lai Y, Zhou X, Weng J. TCR-like antibodies in cancer immunotherapy. J Hematol Oncol 2019;12(1):99.
[37] Courtney AH, Lo WL, Weiss A. TCR signaling: mechanisms of initiation and propagation. Trends Biochem Sci 2018;43(2):108–23.
[38] Hwang JR, Byeon Y, Kim D, Park SG. Recent insights of T cell receptor-mediated signaling pathways for T cell activation and development. Exp Mol Med 2020;52(5):750–61.
[39] Eshhar Z, Waks T, Gross G, Schindler DG. Specific activation and targeting of cytotoxic lymphocytes through chimeric single chains consisting of antibody-binding domains and the gamma or zeta subunits of the immunoglobulin and T-cell receptors. Proc Natl Acad Sci U S A 1993;90(2):720–4.
[40] Sadelain M, Brentjens R, Rivière I. The promise and potential pitfalls of chimeric antigen receptors. Curr Opin Immunol 2009;21(2):215–23.
[41] Krause A, Guo HF, Latouche JB, Tan C, Cheung NK, Sadelain M. Antigen-dependent CD28 signaling selectively enhances survival and proliferation in genetically modified activated human primary T lymphocytes. J Exp Med 1998;188(4):619–26.
[42] Brentjens RJ, Santos E, Nikhamin Y, Yeh R, Matsushita M, La Perle K, et al. Genetically targeted T cells eradicate systemic acute lymphoblastic leukemia xenografts. Clin Cancer Res 2007;13(18 Pt 1):5426–35.
[43] Sadelain M, Brentjens R, Rivière I. The basic principles of chimeric antigen receptor design. Cancer Discov 2013;3(4):388–98.
[44] Guedan S, Posey Jr AD, Shaw C, Wing A, Da T, Patel PR, et al. Enhancing CAR T cell persistence through ICOS and 4-1BB costimulation. JCI. Insight 2018;3(1).
[45] Wu L, Wei Q, Brzostek J, Gascoigne NRJ. Signaling from T cell receptors (TCRs) and chimeric antigen receptors (CARs) on T cells. Cell Mol Immunol 2020;17(6):600–12.
[46] Davenport AJ, Cross RS, Watson KA, Liao Y, Shi W, Prince HM, et al. Chimeric antigen receptor T cells form nonclassical and potent immune synapses driving rapid cytotoxicity. Proc Natl Acad Sci U S A 2018;115(9):E2068–e76.
[47] Watanabe K, Terakura S, Martens AC, van Meerten T, Uchiyama S, Imai M, et al. Target antigen density governs the efficacy of anti-CD20-CD28-CD3 ζ chimeric antigen receptor-modified effector CD8+ T cells. J Immunol 2015;194(3):911–20.
[48] Lu Y-C, Yao X, Crystal JS, Li YF, El-Gamil M, Gross C, et al. Efficient identification of mutated cancer antigens recognized by T cells associated with durable tumor regressions. Clin Cancer Res 2014;20(13):3401–10.
[49] Maby P, Tougeron D, Hamieh M, Mlecnik B, Kora H, Bindea G, et al. Correlation between density of CD8+ T-cell infiltrate in microsatellite unstable colorectal cancers and frameshift mutations: a rationale for personalized immunotherapy. Cancer Res 2015;75(17):3446–55.
[50] Brown SD, Warren RL, Gibb EA, Martin SD, Spinelli JJ, Nelson BH, Holt RA. Neo-antigens predicted by tumor genome meta-analysis correlate with increased patient survival. Genome Res 2014;24(5):743–50.
[51] Leko V, Rosenberg SA. Identifying and targeting human tumor antigens for T cell-based immunotherapy of solid tumors. Cancer Cell 2020;38(4):454–72.
[52] Cherkasova E, Weisman Q, Childs RW. Endogenous retroviruses as targets for antitumor immunity in renal cell cancer and other tumors. Front. Oncologia 2013;3:243.
[53] Mesri EA, Feitelson MA, Munger K. Human viral oncogenesis: a cancer hallmarks analysis. Cell Host Microbe 2014;15(3):266–82.

[54] Doran SL, Stevanović S, Adhikary S, Gartner JJ, Jia L, Kwong MLM, et al. T-cell receptor gene therapy for human papillomavirus-associated epithelial cancers: a first-in-human, phase I/II study. J Clin Oncol 2019;37(30): 2759–68.
[55] Nagarsheth NB, Norberg SM, Sinkoe AL, Adhikary S, Meyer TJ, Lack JB, et al. TCR-engineered T cells targeting E7 for patients with metastatic HPV-associated epithelial cancers. Nat Med 2021;27(3):419–25.
[56] Nardiello T, Jungbluth AA, Mei A, Diliberto M, Huang X, Dabrowski A, et al. MAGE-A inhibits apoptosis in proliferating myeloma cells through repression of Bax and maintenance of survivin. Clin Cancer Res 2011;17 (13):4309–19.
[57] Yang F, Zhou X, Miao X, Zhang T, Hang X, Tie R, et al. MAGEC2, an epithelial-mesenchymal transition inducer, is associated with breast cancer metastasis. Breast Cancer Res Treat 2014;145(1):23–32.
[58] Lurquin C, Lethé B, De Plaen E, Corbière V, Théate I, van Baren N, et al. Contrasting frequencies of antitumor and anti-vaccine T cells in metastases of a melanoma patient vaccinated with a MAGE tumor antigen. J Exp Med 2005;201(2):249–57.
[59] Kunert A, van Brakel M, van Steenbergen-Langeveld S, da Silva M, Coulie PG, Lamers C, et al. MAGE-C2–specific TCRs combined with epigenetic drug-enhanced antigenicity yield robust and tumor-selective T cell responses. J Immunol 2016;197(6):2541–52.
[60] Chou J, Voong LN, Mortales CL, Towlerton AMH, Pollack SM, Chen X, et al. Epigenetic modulation to enable antigen-specific T-cell therapy of colorectal cancer. J Immunother 2012;35(2):131–41.
[61] Robbins PF, Kassim SH, Tran TLN, Crystal JS, Morgan RA, Feldman SA, et al. A pilot trial using lymphocytes genetically engineered with an NY-ESO-1–reactive T-cell receptor: Long-term follow-up and correlates with response. Clin Cancer Res 2015;21(5):1019–27.
[62] Parkhurst MR, Yang JC, Langan RC, Dudley ME, Nathan D-AN, Feldman SA, et al. T cells targeting carcinoembryonic antigen can mediate regression of metastatic colorectal cancer but induce severe transient colitis. Mol Ther 2011;19(3):620–6.
[63] Rohaan MW, Gomez-Eerland R, van den Berg JH, Geukes Foppen MH, van Zon M, Raud B, et al. MART-1 TCR gene-modified peripheral blood T cells for the treatment of metastatic melanoma: a phase I/IIa clinical trial. Immunooncol Technol 2022;15, 100089.
[64] Johnson LA, Morgan RA, Dudley ME, Cassard L, Yang JC, Hughes MS, et al. Gene therapy with human and mouse T-cell receptors mediates cancer regression and targets normal tissues expressing cognate antigen. Blood 2009;114(3):535–46.
[65] Zhu W, Peng Y, Wang L, Hong Y, Jiang X, Li Q, et al. Identification of α-fetoprotein-specific T-cell receptors for hepatocellular carcinoma immunotherapy. Hepatology 2018;68(2):574–89.
[66] Li L-P, Lampert JC, Chen X, Leitao C, Popović J, Müller W, Blankenstein T. Transgenic mice with a diverse human T cell antigen receptor repertoire. Nat Med 2010;16(9):1029–34.
[67] Obenaus M, Leitão C, Leisegang M, Chen X, Gavvovidis I, van der Bruggen P, et al. Identification of human T-cell receptors with optimal affinity to cancer antigens using antigen-negative humanized mice. Nat Biotechnol 2015;33(4):402–7.
[68] Wälchli S, Løset GÅ, Kumari S, Nergård Johansen J, Yang W, Sandlie I, Olweus J. A practical approach to T-cell receptor cloning and expression. PloS One 2011;6(11), e27930.
[69] Cai L, Caraballo Galva LD, Peng Y, Luo X, Zhu W, Yao Y, et al. Preclinical studies of the off-target reactivity of AFP(158)-specific TCR engineered T cells. Front Immunol 2020;11:607.
[70] Luo X, Cui H, Cai L, Zhu W, Yang WC, Patrick M, et al. Selection of a clinical Lead TCR targeting alpha-fetoprotein-positive liver Cancer based on a balance of risk and benefit. Front Immunol 2020;11:623.
[71] Butler MO, Friedlander P, Milstein MI, Mooney MM, Metzler G, Murray AP, et al. Establishment of antitumor memory in humans using in vitro-educated CD8+ T cells. Sci Transl Med 2011;3(80):80ra34.
[72] Long J, Chen X, He M, Ou S, Zhao Y, Yan Q, et al. HLA-class II restricted TCR targeting human papillomavirus type 18 E7 induces solid tumor remission in mice. Nat Commun 2024;15(1):2271.
[73] Campillo-Davo D, Flumens D, Lion E. The quest for the best: how TCR affinity, avidity, and functional avidity affect TCR-engineered T-cell antitumor responses. Cells 2020;9(7).
[74] Govers C, Sebestyén Z, Coccoris M, Willemsen RA, Debets R. T cell receptor gene therapy: strategies for optimizing transgenic TCR pairing. Trends Mol Med 2010;16(2):77–87.
[75] Thomas S, Mohammed F, Reijmers RM, Woolston A, Stauss T, Kennedy A, et al. Framework engineering to produce dominant T cell receptors with enhanced antigen-specific function. Nat Commun 2019;10(1). 4451 p.

[76] Tian S, Maile R, Collins EJ, Frelinger JA. CD8⁺ T cell activation is governed by TCR-peptide/MHC affinity, not dissociation rate. J Immunol 2007;179(5):2952–60.
[77] Zhong S, Malecek K, Johnson LA, Yu Z, Vega-Saenz de Miera E, Darvishian F, et al. T-cell receptor affinity and avidity defines antitumor response and autoimmunity in T-cell immunotherapy. Proc Natl Acad Sci U S A 2013;110(17):6973–8.
[78] Robbins PF, Li YF, El-Gamil M, Zhao Y, Wargo JA, Zheng Z, et al. Single and dual amino acid substitutions in TCR CDRs can enhance antigen-specific T cell functions. J Immunol 2008;180(9):6116–31.
[79] Raman MCC, Rizkallah PJ, Simmons R, Donnellan Z, Dukes J, Bossi G, et al. Direct molecular mimicry enables off-target cardiovascular toxicity by an enhanced affinity TCR designed for cancer immunotherapy. Sci Rep 2016;6, 18851.
[80] Morgan RA, Chinnasamy N, Abate-Daga D, Gros A, Robbins PF, Zheng Z, et al. Cancer regression and neurological toxicity following anti-MAGE-A3 TCR gene therapy. J Immunother 2013;36(2):133–51.
[81] Linette GP, Stadtmauer EA, Maus MV, Rapoport AP, Levine BL, Emery L, et al. Cardiovascular toxicity and titin cross-reactivity of affinity-enhanced T cells in myeloma and melanoma. Blood 2013;122(6):863–71.
[82] Combadière B, e Sousa CR, Trageser C, Zheng L-X, Kim CR, Lenardo MJ. Differential TCR signaling regulates apoptosis and immunopathology during antigen responses in vivo. Immunity 1998;9(3):305–13.
[83] Govers C, Sebestyén Z, Roszik J, van Brakel M, Berrevoets C, Szöőr Á, et al. TCRs genetically linked to CD28 and CD3ε do not Mispair with endogenous TCR chains and mediate enhanced T cell persistence and anti-melanoma activity. J Immunol 2014;193(10):5315–26.
[84] Miyao K, Terakura S, Okuno S, Julamanee J, Watanabe K, Hamana H, et al. Introduction of genetically modified CD3ζ improves proliferation and persistence of antigen-specific CTLs. Cancer Immunol Res 2018;6(6):733–44.
[85] Sakai T, Terakura S, Miyao K, Okuno S, Adachi Y, Umemura K, et al. Artificial T cell adaptor molecule-transduced TCR-T cells demonstrated improved proliferation only when transduced in a higher intensity. Mol Ther Oncolytics 2020;18:613–22.
[86] Casucci M, Hawkins RE, Dotti G, Bondanza A. Overcoming the toxicity hurdles of genetically targeted T cells. Cancer Immunol Immunother 2015;64(1):123–30.
[87] Cameron BJ, Gerry AB, Dukes J, Harper JV, Kannan V, Bianchi FC, et al. Identification of a titin-derived HLA-A1-presented peptide as a cross-reactive target for engineered MAGE A3-directed T cells. Sci Transl Med 2013;5(197). 197ra03-ra03.
[88] Shafer P, Kelly LM, Hoyos V. Cancer therapy with TCR-engineered T cells: current strategies, challenges, and prospects. Front Immunol 2022;13, 835762.
[89] Border EC, Sanderson JP, Weissensteiner T, Gerry AB, Pumphrey NJ. Affinity-enhanced T-cell receptors for adoptive T-cell therapy targeting MAGE-A10: strategy for selection of an optimal candidate. Onco Targets Ther 2019;8(2), e1532759.
[90] Maus MV, Plotkin J, Jakka G, Stewart-Jones G, Rivière I, Merghoub T, et al. An MHC-restricted antibody-based chimeric antigen receptor requires TCR-like affinity to maintain antigen specificity. Mol Ther Oncolytics 2016;3:1–9.
[91] Thakkar N, Bailey-Kellogg C. Balancing sensitivity and specificity in distinguishing TCR groups by CDR sequence similarity. BMC Bioinformatics 2019;20(1):241.
[92] Zhang H, Liu L, Zhang J, Chen J, Ye J, Shukla S, et al. Investigation of antigen-specific T-cell receptor clusters in human cancers. Clin Cancer Res 2020;26(6):1359–71.
[93] Cohen CJ, Li YF, El-Gamil M, Robbins PF, Rosenberg SA, Morgan RA. Enhanced antitumor activity of T cells engineered to express T-cell receptors with a second disulfide bond. Cancer Res 2007;67(8):3898–903.
[94] Cohen CJ, Zhao Y, Zheng Z, Rosenberg SA, Morgan RA. Enhanced antitumor activity of murine-human hybrid T-cell receptor (TCR) in human lymphocytes is associated with improved pairing and TCR/CD3 stability. Cancer Res 2006;66(17):8878–86.
[95] Bialer G, Horovitz-Fried M, Ya'acobi S, Morgan RA, Cohen CJ. Selected murine residues endow human TCR with enhanced tumor recognition. J Immunol 2010;184(11):6232–41.
[96] Legut M, Dolton G, Mian AA, Ottmann OG, Sewell AK. CRISPR-mediated TCR replacement generates superior anticancer transgenic T cells. Blood 2018;131(3):311–22.
[97] Provasi E, Genovese P, Lombardo A, Magnani Z, Liu P-Q, Reik A, et al. Editing T cell specificity towards leukemia by zinc finger nucleases and lentiviral gene transfer. Nat Med 2012;18(5):807–15.
[98] Voss RH, Willemsen RA, Kuball J, Grabowski M, Engel R, Intan RS, et al. Molecular design of the Calphabeta interface favors specific pairing of introduced TCRalphabeta in human T cells. J Immunol 2008;180(1):391–401.

[99] Shilyansky J, Yang JC, Custer MC, Spiess P, Mixon A, Cole DJ, et al. Identification of a T-cell receptor from a therapeutic murine T-cell clone. J Immunother 1997;20(4):247–55.
[100] Haga-Friedman A, Horovitz-Fried M, Cohen CJ. Incorporation of transmembrane hydrophobic mutations in the TCR enhance its surface expression and T cell functional avidity. J Immunol 2012;188(11):5538–46.
[101] Sebestyén Z, Schooten E, Sals T, Zaldivar I, San José E, Alarcón B, et al. Human TCR that incorporate CD3ζ induce highly preferred pairing between TCRα and β chains following gene transfer. J Immunol 2008;180 (11):7736–46.
[102] Aggen DH, Chervin AS, Schmitt TM, Engels B, Stone JD, Richman SA, et al. Single-chain VαVβ T-cell receptors function without mispairing with endogenous TCR chains. Gene Ther 2012;19(4):365–74.
[103] Paulson KG, Voillet V, McAfee MS, Hunter DS, Wagener FD, Perdicchio M, et al. Acquired cancer resistance to combination immunotherapy from transcriptional loss of class I HLA. Nat Commun 2018;9(1):3868.
[104] Lo W, Parkhurst M, Robbins PF, Tran E, Lu YC, Jia L, et al. Immunologic recognition of a shared p53 mutated Neoantigen in a patient with metastatic colorectal Cancer. Cancer Immunol Res 2019;7(4):534–43.
[105] Kaluza KM, Thompson JM, Kottke TJ, Flynn Gilmer HC, Knutson DL, Vile RG. Adoptive T cell therapy promotes the emergence of genomically altered tumor escape variants. Int J Cancer 2012;131(4):844–54.
[106] Blank CU, Haining WN, Held W, Hogan PG, Kallies A, Lugli E, et al. Defining 'T cell exhaustion'. Nat Rev Immunol 2019;19(11):665–74.
[107] McLane LM, Abdel-Hakeem MS, Wherry EJ. CD8 T cell exhaustion during chronic viral infection and Cancer. Annu Rev Immunol 2019;37(1):457–95.
[108] Wherry EJ. T cell exhaustion. Nat Immunol 2011;12(6):492–9.
[109] Wherry EJ, Kurachi M. Molecular and cellular insights into T cell exhaustion. Nat Rev Immunol 2015;15 (8):486–99.
[110] Lynn RC, Weber EW, Sotillo E, Gennert D, Xu P, Good Z, et al. C-Jun overexpression in CAR T cells induces exhaustion resistance. Nature 2019;576(7786):293–300.
[111] Sen DR, Kaminski J, Barnitz RA, Kurachi M, Gerdemann U, Yates KB, et al. The epigenetic landscape of T cell exhaustion. Science 2016;354(6316):1165–9.
[112] Dolina JS, Van Braeckel-Budimir N, Thomas GD, Salek-Ardakani S. CD8+ T cell exhaustion in Cancer. Front Immunol 2021;12.
[113] Zhang L, Kerkar SP, Yu Z, Zheng Z, Yang S, Restifo NP, et al. Improving adoptive T cell therapy by targeting and controlling IL-12 expression to the tumor environment. Mol Ther 2011;19(4):751–9.
[114] Drakes DJ, Rafiq S, Purdon TJ, Lopez AV, Chandran SS, Klebanoff CA, Brentjens RJ. Optimization of T-cell Receptor–Modified T Cells for Cancer Therapy. Cancer Immunol Res 2020;8(6):743–55.
[115] Veatch J, Paulson K, Asano Y, Martin L, Lee B, Hall E, et al. Merkel polyoma virus specific T-cell receptor transgenic T-cell therapy in PD-1 inhibitor refractory Merkel cell carcinoma. J Clin Oncol 2022;40:9549.
[116] Chapuis AG, Roberts IM, Thompson JA, Margolin KA, Bhatia S, Lee SM, et al. T-cell therapy using Interleukin-21-primed cytotoxic T-cell lymphocytes combined with cytotoxic T-cell lymphocyte Antigen-4 blockade results in Long-term cell persistence and durable tumor regression. J Clin Oncol 2016;34(31):3787–95.
[117] Klebanoff CA, Gattinoni L, Restifo NP. Sorting through subsets: which T-cell populations mediate highly effective adoptive immunotherapy? J Immunother 2012;35(9):651–60.
[118] Gattinoni L, Lugli E, Ji Y, Pos Z, Paulos CM, Quigley MF, et al. A human memory T cell subset with stem cell-like properties. Nat Med 2011;17(10):1290–7.
[119] Berger C, Jensen MC, Lansdorp PM, Gough M, Elliott C, Riddell SR. Adoptive transfer of effector CD8+ T cells derived from central memory cells establishes persistent T cell memory in primates. J Clin Invest 2008;118(1): 294–305.
[120] plc. AT. Adaptimmune Announces U.S. FDA Acceptance of Biologics License Application for Afami-cel for the Treatment of Advanced Synovial Sarcoma with Priority Review: Adaptimmune Therapeutics plc.; 2024 [Available from: https://www.adaptimmune.com/investors-and-media/news-center/press-releases/detail/260/adaptimmune-announces-u-s-fda-acceptance-of-biologics.
[121] Depil S, Duchateau P, Grupp SA, Mufti G, Poirot L. 'Off-the-shelf' allogeneic CAR T cells: development and challenges. Nat Rev Drug Discov 2020;19(3):185–99.
[122] Cieri N, Camisa B, Cocchiarella F, Forcato M, Oliveira G, Provasi E, et al. IL-7 and IL-15 instruct the generation of human memory stem T cells from naive precursors. Blood 2013;121(4):573–84.

CHAPTER

7

The combination of local treatment and immunotherapy

Bo Zhu, Jianguo Sun, Qingzhu Jia, Zhi Xu, Yongdong Feng, Luping Zhang, Lingyou Sun, Yaxian Qi, Bin Wang, Yu Liu, Lei Wang, Zheng Jin, Yuxuan Wei, and Shimeng Liang

Xinqiao Hospital, Army Medical University, Chongqing, China

Abstract

Local antitumor therapies have shown promising local and systemic immune effects, prompting clinical trials to evaluate their combination with immunotherapy. Interim data suggest improved median survival, objective response rate (ORR), and progression-free survival (PFS) compared to local therapy alone, indicating a synergistic potential. However, further clinical evidence and careful consideration of potential treatment-related adverse events (TRAEs) are crucial in developing these strategies. Given the high incidence of complications associated with both therapies, novel approaches to minimize side effects are urgently required. In hepatocellular carcinoma (HCC) treatment, reducing side effects is essential to preserving liver function. Multifunctional therapeutic carriers and effective local delivery routes are key to ensuring safe and effective therapy combinations. Local therapy-mediated immunotherapy drug delivery and controlled release have shown effectiveness in overcoming resistance and off-target effects. The multifunctionality of carriers in imaging and therapeutic delivery holds potential for improving procedures. Enhancing therapeutic efficacy requires developing delivery routes for immunotherapy or loaded carriers. Evaluating local administration of immunotherapy compared to systemic administration is underway, leveraging hepatic artery and percutaneous intra-tumoral routes in HCC. Optimal sequencing and intervals of combination therapies need thorough investigation. Identifying key biomarkers to assess immune and therapeutic responses is critical. Integrating multifunctional carriers and image-guided local delivery techniques into novel combination strategies is necessary for safe and effective cancer treatment.

Abbreviations

AE	abscopal effect
APCs	antigen-presenting cells
ATP	adenosine-5-triphosphate
BH	boiling histotripsy

https://doi.org/10.1016/B978-0-443-13827-0.00003-2

C3a	complement component 3a
CALR	calreticulin
CCs	cytokines and chemokines
CD11c+ DCs	CD11c+ dendritic cells
CD47	cluster of differentiation 47
CE	cavitation effect
CRT	calreticulin
CTLA-4	cytotoxic T-lymphocyte associated protein 4
CTLs	cytotoxic T lymphocytes
DAMPs	damage-associated molecular patterns
DCs	dendritic cells
EPR	enhanced permeability and retention
FUS	focused ultrasound
G-CSF	granulocyte colony-stimulating factor
HIFU	high-intensity focused ultrasound
HIFUa	high-intensity focused ultrasound ablation
HMGB1	high-mobility group box 1
HSP70	heat shock protein 70
HSPs	heat shock proteins
ICAM-1	intercellular adhesion molecule-1
ICD	immunogenic cell death
ICIs	immune checkpoint inhibitors
IFN-α	interferon alpha
IFN-γ	interferon gamma
IGRT	image-guided radiation therapy
IL-1	interleukin-1
IMRT	intensity-modulated radiation therapy
IR	ionizing radiation
iRT	immunotherapy
I-SABR	intensified stereotactic ablative radiotherapy
ISV	in situ vaccination
LDR	low-dose radiation ()
LF-HI-LDC	low frequency, high intensity, low duty cycle
LIFU	low-intensity focused ultrasound
L-selectin	low-density lipoprotein
MDSCs	myeloid-derived suppressor cells
MHC-I	major histocompatibility complex I
MHCII	major histocompatibility complex class II
MRI	magnetic resonance imaging
MWA	microwave ablation
na-FUS	nonablation FUS
NF-κB	nuclear factor kappa B
NK	natural killer cells
NSCLC	nonsmall cell lung cancer
NTA	nonthermal ablation
PD-1	programmed death-1
PD-L1	programmed death ligand-1
PDT	photodynamic therapy
PFS	progression-free survival
RE	radscopal effect
RFA	radiofrequency ablation
ROS	reactive oxygen species
RT	radiotherapy
SABR	stereotactic ablative radiotherapy
SBRT	stereotactic body radiation therapy

SCC	squamous cell carcinoma
SR	surgical resection
TAAs	tumor-associated antigens
TACE	transcatheter arterial chemoembolization
Th1	T helper 1
Th2	T helper 2
TLRs	toll-like receptors
TT	thermotherapy
US	ultrasound

Conflict of interest

No potential conflict of interest were disclosed.

Introduction

Local treatment techniques for tumors encompass various methods, such as radiotherapy, radiofrequency ablation (RFA; including microwave ablation), photodynamic therapy, thermotherapy, cryotherapy, and transcatheter arterial chemoembolization (TACE), among others.

Radiotherapy stands as a cornerstone treatment for cancer. It exerts its effects on biomaterials through the absorption of energy from X-rays, gamma rays, or charged particles, which may ionize or excite atoms, leading to a cascade of effects and subsequent biological consequences. By impairing cellular proliferation, radiotherapy successfully achieves the objective of eradicating tumor cells. It induces various forms of tumor cell death, releasing proinflammatory cytokines, chemokines, tumor antigens, and other hazardous signals. Consequently, radiation can potentially enhance tumor immunogenicity through this mechanism [1].

RFA and microwave ablation (MWA) are hyperthermic procedures that utilize energy to heat the tissue to a minimum of 60 degrees Celsius, achieving optimal effectiveness. These modalities have been clinically validated and are commonly employed for ablation [2]. Clinical experiments have shown that RFA treatment can enhance the body's immune response, thereby eradicating tumor cells [3].

Photodynamic therapy (PDT) is a contemporary and noninvasive therapeutic approach used for treating both nononcological diseases and various types and locations of cancers. Photocytotoxic reactions occur exclusively within pathological tissues, specifically in the area where the photosensitizer is distributed, enabling precise and selective destruction. PDT has been reported to stimulate antitumor immune responses in laboratory small animal cancer models with functional immune systems [4].

Thermotherapy refers to the application of physical energy to heat either the entire human body or a specific area, elevating the temperature of tumor tissue to an effective therapeutic level and maintaining it for a specific duration. Hyperthermia modulates the immune state of the tumor microenvironment by delivering danger signals from heat shock proteins (HSPs) and subsequently activating the immune system [5].

Transcatheter arterial chemoembolization (TACE) is an antitumor therapy based on the principle that hepatocellular carcinoma (HCC) prioritizes blood supply from the hepatic artery over the surrounding liver parenchyma while minimizing damage to healthy liver tissue.

Studies have revealed that tumor ischemia and hypoxia upregulate various molecular factors, thereby inhibiting cell apoptosis and stimulating tumor growth [6].

Local and systemic immune response to local therapies

Radiotherapy

Introduction of radiotherapy and its uses in cancer treatment

The history of radiotherapy (RT) can be traced back to the discovery of X-rays by Roentgen in 1895, which were first used in cancer treatment in 1896. From the 1930s to the 1970s, radiotherapy techniques advanced rapidly alongside the development of physics and machinery. Subsequently, breakthroughs in precision were achieved through the development of computers and imaging technology. Radiotherapy has been extensively employed in the treatment of cancer patients. Looking ahead, FLASH RT, proton RT, and carbon ion RT hold promise for benefiting more patients with reduced toxicities and improved survival outcomes.

Radiotherapy plays a significant role in the treatment of both benign and malignant diseases, with up to half of all cancer patients undergoing this treatment. It can be administered as a standalone treatment or in combination with chemotherapy, surgery, and immunotherapy. However, radiotherapy also induces damage to normal tissues and releases immunosuppressive factors. As a result, the focus of radiotherapy has shifted toward minimizing side effects, particularly the late effects that can significantly impact a patient's quality of life [7].

Mainstream radiotherapy techniques currently in use include external-beam radiotherapy, brachytherapy, and the use of protons and heavy ions. External-beam radiotherapy involves delivering radiation from outside the patient and can be administered using highly conformal techniques such as stereotactic body radiation therapy (SBRT) or stereotactic ablative radiotherapy (SABR) [8]. These techniques have evolved over time, with advancements like intensity-modulated radiation therapy (IMRT) and image-guided radiation therapy (IGRT). Ongoing research is also exploring the combination of radiotherapy with immunotherapy to elicit a systemic immune response against cancer. Additionally, low-dose radiation is being investigated as a potential tool to reprogram the tumor microenvironment and enhance the occurrence of the abscopal effect.

Immunological effects of radiotherapy

Radiation primarily targets the DNA within tumor cells, resulting in lethal effects through direct DNA damage caused by radiation. This DNA damage directly affects cancer cells, leading to their demise and activating the cGAS/STING pathway. Additionally, this effect can occur indirectly through the generation of free radicals that initiate chemical reactions with various components within the cells.

Ionizing radiation exerts dual effects on cancer cells: direct DNA damage leading to cell death and modulation of the tumor microenvironment by activating the immune system through immunogenic cell death (ICD).

After cell death, proinflammatory mediators are released, including tumor-associated antigens (antigenic), damage-associated molecular patterns (DAMPs), calreticulin (CALR) expression, and major histocompatibility complex I (MHC-I) surface expression. Toll-like receptors (TLRs) expressed on the surface of dendritic cells (DCs) recognize DAMPs and are responsible for their activation and maturation. Tumor-infiltrating DCs are associated with either a favorable or unfavorable prognosis depending on the cancer type. Although they seem to be quite radioresistant, radiation may impair DCs' functionality, potentially altering the balance between T cell activation and tolerance mediated by DCs. Adenosine-5-triphosphate (ATP) also contributes to tumor growth and myeloid-derived suppressor cells (MDSCs) with immunosuppressive properties induced by radiation.

Radiation also enhances tumor immunogenicity by facilitating the presentation of tumor-specific antigens to T lymphocytes. These processes initiate both adaptive and innate immune responses, contributing to systemic antitumorigenic effects. Additionally, radiation exposure to a tumor induces specific forms of tumor cell death, prompting the release of proinflammatory cytokines, chemokines, tumor antigens, and danger signals. This leads to an increase in the production of molecules involved in cellular adhesion, facilitating the recruitment of antitumor T cells to the corresponding sites. The dead and stressed cells release a variety of substances that give ionizing radiation either immunosuppressive or immune-stimulating properties, which may enhance tumor immunogenicity.

Radiation can induce the generation of a substantial number of tumoral neoantigens, which are subsequently presented to T lymphocytes. Consequently, radiation has the potential to trigger both adaptive and innate immune responses, leading to systemic antitumorigenic effects both within and beyond the irradiation field.

The irradiated sites cause a reprogramming of the tumor microenvironment, promoting molecular and gene-level alterations in nonirradiated sites. Ionizing radiation (IR) can produce the abscopal effect, a phenomenon in which radiotherapy induces a systemic immune response against cancer. This effect occurs when radiation kills tumor cells and initiates an immune response that also targets tumor cells in other parts of the body. The abscopal effect has been well-known since its initial description and has fundamentally changed the perception of radiotherapy from a simple local treatment. This response arises from the immunogenic cell death of tumor cells caused by IR, leading to the release of DAMPs and neoantigens, which in turn enhances antigen presentation by dendritic cells (DCs) and their maturation. As a result, cytotoxic T lymphocytes (CTLs) and natural killer (NK) cells are activated to recognize and attack both primary and abscopal tumor cells. The abscopal effect is also influenced by cytokines and chemokines, as evidenced by observed changes in plasma levels after radiation treatment (Fig. 1). Moreover, low-dose radiation can stimulate the abscopal tumor stroma throughout the body, leading to abscopal tumor regression [9]. It is now widely accepted that radiotherapy can elicit a systemic immune response, providing strong support for combining radiotherapy with immunotherapy.

(1) Radiotherapy directly induces DNA damage in tumor cells, consequently activating the cGAS/STING pathway. Additionally, radiotherapy indirectly generates free radicals that initiate chemical reactions with cellular components.
(2) Radiotherapy triggers immunogenic cell death (ICD) and the release of inflammatory factors, including tumor-associated antigens (antigenic), damage-associated molecular patterns (DAMPs), calreticulin (CALR) expression, and major histocompatibility complex I (MHC-I) surface expression. This upregulation subsequently activates dendritic cells (DCs).
(3) The activated DCs migrate into the lymph nodes to present antigens.
(4) Tumor-specific T cells are then primed and undergo proliferation.

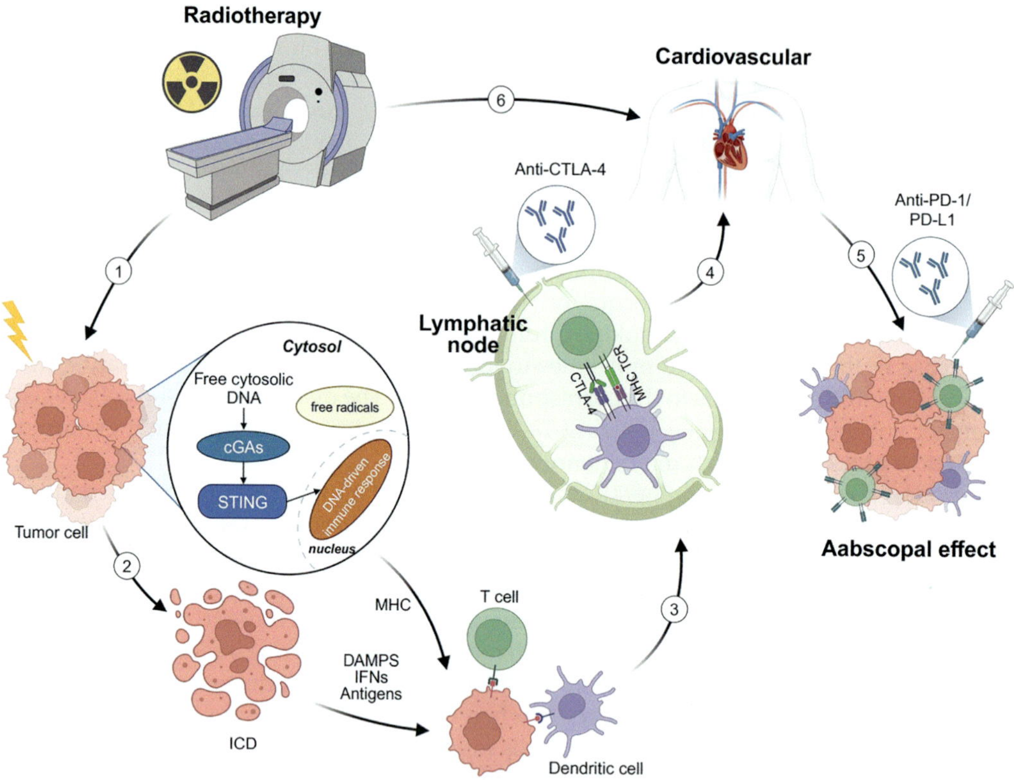

FIG. 1 Immunological and abscopal effect of radiotherapy.

(5) These T cells initiate an attack on tumor cells within the radiation field and also exhibit an abscopal effect by attacking tumors outside the radiation field.
(6) However, radiotherapy could also induce damage to normal tissues and release immunosuppressive factors.

Clinical application of radiotherapy in combinational therapies

Administering immune checkpoint inhibitors (ICIs), such as PD-1/PD-L1 inhibitors or CTLA-4 inhibitors, can counteract the immunosuppressive effects caused by preexisting tumor cells or radiotherapy within the radiation field, thereby enhancing the antitumor immune response. Ongoing clinical trials investigating the combination of checkpoint inhibitors and radiotherapy (iRT) aim to evaluate the abscopal effect in patients with advanced disease. Various vaccines and immune checkpoint inhibitors, such as anti-PD-1, anti-PD-L1, and anti-CTLA-4 antibodies, have been identified and tested in combination with radiotherapy delivered with SBRT. These trials encompass different tumor sites and disease stages, and patients undergo careful monitoring of their immune function.

Clinical trials combining radiotherapy and immunotherapy have shown promising results in enhancing overall survival and disease control by inducing an immune response. Studies have demonstrated that radiotherapy can provoke an in situ vaccination by destroying tumor cells and eliciting a systemic immune response, resulting in an abscopal effect that supports iRT. Clinical trials have proven that iRT's efficacy is significantly superior to mono-immunotherapy, increasing the out-of-field response rate of immunotherapy alone from 19.7% to 41.7%. However, several issues with iRT remain unresolved, including the optimal combination modality, timing for iRT, and how to expand the occurrence of the abscopal effect. Ongoing clinical trials are expected to shed light on these issues. A new concept called the "radscopal effect" has been proposed, referring to the use of low-dose radiation to reprogram the tumor microenvironment and amplify the occurrence of the abscopal effect. Therefore, future challenges include exploring the use of low-dose radiation to amplify the occurrence of the abscopal effect, identifying optimal combination schedules of agents, sequence, dose, fractionation, and irradiated sites, and studying how to broaden the beneficial spectrum of immunotherapy and appropriately select patients.

Current clinical trials are investigating the combination of checkpoint inhibitors and radiotherapy to assess the abscopal effect in patients with advanced disease. Various vaccines and immune checkpoint inhibitors, including anti-PD-1, anti-PD-L1, and anti-CTLA-4 antibodies, have been identified and examined in combination with radiotherapy using SBRT. These trials encompass diverse tumor sites and disease stages, and patients undergo meticulous monitoring of their immune function.

iRT demonstrates enhanced efficacy against micrometastases that are not visually detectable. Meanwhile, visible primary cancer foci are targeted using SABR, promoting the release of tumor-specific antigens that stimulate the body's anticancer immune response. Consequently, the combination of radiotherapy and immunotherapy exhibits a synergistic effect, potentially doubling the effectiveness of radiotherapy. A clinical trial by Joe Y. Chang's team found that I-SABR significantly improved 4-year event-free survival compared to the SABR-only group. Additionally, the study found that I-SABR was well-tolerated and may be a treatment option for patients with early stage nonsmall cell lung cancer [10]. Moreover, the safety and systemic response of combining radiotherapy with checkpoint inhibitors have been assessed in various other cancers, such as prostate cancer, glioblastoma, and cutaneous squamous cell carcinoma, yielding promising results.

Focused ultrasound (FUS)

Ultrasound (US), functioning as a mechanical sound wave, operates at frequencies greater than 20 kHz, making it undetectable by human ears [11]. In medical applications, varying ultrasound energy levels serve different purposes: low energy for diagnostic imaging and high energy for therapeutic use, specifically focused ultrasound (FUS) [12].

Focused ultrasound represents a rapidly advancing, noninvasive therapeutic technology with the potential to revolutionize the treatment of numerous medical conditions. This technique utilizes ultrasonic energy to precisely target deep tissues in the body without the need for incisions or radiation. Its fundamental principle is analogous to using a magnifying glass to focus sunlight onto a single point, resulting in burning a hole in a leaf. Focused ultrasound employs an acoustic lens to concentrate multiple intersecting beams of ultrasound, which are then directed with extreme accuracy and precision at a specific target deep within the body. Depending on the ultrasound parameters and lens design, the target diameter can range from 10×16mm to 1×1.5 mm. Currently, focused ultrasound is being researched for over 170 diseases worldwide and has been applied in the treatment of various conditions, including essential tremor, Parkinson's disease, uterine fibroids, bone metastases, osteoid osteoma, and certain malignant tumors, among others.

Focused ultrasound exhibits two main biological effects: thermal effects and nonthermal effects. The most well-known biological effect of FUS is the thermal effect, which occurs at longer duty cycles and moderate to high intensity [13]. The thermal effect includes both high temperature, inducing coagulative necrosis, and mild temperature, resulting in thermal stress or thermotherapy. High temperature is typically caused by thermal ablation, with the energy density increasing upon ultrasonic beam absorption. In such cases, the focus temperature can rapidly rise to 56°C, leading to protein denaturation, irreversible cell damage, and tumor cell necrosis. Consequently, tumor tissues undergo coagulation necrosis [14,15]. On the other hand, mild temperature can disrupt cell cycles, damage DNA and proteins, and induce cell death or apoptosis [14].

The nonthermal effects of FUS primarily include mechanical effects and cavitation effects [16]. As the ultrasound travels through the tissue, molecular cell particles rotate, vibrate, and rub, thereby altering the cell volume and membrane permeability and resulting in small eddying of fluids and cell plasma flow. These effects strengthen lymph and blood circulation and influence protein synthesis and cellular secretions [17]. Additionally, ultrasound can generate gas-filled voids within tissues and fluids, and when these gas bubbles collapse, a significant amount of energy is released in a flash. This phenomenon is known as the cavitation effect, typically occurring at low frequency, high intensity, and low duty cycle [18].

The different forms of FUS

There are three forms of FUS therapy: thermal ablation, nonthermal ablation, and nonablation (Fig. 2). Thermal ablation, particularly High-Intensity Focused Ultrasound ablation (HIFUa) or High-Intensity Focused Ultrasound (HIFU), represents the most extensively

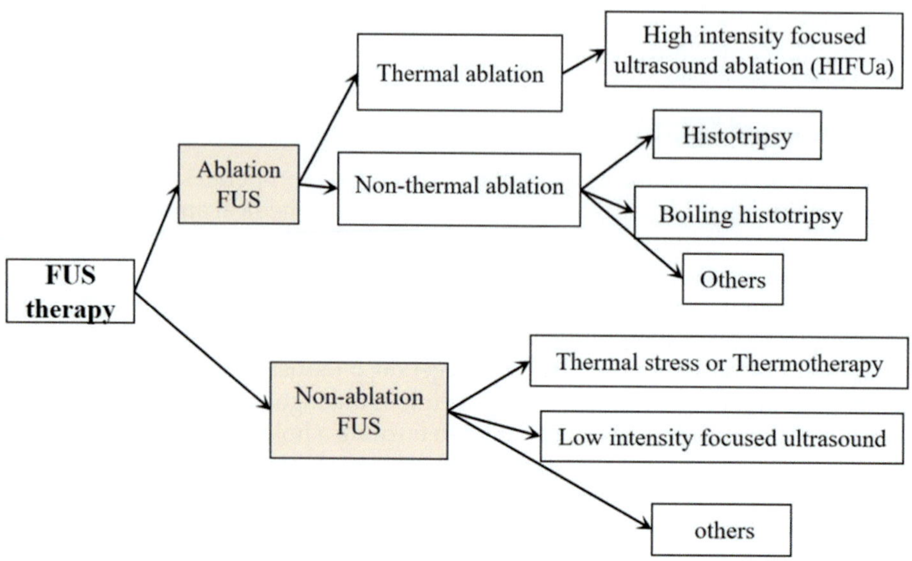

FIG. 2 Different treatment modes of FUS.

used FUS modality in medicine. Nonthermal ablation models of FUS include boiling histotripsy (BH), histotripsy, and other techniques. Additionally, two commonly employed nonablation FUS therapies are low-intensity focused ultrasound (LIFU) and thermotherapy.

Thermal ablation

High-Intensity Focused Ultrasound ablation (HIFUa) is a noninvasive therapy that minimizes damage to surrounding normal tissue, promoting faster patient recovery. In clinical malignant tumor therapy, HIFUa stands as a major treatment modality of FUS. By inducing high temperatures above 56°C, HIFUa can rapidly ablate tumor tissues into coagulative necrosis. In the focal area, tissues experience apoptosis or thermal stress with mild temperatures. This thermal ablation effectively reduces the tumor burden and releases multiple tumor antigens along with various bioactive molecules, such as damage-associated molecular patterns (DAMPs) and inflammation [19,20]. However, due to the potential denaturation of antigens and cytokines, as well as a reduction in tumor blood flow caused by endothelial cell swelling due to high temperature, HIFUa has not proven sufficient to induce long-lasting systemic immunity [21,22]. To significantly enhance therapy efficacy, HIFUa can be combined with chemotherapy, radiotherapy, and immunotherapy [23–25]. Thermal ablation has been successfully applied to treat various malignant tumors worldwide, including prostate cancer, liver cancer, breast cancer, pancreatic cancer, soft tissue cancer, bone sarcoma, and kidney tumors.

Nonthermal ablation

In recent years, nonthermal ablation of FUS has gained considerable interest due to its ability to mechanically fractionate target tissue without causing thermal damage [26]. With high precision, the tissues within the treatment focal area are mechanically ablated into acellular debris [27,28]. Outside the focal area, tissues experience minimal damage without thermal destruction [29,30]. The complete fragmentation of tumor tissue results in emulsified acellular debris that can be rapidly drained and/or absorbed through the normal physiologic healing process [31,32]. Nonthermal ablation of FUS shows great potential in inducing antitumor immune effects. However, it is important to acknowledge some drawbacks, such as the difficulty in eliminating tumors with a high tumor burden. Therefore, nonthermal ablation is still in the preclinical stage.

Nonablation FUS (na-FUS)

Nonablation FUS (na-FUS) does not directly induce tissue necrosis and primarily leads to nonthermal effects. Thermotherapy using FUS is rarely applied in tumor treatment. Low-intensity focused ultrasound (LIFU) is the main treatment form of na-FUS used in clinical practice. LIFU's mechanical effects include tissue disruption, inflammation, neuromodulation, enhanced permeability, or drug delivery. LIFU can directly enhance drug delivery or be combined with artificial agents like nanoparticles or microbubbles [33]. Furthermore, LIFU can be combined with acoustic droplet vaporization to noninvasively generate gas emboli within the tumor vasculature, which can be utilized to reduce tumor volume and be combined with other therapies [34].

Photodynamic therapy

Photodynamic therapy (PDT) presents a novel and minimally invasive treatment option that can be applied to multiple types of cancer and noncancerous diseases, as compared to traditional clinical treatments [35–37]. The basis of PDT lies in the photodynamic effect, where photosensitizers are driven by the permeability and retention effect (EPR) to accumulate in the tumor microenvironment and are subsequently excited by local illumination [38]. The induced photochemical reactions produce abundant reactive oxygen species in the tumor microenvironment, which react with nearby biomacromolecules and ultimately mediate cytotoxic effects on tumor cells through multiple pathways [38].

The mechanism of PDT-induced tumor inhibition involves several aspects. Firstly, the reactive oxygen species produced by PDT mediate apoptosis, necrosis, or autophagy of tumor cells, directly leading to tumor regression. Secondly, PDT destroys the tumor vascular system and damages vascular endothelial cells, resulting in insufficient blood supply to tumor tissues and ultimately leading to "starvation" of tumor cells. These effects are mediated by directly acting on tumor cells or the surrounding microenvironment. Moreover, PDT treatment efficiently activates the immune system. PDT induces cell death by generating ROS, thus activating immunogenic apoptosis and necrosis [38,39]. The dying tumor cells release or expose de novo damage-associated molecular patterns (DAMPs) as danger signals for the innate immune system [40,41]. DAMPs consist of multiple components that trigger immunomodulatory effects, such as inducing dendritic cell maturation or attracting monocytes and neutrophils [42–44].

Neutrophils represent the most abundant leukocytes in the immune system. PDT treatment promotes neutrophil infiltration at tumor sites [45]. In vivo imaging of luminol chemiluminescence also confirms neutrophil infiltration after PDT treatment [46]. Peripheral blood detection found two waves of neutrophilia in mice treated with Photofrin-PDT [47–49]. Neutrophil responses triggered by Photofrin-PDT are essential for successful treatment, and neutrophil activation is associated with therapeutic efficacy [46,50]. The increased neutrophils found in peripheral blood and tumors can be attributed to several reasons. First, PDT promotes the release of IL-1 and granulocyte colony-stimulating factor (G-CSF), responsible for generating and mobilizing neutrophils from the bone marrow [51]. Second, adhesion molecules/ligands involved in neutrophil migration were found upregulated after Photofrin-PDT treatment. Tumor vascular endothelium displayed higher expression of adhesion molecules ICAM-1 and L-selectin, while ICAM-1 ligands were upregulated on neutrophils [52,53]. Third, PDT increases vascular permeability, which may promote neutrophil infiltration [54]. Besides the above factors contributing to neutrophil responses, the complement system is involved in the enhanced neutrophil flux. Blocking the C3a receptor inhibits the second wave of neutrophilia, implying the role of C3 in neutrophilia [47].

Dendritic cells (DCs) are a major subpopulation of antigen-presenting cells [55]. Reports have shown that Photofrin-PDT or ALA-PDT can increase the expression of CRT, HSP70, and HMGB1 in tumors [56–58]. DAMPs can be recognized by antigen-presenting cells (APCs), promoting the presentation of tumor-associated antigens [56]. Antigen-presenting cells engulf tumor-associated antigens (TAAs) and deliver antigenic peptides to effector T cells, thereby eliciting an antitumor adaptive immune response. Photofrin-PDT treatment promotes efficient phagocytosis of dead cells by DCs, resulting in increased IL-12-expressing CD11c + DCs in the tumor-draining lymph nodes [59,60]. Studies have also found that Photofrin-PDT-treated tumor lysates can promote DC maturation, leading to upregulated expression of surface CD86, CD80, and MHCII [58,61]. The effects of PDT on DCs can further enhance T cell proliferation and activation, linking the innate immune system and adaptive immune system.

The above studies indicate that PDT treatment can activate an adaptive immune response. CD8+ T cells are found to be the dominant effector cells in the PDT-mediated adaptive immune response [62,63]. Increased INF-γ secretion from CD4+ T cells and CD8+ T cells is found in the lymph nodes and spleen of Photofrin-PDT-treated mice, indicating enhanced antitumor immunity induced by PDT [64]. The activation of dendritic cells is the major contributor to enhanced T cell activation. Moreover, PDT can stimulate the adaptive antitumor immune response in several ways. Firstly, the enhanced expression of calreticulin (CRT) on the membrane surface of PDT-treated cells provides "eat me" signals to elicit an immune response [65]. Secondly, in addition to cytotoxicity, PDT could increase the release of proinflammatory factors in the tumor microenvironment, including the secretion of antitumor cytokines (IFN-γ, IFN-α) as well as the transcription factor (NF-κB) [66,67]. The response induced by PDT also contributes to systemic tumor responsiveness. PDT treatment not only inhibits primary tumor growth but also simultaneously inhibits lesions outside the primary tumor area [68–70]. Tumor cell lysates generated by PDT treatment could serve as an anticancer vaccine displaying tumor protection [71]. Another study found that PDT treatment could resist the growth of homogeneous tumor cells in immunocompetent mice but not in immunodeficient mice. This indicates that PDT could induce antitumor immune memory [72]. This phenomenon is in accordance with the following study, which discovered memory T lymphocytes in the lymph nodes of mice treated with Photofrin-PDT [73]. These studies highlight the role of adaptive immunity in PDT-induced tumor inhibition.

Based on the above effects of PDT on the tumor microenvironment, combining PDT treatment and immunotherapy shows promise. In colon carcinoma models, the addition of CTLA-4 antibody prior to PDT significantly inhibits tumor growth compared with either treatment alone [74]. CD8 depletion experiments confirmed that the combinational effect depended on CD8+ T cells. PD-L1 expression is reported to be associated with the response rates of PD-1/PD-L1 blockade [75]. It was found that PDT treatment could induce PD-L1 expression in human xenograft tumors. Combined PDT with anti-PD-1/PD-L1 antibody displayed a strong antitumor effect, inhibiting primary tumor growth, preventing lung metastases, and prolonging survival [76]. Analysis of the tumor microenvironment found an increased ratio of CD8+ T cells/Tregs, indicating enhanced antitumor immunity. The above results are based on preclinical models. In a case report, a 62-year-old patient diagnosed with advanced SCC of the mouth floor received surgery but progressed later. However, combined PDT and anti-PD-1 therapy resulted in regression [77]. These results suggest that PDT treatment combined with immunotherapy may be a promising strategy to overcome treatment resistance.

Thermoablation and cryoablation

The primary treatment for early stage lung cancer is surgical resection, while comprehensive treatment is preferred for advanced lung cancer. In recent years, minimally invasive interventional therapies, such as CT or MRI-guided tumor ablation techniques, have rapidly advanced. Ablation therapy techniques offer precise guidance, less invasiveness, fewer complications, significant efficacy, and rapid recovery, making great progress in treating lung cancer, liver cancer, thyroid cancer, prostate cancer, and pancreatic cancer [78,79]. Currently, commonly used ablation treatments in clinical practice include RFA, microwave ablation (MWA), high-intensity focused ultrasound (HIFU), and cryoablation. RFA, MWA, and HIFU use thermal therapy principles, while cryoablation is a type of low-temperature therapy that causes tissue damage through the freeze-thaw process. Thermal ablation involves using different energy sources to rapidly heat the target lesion and cause coagulative necrosis of cells and tissues, thereby inactivating the target lesion. Cryoablation inactivates tumors mainly through short-term extracellular icing leading to a solution effect, intracellular icing, intramicrovascular thrombosis, and mediated apoptosis [80].

Mechanism of ablation therapy and antitumor immune response

Ablation therapy not only causes necrosis and apoptosis of tumor tissue but also promotes the release of tumor-derived autoantigens into circulation. This stimulates the host immune system to produce a good antitumor immune effect against primary and metastatic tumors, leading to the regression of the primary tumor and distant tumor metastases. This additional effect is thought to underlie the synergistic antitumor effect observed with immunotherapy [81,82].

Tumor immunotherapy uses antigens, antibodies, vaccines, cells, or viruses to cause or enhance the body's antitumor immune response, controlling or even destroying tumors. Currently, PD-1/PD-L1 inhibitors are the first-line treatment for nonsmall cell lung cancer with PD-1 expression >50% in tumor tissue and no identified therapeutic tumor target. However, the response rate in most clinical trials is only about 20%, which may be related to insufficient release of autoantigens in the tumor microenvironment and less infiltration of TILs [83].

Exploring the link between thermal ablation therapy and immunotherapy

Thermal ablation therapy, as a minimally invasive and precise tumor treatment technique, has become an effective way to treat tumors in clinical practice and is gradually gaining acceptance. Numerous studies have shown that thermal ablation can enhance patients' antitumor immunity and help reduce the risk of tumor recurrence and metastasis. However, single thermal ablation treatment may also lead to a short-term reduction in the body's antitumor immune strength. It is necessary to define the indications for thermal ablation and combine them with appropriate immune adjuvant therapy. The ability of thermal ablation to improve synergy through combined tumor immunotherapy is a hot topic of current research. Although evidence in NSCLS is limited, some studies conducted in malignancies such as liver, kidney, breast, and prostate cancers have shown that thermal ablation can induce altered immunogenicity. Immune agonists enhance the CTL response induced by thermal ablation, effectively inhibiting tumor growth and metastasis. RFA in combination with immunostimulatory OK-432 increases the number of OX-62 relative specific antigens, CD4+, CD8+, DC, INF-γ, and TNF-α cells and mediates tumor suppression [84]. MWA in combination with OK-432 increases CD4+, CD8+ T cell tumor infiltration and elevates Th1/Th2, polarizing the cellular response to the Th1 dominant state [85]. Several Phase I/II clinical trials with approved immune combination thermal ablation are underway, such as NCT02469701, NCT03290677, NCT04049474, NCT02843815, and NCT04201990. It is hoped that promising results will soon guide clinical treatment.

Studies on the body's antitumor immune mechanisms resulting from nonthermal mechanisms of different thermal ablations are limited and require further investigation. The clinical efficacy of microwave thermal ablation is also influenced by the size, number, and location of the patient's lesions, and the changes in immune cell-related subtypes before and after ablation remain unclear.

Cryoablation therapy linked to immunotherapy exploration

Cryoablation is a cryotherapeutic modality for tumor treatment that promotes the release of a large amount of antigens from the tumor, thereby effectively activating T cell recognition of the tumor [86,87]. However, the use of cryoablation alone cannot trigger a strong enough antitumor immunity in clinical practice, thus making it challenging to completely combat tumors with a single treatment modality. In recent years, a series of studies have found that cryoablation can not only control the primary tumor but also enhance the systemic antitumor immune response by producing an "in situ tumor vaccine" in the body using immune activators, thereby inhibiting the recurrence and metastasis of residual tumors [88]. In one study, PD-1 inhibitors were combined with cryoablation for the treatment of renal cell carcinoma in mice, and the combined treatment demonstrated stronger inhibitory effects on the metastatic growth of the tumor compared to administering the drug alone or cryotherapy alone. Furthermore, PD-1 inhibitors significantly enhanced the immune response induced by cryoablation, including an increase in the number of CD8+ TIL and an increase in the expression level of INF-γ [89]. In another study, the effects of cryoablation combined with IL-12 on the antitumor immunity of rat xenoglioblastoma were observed, and cellular immunity was significantly enhanced after cryoablation combined with IL-12 treatment, resulting in gradual tumor regression and a significant increase in subsequent hyperimmune activity, including an increase in IFN-γ expression, CD11c+ dendritic cells, and CD86+ lymphocytes [90]. In a study by Benzon et al., the potential synergistic antitumor effects between tumor cryoablation and immune checkpoint-blocking antibodies were investigated. The study findings indicated that cryoablation combined with anti-CTLA-4 antibody treatment delayed tumor growth or triggered tumor rejection in mice with prostate cancer [91]. This suggests that the anti-CTLA-4 antibody significantly enhances the cryoablation-mediated immune response.

Cryoablation combined with immunotherapy is a treatment technique that has been validated in several clinical trials. In one study, Niu et al. retrospectively analyzed the efficacy of cryoablation combined with immunotherapy in patients with metastatic pancreatic cancer. They found that the progression-free survival (PFS) of patients was significantly longer than that of patients treated with cryoablation alone or chemotherapy after a 4-year follow-up. However, different immunotherapy drugs have varying immune activation effects on the body. For instance, PD-1 antibody can block the tumor suppression effect on T cells and activate the body's immunity. CTLA-4 antibody can eliminate the suppression of CD8+ T cells by Treg. Additionally, CD47 antibody can promote the recognition of tumor cells by macrophages. Thus, from the perspective of clinical applications, the future of cryoablation combined with immunotherapy has a long way to go. There is a need to specify the appropriate freezing rate, determine how to combine the relevant modes of systemic immunotherapy, and the timing of drug administration. Additionally, it is crucial to reveal the changes in the tumor microenvironment after cryoablation treatment, including observing the changes in different immune cell subtypes before and after treatment. These factors are significant for the future of local cryoablation combined with immunotherapy.

In conclusion, further clinical studies are needed to determine whether local thermal ablation and cryoablation combined with immunotherapy can improve the survival time and survival rate of tumor patients. Moreover, the effectiveness of combined tumor immunotherapy in improving the complete ablation rate and determining the most effective local ablation combined with antitumor immunotherapy requires further exploration.

Transcatheter arterial chemoembolization (TACE)

TACE is the recommended first-line treatment for patients with HCC in the intermediate stage, which is characterized by multinodular tumor burden, preserved liver function, and good performance status. The blood supply to liver cancer tissue differs from that of normal liver tissue. More than 90% of the blood supply to liver cancer tissue comes from the hepatic artery, while only a small amount comes from the portal vein, whereas 75% of the blood supply to normal liver tissue comes from the portal vein. Transarterial embolization (TAE) can reduce the blood supply to liver cancer tissue to 5%–8%, having little effect on normal liver tissue. Intra-arterial ablative techniques for the treatment of HCC are illustrated. TAE aims to obstruct the arterial blood supply to the tumor, causing ischemia, while TACE involves the release of antitumoral agents before obstructing the feeding arteries. Two techniques for performing TACE have emerged: conventional TACE, where cytotoxic drugs (commonly doxorubicin, epirubicin, cisplatin, or miriplatin) are delivered to the tumor with lipiodol followed by arterial vessel embolization, and drug-eluting bead TACE (DEB-TACE), where microspheres carrying doxorubicin are delivered, achieving sustained drug delivery, followed by embolization. Transarterial radioembolization (TARE) is the radiotherapeutic analog of TACE, wherein yttrium-90-loaded microspheres are deposited for local β-radiation.

TACE can shape tumor immunity by altering the composition of the tumor microenvironment. According to a study in 2008, 11% of the tumor cells in HCC patients who underwent TACE were apoptotic, and 88% exhibited necrosis [92]. These findings indicated that TACE can induce both immunogenic cell death and nonimmunogenic cell death. The effects on cell death can differ with the types of chemotherapeutics used; for example, doxorubicin, the most commonly used agent in TACE, can induce immunogenic cell death, while cisplatin cannot [93]. However, the changes in the composition or function of immune cells after TACE remain to be elucidated. Some studies have reported alterations in T cell activation levels and subpopulations of immune cells after TACE [94]. The peripheral CD4+/CD8+ T cell ratio was significantly lower in patients with HCC before TACE than in healthy volunteers and increased 1 month after TACE [95,96]. Other studies have shown that TACE increased the frequency of tumor-specific CD4+ T cells and CD8+ cells, and this increase was associated with a better response to TACE therapy [94,97]. Additionally, TACE can significantly increase the CD4/CD8 ratio from 1 to 5 weeks after treatment, while CD8+ T cells only slightly increased [95]. These data suggest that the immune function is restored in HCC after TACE. Furthermore, TACE has been found to significantly increase the peripheral circulation of T helper 17 cells, a proinflammatory T cell, and markedly decrease CD4+ CD25+ Tregs, an antiinflammatory cell, indicating a positive regulatory effect on immune function.

The immunological changes caused by TACE are complex processes, and TACE also alters the cytokine profiles [98]. It has been shown that TACE is associated with an increase in proinflammatory cytokines, such as IL-6 and IL-22, in the first week after TACE, which correlates with the development of hepatitis post-TACE [99]. However, large tumors exhibit an increase in Th-2-associated cytokines (IL-4, IL-5, or suppressive cytokine IL-10), indicating an immune-suppressive environment 2 months post-TACE [99]. High-mobility group box 1 (HMGB1), an immunologic cell death marker, is an inducer of PD-L1 in tumor cells, and its levels are increased by TACE in HCC patients [100,101]. Vascular endothelial growth factor (VEGF) is one of the most important angiogenic factors highly expressed in patients with HCC. It has been found to increase in patients with HCC after TACE treatment [102]. VEGF is also known to be immunosuppressive, as it can increase the recruitment of Tregs, mast cells, monocytes, and myeloid-derived suppressor cells (MDSCs). Additionally, it promotes the accumulation of tumor-infiltrating macrophages and inhibits the development of T cells and the maturation of DCs [103,104]. VEGF has also been reported to modulate the checkpoint expression of CD8+ T cells in HCC [105].

Programmed death-1 (PD-1) is a transmembrane receptor expressed by activated T cells, natural killer cells, B cells, and antigen-presenting cells. Its ligand, programmed death-ligand 1 (PD-L1), is expressed in different types of cells, including cancer cells, which use this mechanism to escape from immunosurveillance. This interaction can trigger energy or death in antitumor cells, resulting in evasion of the immune system by tumor cells [104]. Therefore, PD-L1 plays a key role in tumor-induced immunosuppression. PD-L1 expression in tumor cells was significantly higher in the surgical specimens that received preoperative TACE treatment than in paired pre-TACE biopsies [106]. Similarly, PD-1 and PD-L1 expression in immune cells were also significantly higher in surgical specimens following TACE compared to paired biopsies taken before TACE treatment. TACE increased the expression of PD-1 and PD-L1 in HCC, potentially inhibiting antitumor immunity [106,107]. Approximately 1 week after TACE therapy, immune inhibition becomes increasingly dominant in the tumor microenvironment (TME), as PD-L1 levels continue to increase [108]. TACE also triggers a significant increase in the sTIM-3 level in plasma within the first week posttreatment, suggesting a reactive expansion of TIM-3 expression by T cells as a negative feedback mechanism in response to intense immune stimulation following tumor necrosis.

When compared with systemic therapies, TACE presents a higher likelihood of inducing sudden hypoxia in treated lesions, leading to the production of more hypoxia-related factors that can influence components of cancer immunity in the short term. Therefore, the interpretation of results among TACE-treated patients is challenging, and further research is necessary to better understand the treatment's immunomodulatory effects [109].

Trials investigating local therapy/immunotherapy combinational therapy

Radiotherapy

Since the first case report demonstrating the great efficacy of combining radiotherapy and immunotherapy (iRT) was published in 2012, numerous clinical trials have been conducted to confirm the clinical benefits of iRT. Golden and his colleagues conducted a proof-of-principle trial (NCT02474186) and showed that RT combined with GM-CSF produced objective abscopal responses in some patients with metastatic solid tumors, providing a basis for subsequent clinical trials. In the phase I trial KEYNOTE-001, Shaverdian et al. evaluated the efficacy and safety of pembrolizumab with or without RT in over 1000 advanced NSCLC patients (NCT01295827). The trial reported that patients in the combination group achieved longer PFS and overall survival (OS) with an acceptable safety profile, compared to the monotherapy group. Another clinical study (NCT02221739) conducted by Formenti et al. found that RT combined with CTLA-4 blockade induced an immune response in some patients who had not responded to immunotherapy previously, resulting in improved survival time.

Furthermore, based on a powerful phase III trial (NCT02125461), the median PFS of patients who received durvalumab plus chemoradiotherapy was 16.8 months, significantly longer than the 5.6 months observed in the placebo group. A pooled analysis of the PEMBRO-RT trial and the MDACC trial showed that the median OS was 15.9 months in the combination group versus 7.6 months in the pembrolizumab alone group. Additionally, adding RT to pembrolizumab immunotherapy significantly increased responses and outcomes in patients with metastatic NSCLC. However, the optimal choice of immune checkpoint inhibitors (ICIs) and RT agents, as well as the sequence, dose, fractionation, and irradiated sites to achieve the best synergies, requires further exploration and optimization through research and clinical trials.

Moreover, a randomized phase II trial (NCT03110978) conducted by Joe Y. Chang's team demonstrated a significant improvement in the 4-year event-free survival when immunotherapy was combined with SABR compared to SABR alone. The event-free survival rate at the 4-year mark was 77% in the SABR with immunotherapy group, in contrast to 53% in the SABR alone group. Notably, the SABR group exhibited a 4-year recurrence rate of 36% for any type of recurrence (local, regional, or distant), while the I-SABR group showed a substantially lower recurrence rate of 12% at the 4-year mark. Additionally, the I-SABR group demonstrated a lower incidence of second primary lung cancers, with an observed rate of 3% compared to 8% in the SABR group.

Radiofrequency ablation

Currently, there are few clinical trials exploring combination therapy of RFA with immunotherapy, most of which focus on HCC. In the NIVOLVE trial, the efficacy and safety of nivolumab as an adjuvant therapy with SR or RFA in HCC were assessed. The 1-year recurrence-free survival rate and recurrence-free survival in the NIVOLVE trial were 76.6% and 26.0 months, respectively, with no significant difference between SR and RFA. Similarly, another study suggested that combination therapy with anti-PD-1 plus RFA significantly improved survival time compared to RFA alone in HCC patients, with 32.5% and 10% recurrence-free survival rates, respectively. A proof-of-concept clinical trial reported that the objective response rate was increased from 10% to 24% by adding RFA therapy to advanced HCC patients with a stable or atypical response during PD-1 therapy (NCT03939975).

An early phase III randomized controlled trial conducted in Korea in 2015 demonstrated that patients who underwent SR or RFA for HCC had longer recurrence-free survival (RFS) and overall survival (OS) when adjuvant immunotherapy with activated CIK cells was administered, compared to those without immunotherapy. Subsequently, the same institution conducted a retrospective study to evaluate the efficacy of this combination treatment in real-world clinical practice. During a median follow-up of 28 months, adjuvant immunotherapy in patients who underwent SR or RFA showed a significantly longer RFS. However, a single-arm phase II trial of pembrolizumab in combination with RFA showed that the hypothesized objective response rate of 30% before local therapy was not reached (NCT03753659).

Photodynamic therapy

Although photodynamic therapy (PDT) may induce antitumor immunity, its capacity for tumor elimination is limited, and its efficacy for distant tumors is poor [110]. Combining PDT with immunotherapy can stimulate a strong antitumor immune response to maximize the clinical outcome [111]. However, the effectiveness of this combination mainly relies on preclinical models, and relevant clinical trials are eagerly awaited and necessary.

Thermotherapy/cryotherapy

Combining cryotherapy and immunotherapy has shown promising clinical efficacy in breast cancer and melanoma, as indicated by preclinical and pilot studies [86]. In a pilot trial, Ross et al. assessed the safety and feasibility of cryotherapy combined with pembrolizumab in prostate cancer [112]. The median PFS of the 12 enrolled patients was 14 months without apparent complications from cryotherapy. Another recent pilot study in metastatic renal cell carcinoma (RCC) reported a significant immune cell infiltration with tremelimumab plus cryoablation compared to tremelimumab monotherapy [113]. Both pilot studies demonstrated the feasibility and safety of cryotherapy in combination with immunotherapy. An ongoing phase II trial in metastatic or locally advanced soft tissue sarcoma is evaluating the use of ipilimumab or nivolumab plus cryotherapy, with 3 out of 29 patients achieving partial response (PR) and 7 patients experiencing stable disease (SD) [114] (NCT04118166). Additionally, two clinical trials investigating the combination of nivolumab and ipilimumab with cryotherapy for breast cancer treatment are also in progress (NCT02833233, NCT03546686).

TACE

Transarterial chemoembolization (TACE), an approach that embolizes the hepatic artery in HCC, is a standard treatment option for intermediate-stage HCC. Combining TACE with immunotherapy, especially immune checkpoint inhibitors, has achieved significant breakthroughs in the systemic therapy of HCC. In a phase Ib trial, the combination of TACE and pembrolizumab was found to be safe, with no evidence of synergistic toxicity [114] (NCT03397654). A randomized phase II trial reported that the median PFS and OS were significantly longer in the combination of TACE with low-dose cyclophosphamide and DC vaccination group, compared to the TACE plus low-dose cyclophosphamide group. The objective response rate (ORR) in the former group reached 54%, while in the latter group, it was only 29% [115] (NCT11889464). A recent study on primary HCC reported that the treatment efficacy of RFA and TACE plus CIK cell immunotherapy was significantly better than RFA and TACE alone [116]. The follow-up results showed a mean OS of 42.1 $\pm$ 5.6 months and 37.8 $\pm$ 4.8 months, with a 5-year OS rate of 29.3% and 13.8%, respectively, in each group. However, it should be noted that a recent study on recurrent HCC patients from China showed a similar OS in the TACE plus camrelizumab group compared with the TACE alone group [140]. To assess the efficacy of combinational therapy with TACE in real-world scenarios, several clinical trials are ongoing. The Emerald-1 study is a phase III trial aiming to compare TACE treatment with either placebo, durvalumab, or durvalumab/bevacizumab (NCT03778957). Another TACE-3 study is designed to assess the efficacy between TACE alone and TACE plus nivolumab (NCT04268888) (Table 1).

Summary

Local antitumor therapies have shown promising local and systemic immune effects, prompting several clinical trials to evaluate their combination with immunotherapy. Interim data released from these trials indicate improved median survival, objective response rate

TABLE 1 Clinical trials in local therapy/immunotherapy combination therapies.

Therapy	Clinical trial	Researchers	Objective	Findings
Radiotherapy	NCT02474186	Golden et al.	Validate the efficacy of RT combined with GM-CSF in patients with metastatic solid tumors	RT combined with GM-CSF produced objective abscopal responses in some patients with metastatic solid tumors, indicating feasibility for subsequent trials
	KEYNOTE-001 (NCT01295827)	Shaverdian et al.	Evaluate the efficacy and safety of pembrolizumab with or without RT in advanced NSCLC patients	Combination group achieved longer progression-free survival (PFS) and overall survival (OS) compared to monotherapy group
	NCT02221739	Formenti et al.	Investigate the immune response induced by RT combined with CTLA-4 blockade in previously nonresponsive patients	RT combined with CTLA-4 blockade induced an immune response in previously nonresponsive patients, leading to improved survival
	NCT03110978	Joe Y. Chang's team	Assess the impact of combining immunotherapy with SABR on 4-year event-free survival and recurrence rates	Combination of immunotherapy with SABR improved 4-year event-free survival compared to SABR alone. I-SABR group had lower recurrence rate and incidence of second primary lung cancers
	NCT02125461	Theelen, W., et al.	Evaluate the efficacy of durvalumab plus chemoradiotherapy compared to placebo	Durvalumab plus chemoradiotherapy yielded significantly longer median progression-free survival (PFS) compared to placebo
	Pooled analysis of the PEMBRO-RT trial and the MDACC trial	Theelen, W., et al.	Analyze the outcomes of combining pembrolizumab immunotherapy with RT in metastatic NSCLC patients	Combination of pembrolizumab immunotherapy with RT increased responses and outcomes in metastatic NSCLC patients
Radiofrequency ablation	NIVOLVE trial (HCC)	–	Assess efficacy and safety of nivolumab as adjuvant therapy with SR or RFA in HCC	1-year recurrence-free survival rate (RFSR): 76.6% Recurrence-free survival (RFS): 26.0 months - No significant difference between SR and RFA

TABLE 1 Clinical trials in local therapy/immunotherapy combination therapies—cont'd

Therapy	Clinical trial	Researchers	Objective	Findings
	Study on anti-PD-1 plus RFA (HCC)	Wang, X., et al.	Investigate combination therapy with anti-PD-1 plus RFA in HCC patients	- Improved survival time compared to RFA alone - RFSR rates: 32.5% (combination therapy) and 10% (RFA alone)
	Proof-of-concept trial (HCC)	Lyu, N., et al.	Determine the impact of additional RFA therapy in advanced HCC patients with stable or atypical responses during PD-1 therapy	- Increased objective response rate (ORR) from 10% to 24% with additional RFA therapy
	Early phase III trial (Korea, 2015)	Lee, J.H., et al.	Evaluate adjuvant immunotherapy with activated CIK cells in patients undergoing SR or RFA for HCC	- Longer recurrence-free survival (RFS) and overall survival (OS) compared to without immunotherapy
	Retrospective study (Korea)	Lee, J.H., et al.	Assess the efficacy of adjuvant immunotherapy in real-world clinical practice for patients undergoing SR or RFA for HCC	- Significantly longer recurrence-free survival (RFS) during a median follow-up of 28 months
	Pembrolizumab + RFA phase II trial	–	Investigate the combination of pembrolizumab and RFA in HCC	- Hypothesized ORR of 30% before local therapy was not reached
Photodynamic therapy				
Thermotherapy/ Cryotherapy	pilot trial	Ross et al.	Assess the safety and feasibility of cryotherapy combined with pembrolizumab in prostate cancer	Median progression-free survival (PFS) was 14 months. No apparent complications from cryotherapy were observed
	Pilot Study	Campbell, M.T., et al.	Metastatic renal cell carcinoma (RCC)	Significant immune cell infiltration was observed in patients receiving tremelimumab plus cryoablation compared to tremelimumab monotherapy.
	Ongoing Phase II Trial (NCT04118166)	–	Metastatic or locally advanced soft tissue sarcoma	3 out of 29 patients achieved a partial response (PR) when treated with ipilimumab or nivolumab plus cryotherapy. - 7 patients experienced temporary stable disease (SD).

Continued

TABLE 1 Clinical trials in local therapy/immunotherapy combination therapies—cont'd

Therapy	Clinical trial	Researchers	Objective	Findings
	Ongoing Clinical Trials (NCT02833233, NCT03546686).	–	Investigate the combination of nivolumab and ipilimumab with cryotherapy that is currently in progress for the treatment of breast cancer	–
TACE	Phase Ib Trial (NCT03397654)	Pinato et al.	Assess safety of combining TACE with Pembrolizumab in HCC	No evidence of synergistic toxicity, indicating safety of the combination therapy
	Randomized Phase II Trial (NCT11889464)	–	Evaluate the efficacy of combining TACE with low-dose cyclophosphamide and DC vaccination in HCC	Combination therapy showed significantly longer median progression-free survival (PFS) and overall survival (OS) compared to TACE plus low-dose cyclophosphamide alone. Higher objective response rate (ORR) of 54% in the combination group
	Study on Primary HCC	Ji, Q., et al.	Investigate the efficacy of combining RFA and TACE with CIK cell immunotherapy in primary HCC	Combination therapy demonstrated significantly superior treatment efficacy compared to RFA and TACE alone. Higher mean OS and 5-year OS rates in the combination group
	Study on Recurrent HCC	Guo, Y., et al.	Assess the efficacy of combining TACE with camrelizumab in recurrent HCC	Similar overall survival (OS) was observed in the group receiving TACE plus camrelizumab compared to TACE alone
	Emerald-1 Study (NCT03778957)	–	Conduct a phase III trial comparing different treatment options with TACE in HCC	–
	TACE-3 Trial (NCT04268888)	–	Evaluate the efficacy of TACE alone compared to TACE plus nivolumab	–

(ORR), and PFS compared to local therapy alone, suggesting a synergistic rationale for this combination. However, additional robust clinical evidence is needed to validate these findings further, and careful consideration of potential treatment-related adverse events (TRAEs) is essential in developing synergistic combination strategies.

Given the high percentage of complications associated with both local therapies and immunotherapy, novel approaches that minimize side effects in their combination are urgently required. In the context of HCC treatment, reducing side effects is crucial for preserving liver function during combination therapy. Developing multifunctional therapeutic carriers and effective local delivery routes can play a critical role in ensuring the safe and effective combination of local therapy and immunotherapy. Furthermore, local therapy-mediated immunotherapy drug delivery, controlled release, and immune modulation have shown effectiveness in overcoming immunotherapy resistance, ignorance, and off-target side effects (irAEs). Additionally, the multifunctionality of carriers in imaging and therapeutic delivery holds excellent potential for improving interventional procedures.

To enhance the therapeutic efficacy of combining local therapy and immunotherapy, the development of delivery routes for immunotherapy or loaded carriers is essential. While most immunotherapies have been tested with systemic administration, evaluating local administration of immunotherapy in comparison to systemic administration has recently begun. Leveraging established hepatic artery local administration and percutaneous intra-tumoral administration routes in HCC can facilitate the rapid development and optimization of local immunotherapy and its combination with local therapy. Moreover, the optimal sequencing and interval of local therapies and immunotherapy in the combination also require thorough investigation.

Lastly, the identification of key biomarkers that can assess immune response and therapeutic response to the combination of local therapy and immunotherapy is critical. Given the complexity of the immune-suppressive tumor microenvironment (TME) and anticancer immunity in solid tumors, substantial effort is necessary to integrate multifunctional carriers and image-guided local delivery techniques into novel combination strategies of local therapy and immunotherapy for the safe and effective treatment of various stages.

References

[1] Carvalho HA, Villar RC. Radiotherapy and immune response: the systemic effects of a local treatment. Clinics 2018;73(Suppl 1), e557s.

[2] Izzo F, et al. Radiofrequency ablation and microwave ablation in liver tumors: an update. Oncologist 2019;24 (10):e990–e1005.

[3] Schneider T, et al. Immune response after radiofrequency ablation and surgical resection in nonsmall cell lung cancer. Semin Thorac Cardiovasc Surg 2016;28(2):585–92.

[4] Hamblin MR, Abrahamse H. Factors affecting photodynamic therapy and anti-tumor immune response. Anti Cancer Agents Med Chem 2021;21(2):123–36.

[5] Li Z, et al. Hyperthermia targeting the tumor microenvironment facilitates immune checkpoint inhibitors. Front Immunol 2020;11:595207.

[6] Liapi E, et al. Transcatheter arterial chemoembolization: current technique and future promise. Tech Vasc Interv Radiol 2007;10(1):2–11.

[7] Zhang Z, et al. Radiotherapy combined with immunotherapy: the dawn of cancer treatment. Signal Transduct Target Ther 2022;7(1):258.

[8] Citrin DE. Recent developments in radiotherapy. N Engl J Med 2017;377(11):1065–75.

[9] Rodriguez-Ruiz ME, et al. Immunological mechanisms responsible for radiation-induced abscopal effect. Trends Immunol 2018;39(8):644–55.

[10] Chang JY, et al. Stereotactic ablative radiotherapy with or without immunotherapy for early-stage or isolated lung parenchymal recurrent node-negative non-small-cell lung cancer: an open-label, randomised, phase 2 trial. Lancet 2023;871–81.

[11] O'Brien Jr WD. Ultrasound-biophysics mechanisms. Prog Biophys Mol Biol 2007;93(1–3):212–55.
[12] Miller DL, et al. Overview of therapeutic ultrasound applications and safety considerations. J Ultrasound Med 2012;31(4):623–34.
[13] Baker KG, Robertson VJ, Duck FA. A review of therapeutic ultrasound: biophysical effects. Phys Ther 2001; 81(7):1351–8.
[14] Hoogenboom M, et al. In vivo MR guided boiling histotripsy in a mouse tumor model evaluated by MRI and histopathology. NMR Biomed 2016;29(6):721–31.
[15] Tempany CM, et al. Focused ultrasound surgery in oncology: overview and principles. Radiology 2011; 259(1):39–56.
[16] O'Neill BE, et al. Pulsed high intensity focused ultrasound mediated nanoparticle delivery: mechanisms and efficacy in murine muscle. Ultrasound Med Biol 2009;35(3):416–24.
[17] Watson T. Ultrasound in contemporary physiotherapy practice. Ultrasonics 2008;48(4):321–9.
[18] Wagterveld RM, et al. Visualization of acoustic cavitation effects on suspended calcite crystals. Ultrason Sonochem 2011;18(1):216–25.
[19] Wu F, et al. Expression of tumor antigens and heat-shock protein 70 in breast cancer cells after high-intensity focused ultrasound ablation. Ann Surg Oncol 2007;14(3):1237–42.
[20] Hu Z, et al. Release of endogenous danger signals from HIFU-treated tumor cells and their stimulatory effects on APCs. Biochem Biophys Res Commun 2005;335(1):124–31.
[21] den Brok MH, et al. Synergy between in situ cryoablation and TLR9 stimulation results in a highly effective in vivo dendritic cell vaccine. Cancer Res 2006;66(14):7285–92.
[22] van den Bijgaart RJ, et al. Thermal and mechanical high-intensity focused ultrasound: perspectives on tumor ablation, immune effects and combination strategies. Cancer Immunol Immunother 2017;66(2):247–58.
[23] Eranki A, et al. High-intensity focused ultrasound (HIFU) triggers immune sensitization of refractory murine neuroblastoma to checkpoint inhibitor therapy. Clin Cancer Res 2020;26(5):1152–61.
[24] Bandyopadhyay S, et al. Low-intensity focused ultrasound induces reversal of tumor-induced T cell tolerance and prevents immune escape. J Immunol 2016;196(4):1964–76.
[25] Tao SF, et al. A retrospective case series of high-intensity focused ultrasound (HIFU) in combination with gemcitabine and oxaliplatin (Gemox) on treating elderly middle and advanced pancreatic cancer. Onco Targets Ther 2019;12:9735–45.
[26] Mauri G, et al. Focused ultrasound: tumour ablation and its potential to enhance immunological therapy to cancer. Br J Radiol 2018;91(1083), 20170641.
[27] Maxwell AD, et al. Probability of cavitation for single ultrasound pulses applied to tissues and tissue-mimicking materials. Ultrasound Med Biol 2013;39(3):449–65.
[28] Vlaisavljevich E, et al. Effects of tissue stiffness, ultrasound frequency, and pressure on histotripsy-induced cavitation bubble behavior. Phys Med Biol 2015;60(6):2271–92.
[29] Hall TL, et al. Histotripsy of rabbit renal tissue in vivo: temporal histologic trends. J Endourol 2007;21(10): 1159–66.
[30] Lake AM, et al. Renal ablation by histotripsy—does it spare the collecting system? J Urol 2008;179(3):1150–4.
[31] Wu F. High intensity focused ultrasound ablation and antitumor immune response. J Acoust Soc Am 2013;134 (2):1695–701.
[32] Hoogenboom M, et al. Mechanical high-intensity focused ultrasound destruction of soft tissue: working mechanisms and physiologic effects. Ultrasound Med Biol 2015;41(6):1500–17.
[33] Kwan JJ, et al. Ultrasound-propelled nanocups for drug delivery. Small 2015;11(39):5305–14.
[34] Harmon JS, et al. Minimally invasive gas embolization using acoustic droplet vaporization in a rodent model of hepatocellular carcinoma. Sci Rep 2019;9(1):11040.
[35] van Straten D, et al. Oncologic photodynamic therapy: basic principles, current clinical status and future directions. Cancers (Basel) 2017;9(2):19.
[36] Velez-Montoya R, et al. Current knowledge and trends in age-related macular degeneration: today's and future treatments. Retina 2013;33(8):1487–502.
[37] Reinhard A, et al. Photodynamic therapy as a new treatment modality for inflammatory and infectious conditions. Expert Rev Clin Immunol 2015;11(5):637–57.
[38] Beltran Hernandez I, et al. Preclinical and clinical evidence of immune responses triggered in oncologic photodynamic therapy: clinical recommendations. J Clin Med 2020;9(2):333.

[39] Garg AD, Agostinis P. ER stress, autophagy and immunogenic cell death in photodynamic therapy-induced anti-cancer immune responses. Photochem Photobiol Sci 2014;13(3):474–87.

[40] Garg AD, et al. DAMPs and PDT-mediated photo-oxidative stress: exploring the unknown. Photochem Photobiol Sci 2011;10(5):670–80.

[41] Krysko DV, et al. Immunogenic cell death and DAMPs in cancer therapy. Nat Rev Cancer 2012;12(12):860–75.

[42] Cui J, et al. Mechanisms and pathways of innate immune activation and regulation in health and cancer. Hum Vaccin Immunother 2014;10(11):3270–85.

[43] Garg AD, et al. Hypericin-based photodynamic therapy induces surface exposure of damage-associated molecular patterns like HSP70 and calreticulin. Cancer Immunol Immunother 2012;61(2):215–21.

[44] Yang D, et al. High-mobility group nucleosome-binding protein 1 acts as an alarmin and is critical for lipopolysaccharide-induced immune responses. J Exp Med 2012;209(1):157–71.

[45] Krosl G, Korbelik M, Dougherty GJ. Induction of immune cell infiltration into murine SCCVII tumour by photofrin-based photodynamic therapy. Br J Cancer 1995;71(3):549–55.

[46] Davis RWT, et al. Luminol chemiluminescence reports photodynamic therapy-generated neutrophil activity in vivo and serves as a biomarker of therapeutic efficacy. Photochem Photobiol 2019;95(1):430–8.

[47] Cecic I, Sun J, Korbelik M. Role of complement anaphylatoxin C3a in photodynamic therapy-elicited engagement of host neutrophils and other immune cells. Photochem Photobiol 2006;82(2):558–62.

[48] Cecic I, Parkins CS, Korbelik M. Induction of systemic neutrophil response in mice by photodynamic therapy of solid tumors. Photochem Photobiol 2001;74(5):712–20.

[49] Cecic I, Stott B, Korbelik M. Acute phase response-associated systemic neutrophil mobilization in mice bearing tumors treated by photodynamic therapy. Int Immunopharmacol 2006;6(8):1259–66.

[50] de Vree WJ, et al. Evidence for an important role of neutrophils in the efficacy of photodynamic therapy in vivo. Cancer Res 1996;56(13):2908–11.

[51] de Vree WJ, et al. Role of interleukin 1 and granulocyte colony-stimulating factor in photofrin-based photodynamic therapy of rat rhabdomyosarcoma tumors. Cancer Res 1997;57(13):2555–8.

[52] Cecic I, Korbelik M. Mediators of peripheral blood neutrophilia induced by photodynamic therapy of solid tumors. Cancer Lett 2002;183(1):43–51.

[53] Sun J, et al. Neutrophils as inflammatory and immune effectors in photodynamic therapy-treated mouse SCCVII tumours. Photochem Photobiol Sci 2002;1(9):690–5.

[54] Schacht V, Szeimies RM, Abels C. Photodynamic therapy with 5-aminolevulinic acid induces distinct microcirculatory effects following systemic or topical application. Photochem Photobiol Sci 2006;5(5):452–8.

[55] Galati D, Zanotta S. Dendritic cell and cancer therapy. Int J Mol Sci 2023;24(4):4253.

[56] Korbelik M, Sun J, Cecic I. Photodynamic therapy-induced cell surface expression and release of heat shock proteins: relevance for tumor response. Cancer Res 2005;65(3):1018–26.

[57] Korbelik M, Zhang W, Merchant S. Involvement of damage-associated molecular patterns in tumor response to photodynamic therapy: surface expression of calreticulin and high-mobility group box-1 release. Cancer Immunol Immunother 2011;60(10):1431–7.

[58] Wang X, et al. Stimulation of dendritic cells by DAMPs in ALA-PDT treated SCC tumor cells. Oncotarget 2015; 6(42):44688–702.

[59] Jalili A, et al. Effective photoimmunotherapy of murine colon carcinoma induced by the combination of photodynamic therapy and dendritic cells. Clin Cancer Res 2004;10(13):4498–508.

[60] Hwang HS, et al. Combination of photodynamic therapy (PDT) and anti-tumor immunity in cancer therapy. J Pharm Investig 2018;48(2):143–51.

[61] Etminan N, et al. Heat-shock protein 70-dependent dendritic cell activation by 5-aminolevulinic acid-mediated photodynamic treatment of human glioblastoma spheroids in vitro. Br J Cancer 2011;105(7):961–9.

[62] Korbelik M, Dougherty GJ. Photodynamic therapy-mediated immune response against subcutaneous mouse tumors. Cancer Res 1999;59(8):1941–6.

[63] Korbelik M, Cecic I. Contribution of myeloid and lymphoid host cells to the curative outcome of mouse sarcoma treatment by photodynamic therapy. Cancer Lett 1999;137(1):91–8.

[64] Wachowska M, et al. 5-Aza-2′-deoxycytidine potentiates antitumour immune response induced by photodynamic therapy. Eur J Cancer 2014;50(7):1370–81.

[65] Shaif-Muthana M, et al. Dead or alive: immunogenicity of human melanoma cells when presented by dendritic cells. Cancer Res 2000;60(22):6441–7.

[66] Hatton P, et al. A cervical cord tumour presenting with earache. Br J Hosp Med 1988;39(1):72.
[67] Srivastava P. Roles of heat-shock proteins in innate and adaptive immunity. Nat Rev Immunol 2002;2(3):185–94.
[68] Kabingu E, et al. CD8+ T cell-mediated control of distant tumours following local photodynamic therapy is independent of CD4+ T cells and dependent on natural killer cells. Br J Cancer 2007;96(12):1839–48.
[69] Kabingu E, et al. Enhanced systemic immune reactivity to a basal cell carcinoma associated antigen following photodynamic therapy. Clin Cancer Res 2009;15(13):4460–6.
[70] Thong PS, et al. Photodynamic-therapy-activated immune response against distant untreated tumours in recurrent angiosarcoma. Lancet Oncol 2007;8(10):950–2.
[71] Gollnick SO, Vaughan L, Henderson BW. Generation of effective antitumor vaccines using photodynamic therapy. Cancer Res 2002;62(6):1604–8.
[72] Falk-Mahapatra R, Gollnick SO. Photodynamic therapy and immunity: an update. Photochem Photobiol 2020; 96(3):550–9.
[73] Korbelik M, Sun J, Posakony JJ. Interaction between photodynamic therapy and BCG immunotherapy responsible for the reduced recurrence of treated mouse tumors. Photochem Photobiol 2001;73(4):403–9.
[74] Kleinovink JW, et al. Photodynamic-immune checkpoint therapy eradicates local and distant tumors by CD8(+) T cells. Cancer Immunol Res 2017;5(10):832–8.
[75] Davis AA, Patel VG. The role of PD-L1 expression as a predictive biomarker: an analysis of all US Food and Drug Administration (FDA) approvals of immune checkpoint inhibitors. J Immunother Cancer 2019;7(1):278.
[76] O'Shaughnessy MJ, et al. Systemic antitumor immunity by PD-1/PD-L1 inhibition is potentiated by vascular-targeted photodynamic therapy of primary tumors. Clin Cancer Res 2018;24(3):592–9.
[77] Santos LL, et al. Treatment of head and neck cancer with photodynamic therapy with redaporfin: a clinical case report. Case Rep Oncol 2018;11(3):769–76.
[78] Korpan NN. History of cryosurgery. Springer Vienna; 2001.
[79] Korpan NN. A history of cryosurgery: its development and future. J Am Coll Surg 2007;204(2):314–24.
[80] Erinjeri JP, Clark TWI. Cryoablation: mechanism of action and devices. J Vasc Interv Radiol 2010;21(8):S187–91.
[81] Joe A, et al. Immunotherapy plus cryotherapy: potential augmented abscopal effect for advanced cancers. Front Oncol 2018;8:85.
[82] Staib E, et al. P02.03 microwave ablation enhances tumor-specific immune response in patients with hepatocellular carcinoma. J Immunother Cancer 2020;70:893–907.
[83] Moon EK, Langer CJ, Albelda SM. The era of checkpoint blockade in lung cancer: taking the brakes off the immune system. Ann Am Thorac Soc 2017; AnnalsATS.201702-152FR.
[84] Li L, et al. Microwave ablation combined with OK-432 induces Th1-type response and specific antitumor immunity in a murine model of breast cancer. J Transl Med 2017;15(1):23.
[85] Sheybani ND, et al. Combination of thermally ablative focused ultrasound with gemcitabine controls breast cancer via adaptive immunity. J Immunother Cancer 2020;8(2), e001008.
[86] Yakkala C, et al. Cryoablation and immunotherapy: an enthralling synergy to confront the tumors. Front Immunol 2019;10:2283.
[87] Helmink BA, et al. B cells and tertiary lymphoid structures promote immunotherapy response. Nature 2020;577 (7791):549–55.
[88] Yakkala C, et al. Cryoablation and immunotherapy of cancer. Curr Opin Biotechnol 2020;65:60–4.
[89] Zhu C, et al. PD-1 blockade enhances the anti-tumor immune response induced by cryoablation in a murine model of renal cell carcinoma. Cryobiology 2019;87:86–90.
[90] Waitz R, et al. Potent induction of tumor immunity by combining tumor cryoablation with anti-CTLA-4 therapy. Cancer Res 2012;72(2):430–9.
[91] Benzon, et al. Combining immune check-point blockade and cryoablation in an immunocompetent hormone sensitive murine model of prostate cancer. Prostate Cancer Prostatic Dis 2018;21:126–36.
[92] Lu W, et al. Necrosis and apoptosis in hepatocellular carcinoma following low-dose versus high-dose preoperative chemoembolization. Cardiovasc Intervent Radiol 2008;31(6):1133–40.
[93] Galluzzi L, et al. Immunological effects of conventional chemotherapy and targeted anticancer agents. Cancer Cell 2015;28(6):690–714.
[94] Ayaru L, et al. Unmasking of alpha-fetoprotein-specific CD4(+) T cell responses in hepatocellular carcinoma patients undergoing embolization. J Immunol 2007;178(3):1914–22.
[95] Liao J, et al. Effect of transcatheter arterial chemoembolization on cellular immune function and regulatory T cells in patients with hepatocellular carcinoma. Mol Med Rep 2015;12(4):6065–71.

[96] Huang M, Wang X, Bin H. Effect of transcatheter arterial chemoembolization combined with argon-helium cryosurgery system on the changes of NK cells and T cell subsets in peripheral blood of hepatocellular carcinoma patients. Cell Biochem Biophys 2015;73(3):787–92.

[97] Xiaochen M, et al. The influence of transarterial chemoembolization on serum levels of soluble programed cell death ligand-1 in advanced hepatocellular carcinoma patients. Asia Pac J Clin Oncol 2022;18(5):e515–23.

[98] Ikei S, et al. Changes in IL-6, IL-8, C-reactive protein and pancreatic secretory trypsin inhibitor after transcatheter arterial chemo-embolization therapy for hepato-cellular carcinoma. Cytokine 1992;4(6):581–4.

[99] Kim MJ, et al. Change in inflammatory cytokine profiles after transarterial chemotherapy in patients with hepatocellular carcinoma. Cytokine 2013;64(2):516–22.

[100] Kohles N, et al. Predictive value of immunogenic cell death biomarkers HMGB1, sRAGE, and DNase in liver cancer patients receiving transarterial chemoembolization therapy. Tumour Biol 2012;33(6):2401–9.

[101] Zheng L, et al. Concurrent EGFR-TKI and thoracic radiotherapy as first-line treatment for stage IV non-small cell lung cancer harboring EGFR active mutations. Oncologist 2019;24(8):1031–e612.

[102] Petrillo M, et al. Hypoxia and tumor angiogenesis in the era of hepatocellular carcinoma transarterial loco-regional treatments. Future Oncol 2018;14(28):2957–67.

[103] Zhang W, et al. Blocking the PD-1/PD-L1 axis in dendritic cell-stimulated cytokine-induced killer cells with pembrolizumab enhances their therapeutic effects against hepatocellular carcinoma. J Cancer 2019;10(11): 2578–87.

[104] Cao F, et al. The efficacy of TACE combined with lenvatinib plus sintilimab in unresectable hepatocellular carcinoma: a multicenter retrospective study. Front Oncol 2021;11:783480.

[105] Voron T, et al. VEGF-A modulates expression of inhibitory checkpoints on CD8+ T cells in tumors. J Exp Med 2015;212(2):139–48.

[106] Montasser A, et al. Transarterial chemoembolisation enhances programmed death-1 and programmed death-ligand 1 expression in hepatocellular carcinoma. Histopathology 2021;79(1):36–46.

[107] Ma XL, et al. Elevated soluble programmed death-ligand 1 levels indicate immunosuppression and poor prognosis in hepatocellular carcinoma patients undergoing transcatheter arterial chemoembolization. Clin Chim Acta 2020;511:67–74.

[108] Odagiri N, et al. Early change in the plasma levels of circulating soluble immune checkpoint proteins in patients with unresectable hepatocellular carcinoma treated by lenvatinib or transcatheter arterial chemoembolization. Cancers (Basel) 2020;12(8):2045.

[109] Tampaki M, et al. Association of TIM-3 with BCLC stage, serum PD-L1 detection, and response to transarterial chemoembolization in patients with hepatocellular carcinoma. Cancers (Basel) 2020;12(1):212.

[110] Xu J, et al. Near-infrared-triggered photodynamic therapy with multitasking upconversion nanoparticles in combination with checkpoint blockade for immunotherapy of colorectal cancer. ACS Nano 2017;11(5):4463–74.

[111] Liu Z, et al. Photodynamic immunotherapy of cancers based on nanotechnology: recent advances and future challenges. J Nanobiotechnol 2021;19(1):160.

[112] Ross AE, et al. A pilot trial of pembrolizumab plus prostatic cryotherapy for men with newly diagnosed oligometastatic hormone-sensitive prostate cancer. Prostate Cancer Prostatic Dis 2020;23(1):184–93.

[113] Campbell MT, et al. Pilot study of Tremelimumab with and without cryoablation in patients with metastatic renal cell carcinoma. Nat Commun 2021;12(1):6375.

[114] Fessas P, et al. PETAL protocol: a phase Ib study of pembrolizumab after transarterial chemoembolization in hepatocellular carcinoma. Future Oncol 2023;19(7):499–507.

[115] Ji Q, et al. Effect of RFA and TACE combined with postoperative cytokine-induced killer cell immunotherapy in primary hepatocellular carcinoma. J BUON 2021;26(1):235–42.

[116] Guo Y, et al. Transarterial chemoembolization combined with camrelizumab for recurrent hepatocellular carcinoma. BMC Cancer 2022;22(1):270.

CHAPTER

8

T cell metabolism and cancer immunotherapy

Qianqian Li and Zhuangzhuang Zhang

Guangdong-Hong Kong-Macao University Joint Laboratory of Interventional Medicine, The Fifth Affiliated Hospital, Sun Yat-sen University, Zhuhai, China

Abstract

Immunometabolism is an emerging field that explores how metabolic processes influence immune cell functions and their role in tumor immunotherapy. Recent research has increasingly focused on how T cell metabolism impacts antitumor responses, with promising implications for improving the effectiveness of immunotherapies. By leveraging the metabolic advantages of antitumor immune cells, novel therapeutic strategies are being developed to target the metabolism of effector T cells, potentially enhancing their ability to combat tumors resistant to conventional treatments. Despite these advances, balancing tumor suppression with the maintenance of immune cell activity poses significant challenges. A key area of concern is optimizing checkpoint inhibitors and understanding how metabolic reprogramming affects their efficacy. This chapter emphasizes the urgent need for new drugs that not only enhance antitumor immunity but also exhibit direct cytotoxic effects on tumors.

Additionally, the interaction between different immune cells and their metabolic impacts on the tumor microenvironment (TME) is crucial for developing effective therapies. Investigating the roles of specific metabolic molecules across various cell types will be essential for understanding their collective impact on the TME. The chapter also highlights the importance of considering patient heterogeneity in immunotherapy. Factors such as diet and obesity can alter systemic metabolism, which in turn affects treatment outcomes. Variations in metabolic dependencies among different tumor types, patients, and even regions within the same tumor further complicate the development of personalized immunotherapeutic strategies. Addressing these complexities will be key to advancing tailored treatments and improving clinical outcomes.

Abbreviations

2-DG	2-deoxyglucose
2-HG	2-hydroxyglutarate
ACAT-1	acyl-CoA cholesterol acyltransferase 1
ACAT-1	acyl-CoA: cholesterol acyltransferase 1
ACAT-2	acyl-CoA cholesterol acyltransferase 2

https://doi.org/10.1016/B978-0-443-13827-0.00006-8

ACC1	acetyl-CoA carboxylase 1
ACT	adoptive cell transfer
AKT	protein kinase B
AMPK	AMP-activated protein kinase
ASCT2	alanine, serine, cysteine transporter 2
ATP	adenosine triphosphate
CD4+	cluster of differentiation 4 positive
CD8+	cluster of differentiation 8 positive
CPT1A	carnitine palmitoyltransferase 1A
CTLA-4	cytotoxic T-lymphocyte antigen 4
CTLs	cytotoxic T lymphocytes
ECAR	extracellular acidification rate
FAO	fatty acid oxidation
FAS	fatty acid synthesis
FASN	fatty acid synthase
GITR	glucocorticoid-induced tumor necrosis factor receptor
GLUT1	glucose transporter 1
HBP	hexosamine biosynthetic pathway
HIF-1α	hypoxia-inducible factor 1 alpha
HK2	hexokinase 2
ICB	immune checkpoint blockade
ICOS	inducible T cell co-stimulator
IDO1	indoleamine 2,3-dioxygenase 1
LAG-3	lymphocyte activation gene 3
LAT1	L-type amino acid transporter 1
LKB1	liver kinase B1
LXR	liver X receptor
MDSC	myeloid-derived suppressor cell
mTOR	mammalian target of rapamycin
mTORC1	mammalian target of rapamycin complex 1
NK	natural killer
OCR	oxygen consumption rate
Otub1	OTU domain-containing ubiquitin aldehyde-binding protein 1
OXPHOS	oxidative phosphorylation
PCK1	pyruvate carboxykinase 1
PD-1	programmed cell death protein 1
PD-L1	programmed death-ligand 1
PFKFB3	6-phosphofructo-2-kinase/fructose-2,6-biphosphatase 3
PGC1α	peroxisome proliferator-activated receptor gamma coactivator 1 Alpha
PI3K	phosphoinositide 3-kinase
PI3K	phosphoinositide 3-kinase
PKM1	pyruvate kinase muscle isozyme 1
PPAR	peroxisome proliferator-activated receptor
PPARs	peroxisome proliferator-activated receptors
SERBP-1	sterol regulatory element-binding protein 1
SLC1A5	solute carrier family 1 member 5
SLC38A1	solute carrier family 38 member 1
SLC38A2	solute carrier family 38 member 2
SLC7A5	solute carrier family 7 member 5
SNAT1	sodium-dependent neutral amino acid transporter 1
SNAT2	sodium-dependent neutral amino acid transporter 2
SREBP	sterol regulatory element-binding protein
STAT5	signal transducer and activator of transcription 5
TAA	tumor-associated antigen
TAM	tumor-associated macrophages

TCR	T cell receptor
TDO2	tryptophan 2,3-dioxygenase 2
Teff	effector T cells
TIM-3	T cell immunoglobulin and mucin domain 3
Tm	memory T cells
TME	tumor microenvironment
TRAF3IP3	TRAF3 interacting protein 3
Treg	regulatory T cells
TSA	tumor-specific antigen
UDPGlcNAc	uridine diphospho-*N*-acetylglucosamine
XBP1	X-box binding protein 1

Conflict of interest

No potential conflict of interest were disclosed.

Introduction

The immune system contains a variety of immune cells, such as macrophages, neutrophils, monocytes, eosinophils, basophils, lymphocytes, and natural killer cells. Of the immune cells with different functions, T cell is the main cell type that exerts an antitumor effect in the adaptive immunity phase [1]. Energy generated from metabolizing nutrients to produce ATP is critical for survival and normal function of cells. Like other cells, the metabolic pathways of T cells are influenced by the availability of nutrients such as glucose, glutamine, and fatty acids, which determine the activation and function of immune cells [2]. T cells are divided into various subtypes that are the main immune component of the tumor microenvironment (TME), and activated T cells directly or indirectly kill tumors via secreting a variety of factors. Naive T cells undergo metabolic reprogramming during proliferation, differentiation, and execution of effector functions. Activated T cells increase glutamine catabolism and decrease fatty acid oxidation to supply energy for cell growth, proliferation, differentiation, and cytokine secretion [3]. Thus, different types of T cells in TME are metabolically reprogrammed to perform their functions.

With the success of the clinical use of immune checkpoint inhibitors and the tremendous advances made in adoptive cell transfer therapy, tumor immunotherapy has become an important tool in oncology treatment. A sustained response of effector T cells to induce a durable clinical response is required for these anticancer immunotherapy strategies [4,5]. These T cells, either endogenous or transmigratory T cells, must infiltrate and grow within TME. Hypoxia, low pH, suppressive metabolites, and low availability of nutrients in TME prevent T cells from achieving optimal effector function. Currently, the metabolic limitations of effector T cells in tumors are still not addressed in therapeutic interventions [6,7]. Thus, understanding the metabolic reprogramming of T cells in TME could help improve immunotherapy for tumors.

In this chapter, we review the metabolic reprogramming of T cells in TME and attempt to provide new insights for tumor immunotherapy by targeting T cell metabolism.

Metabolic abnormalities in T cells

Human immunity is conferred by two corporative systems, the innate and adaptive immune systems, protecting our body from outside invaders. The innate and adaptive immune

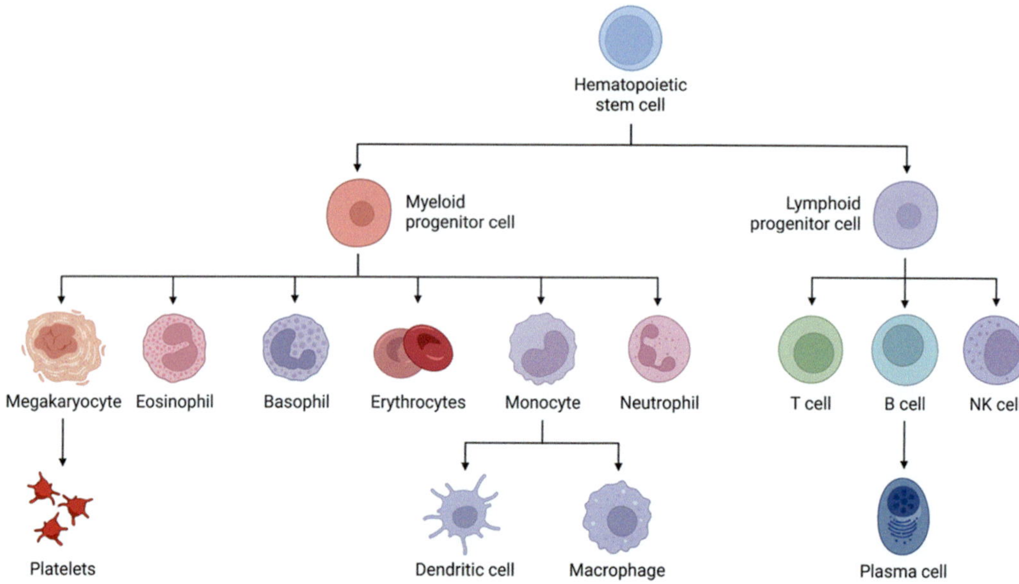

FIG. 1 Immune cell lineage is differentiated from hematopoietic stem cell.

systems play a critical role in antitumor effect, promoting the unprecedented development of immunotherapy for tumors. The immune system is made up of many kinds of immune cells, including T cells, B cells, NK cells, dendritic cells (DCs), and microphages (Fig. 1). The immune cells in TME undergo metabolic reprogramming in the processes of proliferation, differentiation, and execution of effector functions, which are closely related to the antitumor immune response.

T cells, as the main group of cellular immunity, play an important role in antitumor immunity. Depending on different surface antigens, T cells can be classified into two groups, CD4+ and CD8+ T cells. The initial CD4+ T cells can be further differentiated into helper T cells (Th) and regulatory T cells (Treg), while the initial CD8+ T cells can be differentiated into cytotoxic T lymphocytes (CTLs), responding to cellular, molecular, and other stimuli in TME [8]. Different T cells initiate distinct immune functions through various mechanisms. For example, different subpopulations of Th cells regulate the immune functions of CTL and macrophages by secreting various cytokines [9]. Tregs suppress immune response by directly contacting or secreting inhibitory cytokines [10]. CTLs mainly kill target cells by secreting perforin, granzyme, and other substances or inducing apoptosis of target cells through apoptosis-related factor ligands and TNF-α [11]. Memory T cells (Tm) are derived from initial T cells or Th, Treg, and CTL cells. A rapid secondary immune response to the same antigen can be performed by Tm cells [12].

The metabolic responses of tumor cells and immune cells can be regulated by nutrients and metabolites in TME [13]. A high level of metabolism in tumor cells significantly influences the nutrition composition in TME and may have a critical impact on immune responses. Both the high level of metabolic activity of cancer cells and the disordered vascular system in TME lead to nutrient deficiencies and hypoxic microenvironment, establishing metabolic competition between cancer cells and infiltrating immune cells. Because of the competition for nutrients between tumor cells and T cells, the metabolic adaptation of cells to TME is a key factor in maintaining cell function.

Different cell functions can be regulated by altered T cell metabolism [14]. In addition to directly inhibiting immunity by modulating ligands, tumor cells also create a microenvironment that is metabolically disadvantageous for effector lymphocytes. Tumors consume nutrients and produce wastes, such as lactate or kynurenine, to inhibit T cell function. The inhibitory metabolism within T cells may impair the immune activity of effector T cells while promoting inhibitory Tregs and is an obstacle for effective immunotherapy. A thorough understanding of the impact of TME on the immune system will continually improve the efficacy of immunotherapy for cancer patients.

Glucose metabolism in T cells

Cell population consisting of CD4+ T cells and CD8+ T cells is critical for antitumor immune response. T cells experience metabolic reprogramming during their activation, leading to distinct function fates. Naive T cells can proliferate and be metabolically reprogramed after recognizing signals presented by antigen and costimulator [15,16]. Several studies demonstrated that upregulation of aerobic glycolysis was a sign of T cell activation [16]. Additionally, it was shown that an increase in tricarboxylic acid (TCA) cycle and upregulation of oxidative phosphorylation (OXPHOS) played key roles in the activation of CD4+ and CD8 + T as well [17,18]. TCA cycle is upregulated within 24 h after T cell activation; however, aerobic glycolysis can be upregulated within 6 h upon T cell activation. Compared with OXPHOS, glycolysis produces ATP faster and provides more energy for T cell activation. Therefore, T cells in tumors prefer glycolysis to support their rapid growth and differentiation upon the initiation of TCR and costimulatory signals [17]. During this process, naive T cells shift to effector T cells (Teffs) that can effectively kill tumor cells. On the contrary, Treg and CD8+ Tm cells rely on OXOPHOS for survival. OXOPHOS usually consumes a large amount of oxygen, providing energy for the activation of Tm cells upon re-stimulation [19,20].

Different signaling pathways are involved in the metabolic processes of T cell [21]. Activation of PI3K/AKT/mTOR signaling pathway increases the expression of nutrient transporter proteins to further promote glycolysis upon T cell activation [22] (Fig. 2A). In addition, AMP-activated protein kinase K (AMPK) can inhibit T cell activity by suppressing the activation of mTORC1. Inhibitors of the PI3K/AKT pathway can reduce the induction of 6-phosphofructo-2-kinase/fructose-2,6-biphosphatase 3 (PFKFB3), as well as the production of 2,6-fructose diphosphate and lactate [23]. mTOR also regulates glucose metabolism in Treg. A previous study found that activation of T cell receptor (TCR) promoted assembly and activation of mTORC1 on the lysosomal membrane of Treg, while TRAF3 interacting protein 3 (TRAF3IP3), a transmembrane molecule located in lysosomal membrane, inhibited mTORC1 activity and thus glycolysis within Treg [20]. Moreover, OTU domain-containing ubiquitin aldehyde-binding protein 1 (Otub1) regulates activation of T cells by inhibiting ubiquitination of AKT [24].

In addition to the PI3K/AKT/mTOR signaling pathway, Myc and nuclear receptor superfamily also play important roles in glucose metabolism of T cells. Wang et al. and Finlay et al. found that the transcriptional activities of Myc and hypoxia-inducible factor-1α (HIF-1α) were upregulated in response to T cell activation and thus promoted metabolic reprogramming [25,26]. Notably, the activity of HIF-1α is induced in response to T cell activation under

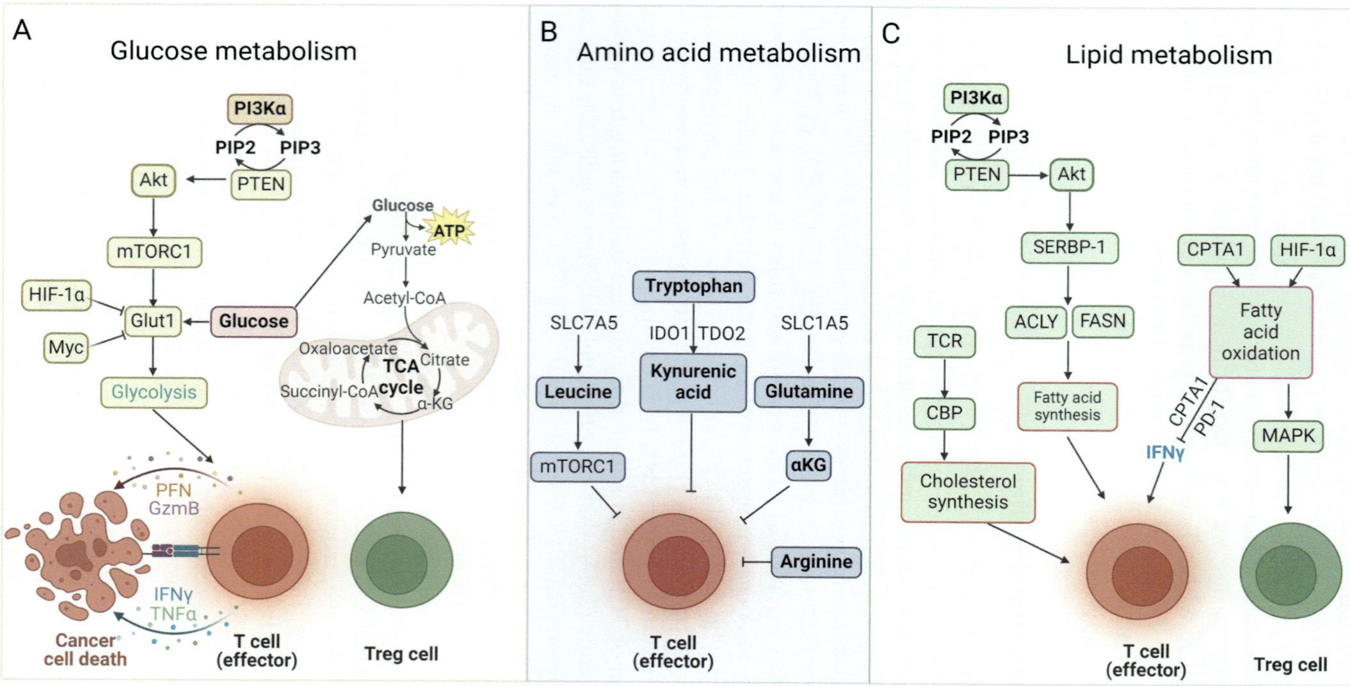

FIG. 2 Metabolisms in T cell. (A) Glucose metabolism in T cell. (B) Amino acid metabolism in T cell. (C) Lipid metabolism in T cell.

hypoxic conditions. In this process, Myc and HIF-1α upregulated the expression of enzymes related to glycolysis, such as pyruvate kinase muscle isozyme 1 (PKM1) and hexokinase 2 (HK2). An increased metabolism rate of activated T cells leads to an increase in glucose transporter 1 (GLUT1) [27]. GLUT1-deficient CD4+ T cells took up significantly less glucose, thereby inhibiting proliferation of CD4+ Teff cells. However, the proliferation and effector functions of GLUT1-deficient CD8+ Teff cells were not affected, suggesting that there may be a compensatory expression of other glucose transporter proteins [28]. Unlike CD4+ Teff cells, Treg cells generated energy via fatty acid oxidation (FAO) to survive and exerted immunosuppressive effects under low glucose conditions [29].

The proximal products of glycolysis are also involved in the activation and effector functions of T cells. Pentose phosphate pathway (PPP) metabolizes glucose 6-phosphate to NADPH and 5-ribose. Wang et al. found a significant increase in glucose metabolization by the PPP in activated CD4+ T cell [25]. NADPH was mainly derived from the PPP and essential for the synthesis of fatty acid and plasma membrane in activated CD8+ T cell [30,31]. Moreover, NADPH is also essential for REDOX homeostasis in proliferating mammalian cells. Uridine diphospho-*N*-acetylglucosamine (UDPGlcNAc), a product of hexosamine biosynthetic pathway (HBP), is a substrate to regulate T cell function through post-translational modification of proteins [32]. For example, O-GlcNAcylation of Myc prevents glycogen synthase kinase 3β-mediated protein degradation, thereby maintaining Myc-induced T cell metabolism. In addition, glucose carbon that has not been metabolized through the glycolytic pathway is important for TCA cycle in Teff cells. For example, citrate can be exported into the cytoplasm to regenerate acetyl coenzyme A for lipid and cholesterol synthesis, both of which are essential for biofilm synthesis in proliferating Teff cells [33]. In summary, specific glycolytic pathways, including aerobic glycolysis, PPP, HBP, and TCA cycle, are upregulated in Teff cells to support rapid proliferation and antitumor effects of these cells.

Limitation of glucose in TME significantly affects the antitumor immune response of T cells. Qiu et al. found that tumor cells suppressed the functions of tumor-infiltrating Teff cells by competing for uptake of glucose in the presence of sufficient tumor antigen [34]. The glycolytic activity of tumor cells can limit the consumption of glucose by tumor-infiltrating Teff cells, thereby inducing exhaustion of Teff cells and immune escape [35]. Blagih et al. found that a low level of glucose in growth medium suppressed the extracellular acidification rate (ECAR), increased the oxygen consumption rate (OCR), diminished mTOR signaling, and inhibited the differentiation and antitumor effects of Teff cells [36]. The mTOR-deficient T cells showed normal activation and IL-2 production after initial stimulation but failed to differentiate into Teff cells and favored the differentiation into Treg cells [37]. However, Sukumar et al. reported that enhancing glycolytic metabolism severely impaired the formation of long-term CD8+ Tm cells, whereas activation of CD8+ T cells in the presence of the glycolytic inhibitor 2-deoxyglucose enhanced the generation of Tm cells and antitumor immune function [38]. Thus, the effect of glycolytic metabolism on antitumor responses may be dependent on different T cell subsets.

Amino acid metabolism in T cells

The proliferation of activated T cells relies heavily on amino acid metabolism for the synthesis of protein and nucleotide [39] (Fig. 2B). The expressions of amino acid transport proteins, including SLC7A5 (also known as LAT1), SLC38A1 (also known as SNAT1), SLC38A2 (also known as SNAT2), and SLC1A5 (also known as ASCT2), are upregulated in activated T cells compared to naive T cells [40]. The essential amino acid leucine is essential for the differentiation and function of T cells. Sinclair et al. found that activated T cells took up leucine and activated mTORC1 signaling by upregulating expression of SLC7A5, the gene encoding the transporter protein of leucine [41]. In the absence of SLC7A5, CD4+ T cells fail to differentiate into Teff cells, and the availability of CD8+ T cells differentiated into Teff cells is severely impaired. However, the differentiation of CD4+ T cells is not affected. These differences may be associated with the different metabolisms of T cell subsets. Arginine is a precursor for the biosynthesis of many macromolecules, including proteins, creatine, and polyamines. Arginine plays an important role in T cell proliferation, as arginine deprivation can lead to cell cycle arrest [42]. Geiger et al. showed that activated T cells can rapidly metabolize arginine, and supplementation with exogenous arginine improved adaptation of T cells and increased generation of Tm cell [43]. Tryptophan is essential for protein synthesis and other vital metabolic activities in living organisms. Tryptophan degradation was mainly induced by the conversion of tryptophan to kynurenic acid by indoleamine 2,3-dioxygenase 1 (IDO1) and tryptophan 2,3-dioxygenase 2 (TDO2) [44]. Cancer cells, tumor-associated macrophages (TAM), myeloid-derived suppressor cells (MDSCs), suppressor dendritic cells, and cancer-associated fibroblasts can deplete tryptophan through IDO, which is highly expressed within TME. Reducing the availability of tryptophan in TME inhibits the antitumor effects of Teff cells. T cell activation is extremely sensitive to tryptophan concentrations in the peripheral environment, and apoptosis of T cells is triggered by deficiency of tryptophan upon heavy utilization by tumor cells in TME [45].

Glutamine metabolism provides carbon and nitrogen for the synthesis of amino acids, nucleic acids, and lipids, which is an integral part of anabolism. Nakaya et al. found that glutamine transporter protein SLC1A5 was expressed at a low level in initial T cells but was significantly upregulated during T cell activation to maintain a high demand for glutamine [46]. Carr et al. showed that restricting glutamine in the medium inhibited the proliferation and effector function of CD8+ Teff cell [47]. When the availability of glutamine was restricted, activated CD8+ T cells tended to differentiate toward a memory cell phenotype [48]. α-Ketoglutarate (αKG), a glutamine-derived metabolite participating in the mitochondrial citric acid cycle, can act as a metabolic regulator of CD4+ T cell differentiation. The limited availability of extracellular glutamine leads to a decrease in the amount of intracellular αKG, thus shifting the balance between Teff and Treg cell production toward a Treg phenotype [49,50].

Lipid metabolism in T cells

Lipid metabolism is mainly consisted of fatty acid oxidation (FAO) and cholesterol metabolism. Under hypoxic conditions, HIF-1α induces pyruvate to leave the mitochondria, allowing Tregs to rely on fatty acids for mitochondrial metabolism in hypoxic tumors. Thus, FAO is essential for Treg metabolism in cancer [51]. Furthermore, Teffs can obtain fatty acids from the microenvironment, whereas Tmem cells only use carbon generated from glucose metabolism to synthesize fatty acids [52]. Lipid metabolism is also important for maintaining the balance between Teffs and Tregs [27,53]. FAO inhibits Teff cell activation and suppresses the expression of PD-1 and carnitine palmitoyltransferase 1A (CPT1A), thereby attenuating the secretion of interferon-γ (IFN-γ). Conversely, FAO-related genes are upregulated in Treg cells to increase FAO levels, thus promoting the generation of Treg cell [54]. FAO plays an important role in the generation and functional maintenance of Tmem cells. Saibil et al.

showed that FAO was the basis for the energy metabolism of Tmem in response to antigenic stimuli and contributed to the maintenance of normal Tmem mitochondrial function and long-term cell survival [55].

Fatty acid synthesis (FAS) is primarily used to produce key lipid cellular structures such as cell membranes required for cell proliferation, while FAO primarily provides ATP and produces metabolic intermediates with important physiological functions. Sterol regulatory element-binding protein (SERBP)-1 is activated by the PI3K/AKT signaling pathway, and then ATP citrate lyase (ACLY) and fatty acid synthase (FASN) are upregulated to promote fatty acid synthesis [56]. Activated T cells are largely dependent on FAS [57], whereas naive T cells and memory T cells maintain essential functions such as the functional integrity of membranes by increasing FAO [19] (Fig. 2C). FAO can promote the generation of Treg cells through activation of MAPK signaling pathway [58]. Tregs are important for immune homeostasis. Treg cells maintain immunosuppression of tumor-associated macrophages and orchestrate the tumor-associated immunosuppressive microenvironment by promoting SERBP-1-dependent lipid metabolism, inhibiting IFN-γ produced by CD8+ T cells [59]. In the absence of SREBP signaling, metabolic reprogramming of activated CD8+ T cells was significantly inhibited. In addition, FAS in T cells is highly dependent on acetyl-CoA carboxylase 1 (ACC1) to initiate the production of long-chain fatty acids. Lee et al. showed that the loss function of ACC1 may disrupt the homeostasis of resting CD8+ T cells and limit the survival of proliferating CD8+ T cells [60]. In contrast, Treg cells are dependent on exogenous fatty acids, and therefore ACC1 deficiency does not affect the differentiation and function of Treg cell [61]. Immunosuppression in Treg cell-driven TME is a key factor contributing to immune tolerance [62]. Peroxisome proliferator-activated receptors (PPARs) also regulate lipid metabolism, and the high activation of PPARs is associated with immunosuppression. Increased PPAR-γ activity inhibits lipolysis and OXPHOS of T cells but promotes differentiation of Tregs [63].

Cholesterol is involved in maintaining cell membrane homeostasis and is a raw material for the synthesis of vitamin D, bile acids, and steroid hormones. Studies have shown that cholesterol can affect the metabolism of immune cells in TME. When T cells are activated, TCR activation promotes cholesterol synthesis by affecting the transcription of key enzymes in the cholesterol biosynthesis pathway (CBP). Meanwhile, liver X receptor (LXR), SERBP-2, and acyl coenzyme A acyltransferase (ACAT)-1 play key regulatory roles in maintaining intracellular cholesterol stability [64]. Earlier studies have shown that cholesterol levels of both whole cell and plasma membrane are significantly elevated in activated CD8+ T cells. Lipid mediators and cytokines promote T cell migration, proliferation, and differentiation upon T cell activation [48,65]. ACAT-1 and ACAT-2 are two cholesterol esterases that convert free cholesterol into cholesteryl esters for storage. ACAT-1 is predominantly expressed in CD8+ T cells and upregulated upon CD8+ T cell activation. In addition, ACAT-1 deficiency inhibits cholesterol esterification but promotes cholesterol biosynthesis, which can upregulate cholesterol levels in the plasma membrane of CD8+ T cells, ultimately enhancing TCR aggregation and activation of signaling [66]. However, several studies have found that cholesterol or its derivatives can inhibit the differentiation of T cells and their anticancer response [67]. Furthermore, a high level of cholesterol in tumor-infiltrating lymphocytes (TIL) upregulates the expression of the endoplasmic reticulum stress receptor XBP1, which further promotes the expression of immune checkpoints and suppresses T cell function [68]. Unlike CD8+ T cells cultured in vitro, the antitumor activity of TIL is usually limited due to the expression of many inhibitory receptors [69]. Therefore, studies on the effects of intracellular cholesterol on T cell function need to focus on T cell types.

Metabolic abnormalities in T cells impair efficacy of cancer immunotherapy

Immunotherapies in cancers

Multigenerational efforts to provide better treatment for cancer patients have led to marked improvements in therapeutic regimens and clinical outcomes. The use of chemotherapy, radiotherapy, and surgical strategies improved patient prognosis in many types of cancer [70–72], and the advent of targeted therapies directly inhibiting oncogenic signaling pathways has also brought benefits to patients with certain cancers [73]. However, to date, treatment options and outcomes have been limited for patients with metastatic cancers and diseases that are not to be controlled by traditional treatment strategies. Therefore, therapies targeting metastatic cancers or traditionally drug-resistant tumors would be of great benefit to a large number of cancer patients.

Approaches that use the body's immune system to treat cancer, collectively known as cancer immunotherapy, have emerged as a powerful set of approaches to address these critical needs. Plenty of cancer immunotherapy approaches are being investigated, including therapeutic strategies of various cellular mechanisms and targets. Currently, there are five main types of immunotherapies: (1) molecular targeted therapy; (2) immune checkpoint inhibitors (PD-1/L1 and CTLA-4 inhibitors); (3) adoptive cell transfer therapy; (4) cytokine; and (5) vaccines (provenge/cimavax, etc.).

Molecularly targeted therapies are designed to reverse the onset of malignant biology at the molecular level by targeting cellular signaling pathways, oncogenes, and cytokine receptors that may lead to cellular carcinogenesis, thereby controlling the growth of tumor cells or even causing complete tumor regression [74].

Immune checkpoint molecules are regulatory molecules that play a suppressive role in the immune system. They are essential for maintaining self-tolerance, preventing autoimmune responses, and minimizing tissue damage by controlling the timing and intensity of the immune response. The expression of immune checkpoint molecules on immune cells will inhibit the function of immune cells to prevent the body from an effective antitumor immune response and result in immune escape for tumors. The main immune checkpoint molecules associated with tumors are PD1, CTLA4, Tim3, and LAG3. Among which, PD1 and CTLA4 are mostly studied [75].

Adoptive cell transfer therapy (ACT) is an immunotherapeutic approach to transfer specific or nonspecific immune cells with antitumor activity to tumor patients. By which, tumor cells will be killed directly or through stimulating an antitumor immune response. Cellular immunotherapy is the most active area of tumor biotherapy, which is more suitable for patients with a low level of cellular immunity [76].

The cytokines commonly used in antitumor immunotherapy include IL-2, IL-12, INF-γ, and TNF. There are many mechanisms by which cytokines exert their tumor-killing effects, such as upregulating the expression and secretion of surface molecules and receptors of immune cells, enhancing the immune surveillance function, promoting the proliferation and differentiation of T cells and the maturation of CTL, stimulating the production of antibodies by B cells, and increasing the activity of NK cells, etc. to generate antitumor immune responses [77,78].

Tumor vaccine is the most anticipated therapy for cancer. Vaccines have been used to prevent diseases, such as hepatitis B and measles, by activating the immune system. Therefore, we are wondering whether cancer could be prevented by activating the immune system as well. Tumor vaccines are usually tumor cells or fragments and fractions containing tumor-specific antigens (TSA) or tumor-associated antigens (TAA). The tumor vaccine is a form of active immunotherapy. Upon entering the body, these cell fragments activate autoimmunity in patients by inducing a specific immune response to overcome immunosuppression and improve resistance to a specific tumor [79]. A great deal of research is currently underway, and some progress has been made. For example, the HPV vaccine is now well established for the prevention of cervical cancer, and Provenge (Sipuleucel-T) is the world's first vaccine for advanced prostate cancer.

Many cancer immunotherapies are being investigated. ACT and ICB are used to eliminate target cancer cells by the antitumor effect of T cells or by removing the suppressiveness of T cells. Despite the early success of ACT and ICB, ACT and ICB immunotherapy are not effective enough in all cancer types, as most patients are not responsive, thus requiring new strategies to improve immunotherapy outcomes. The underlying biology of T cells and how T cells interact with and are suppressed by TME may be responsible for the failure of ACT and ICB [2].

It has been reported that tumors with high glycolytic capacity can consume glucose in TEM, thereby inhibiting the ability of T cells to maintain synthetic growth signals and produce inflammatory cytokines [80,81]. Increasing the competitiveness of T cells in obtaining nutrients through overexpression of PCK1 [81] or ICB therapies targeting CTLA-4 and PD-1 [80] leads to increases in T cell metabolic activity, anabolic signaling, and response to tumors. Interestingly, expression of PD-L1 in tumor cells may promote glycolysis by the autocrine signaling pathway in tumor cells [80]. Amino acids are important for promoting T cell growth and biosynthesis [39], as well as proliferation and effector properties of T cells [41]; thus, a high rate of amino acid uptake by tumor cells may inhibit antitumor responses of T cell [82]. There is also some pieces of evidence showing that some tumors may have a high rate of fatty acid uptake [83]. Considering that fatty acids have been uptaken at a high rate in effector T cells [52], it may be another environmental nutrient that T cells compete for in TEM. There are also indications that metabolic alterations caused by oncogenic mutations found in some tumors may directly affect T cell function. Tumors expressing mutant forms of IDH produce high levels of 2-HG, and treatment of T cells with 2-hydroxyglutarate (2-HG, 2-hydroxyglutarate) [84] results in reduced immunogenicity of T cell [85]. Tumors with mutant IDH create an immunosuppressive environment by producing a high concentration of 2-HG at the tumor site, which can be overcome by pharmacological inhibition of 2-HG [86].

In addition, immunosuppressive nontumor cells and tumor metabolites in the tumor microenvironment also inhibit antitumor function by regulating T cell metabolism. Tregs have a metabolic advantage for survival and function in TEM [29] and promote the expression of indoleamine 2,3-dioxygenase (IDO) in dendritic cells, leading to tryptophan deficiency [87] and suppressing the immune activity of antitumor T cells [88].

Furthermore, Tregs can express CD39, allowing them to hydrolyze extracellular ATP and reduce the function of the immune response [89]. Other subpopulations of suppressive cell types exist in TEM. For example, myeloid-derived suppressor cells (MDSCs) can reduce the availability of amino acids, including arginine and tryptophan [90], thereby suppressing antitumor responses of T cells. Tumor-derived lactate can generate suppressors that promote the polarization of immunosuppressive macrophages, thereby promoting tumor growth [91] and induction of Treg under polarized conditions in vitro [29]. These results suggest that tumor-derived lactate is one of the mechanisms contributing to the immunosuppressive properties of TME.

In summary, T cells face a hostile tumor environment with fierce competition for nutrients. T cells must be able to overcome these difficulties of the unfavorably metabolic environment to perform the antitumor function. On this occasion, cellular metabolism offers a way to understand the limitations of immunotherapy and may offer the potential to enhance the activity by improving the metabolic adaptations of antitumor T cells. Targeting T cell metabolism is therefore emerging as a direction for the next generation of immunotherapies.

Targeting metabolic abnormalities in T cells to enhance efficacy of immunotherapy

Regulation of T cell metabolism in ACT

ACT includes engineered T cell receptor (TCR) therapy, chimeric antigen receptor (CAR) T cell therapy, natural killer (NK) therapy, and tumor-infiltrating lymphocyte (TIL) therapy. ACT has shown success in treating B-cell acute lymphoblastic leukemia (B-ALL) and B-cell lymphoma, with up to half of the patients achieving durable remission [92]. However, responses in other cases remain poor. To improve the response to these immunotherapies, optimization of T cell metabolism is necessary.

Engineered CAR-T cells offer a particularly promising approach to regulating T cell metabolism. By constructing specific signaling domains in CARs, transmissible T cells may be tailored for maintaining effector activity or persistent memory. FDA-approved CD19-targeting second-generation CAR-Ts contain a costimulatory domain derived from CD28 or 4-1BB structural domain, located between the transmembrane region and the CD3 signaling domain. CD28 costimulation typically promotes glycolysis and effector differentiation by activating the PI3K/AKT/mTORC1 pathway to regulate mitochondrial morphology and function [93–95], whereas 4-1BB activates AMPK to stimulate lipid oxidation and mitochondrial biogenesis and metabolism [96]. 4-1BB also stimulates noncanonical NF-κB signaling to promote cell survival. These metabolic phenotypes of CD28 and 4-1BB CAR are associated with functions of short-term effector T cells and long-term memory T cells, respectively. Given the mechanisms mentioned previously, the introduction of signaling mutations may be a promising approach to increase effector function or survival of CAR-T cells. For example, mutations in the YMNM signaling motif of CD28 were shown to increase CAR-T cell survival and reduce T cell exhaustion [97], thereby enhancing durable tumor control. The metabolic significance of this mutation is still unclear. However, phosphorylation of AKT is maintained, and MAPK and calcium signaling pathways are reduced, suggesting changes in metabolic and mitochondrial regulation. Other costimulatory molecules can also regulate T cell metabolism. They can be integrated into CAR structures or targeted by agonist antibodies. For example, ICOS is a member of the CD28 family that promotes glycolysis and mTORC1 activity in Th cells [98]. GITR agonists increase metabolism of CD8+ TIL [99]. Expression of OX40 in Treg TIL is higher than that of CD4 Tconv TIL, which is associated with glycolysis and enrichment of transcription products related to lipid metabolism. Agonism of OX40 enhances lipid uptake of Treg [100].

In addition to CARs, tumor-derived T cells, after in vitro expansion or transduction with engineered TCRs, can also be optimized for function by modulating metabolism in these cells. These approaches usually target solid tumors by fully taking advantage of tumor specificity. However, there are both advantages and challenges for these means. Although the tumor is specific, T cells obtained from a patient's tumor may already be epigenetically programmed. Manipulation of metabolism is needed to overcome this obstacle. Differentiation of naive T cells is needed for sufficient effector function and lifespan of engineered TCRs to eliminate the tumor. T cells with a low level of glycolytic rates or mitochondrial potential may be generated or selected to obtain a longer lifespan while retaining effector functions [101]. Similarly, targeting other pathways, such as glutamine metabolism, may also promote T cell function in these contexts [102].

ACT has a significant advantage over ICB in that the treatment can be administered during in vitro stimulation and T cell engineering rather than in situ. This approach can alter T cell metabolism and mitochondrial function without affecting other cells and tissues, thus preventing any potential risk of metabolic therapies. This approach also eliminates concerns about the potential toxic effects of targeting the metabolism of other tissues.

Glycolysis can be directly targeted with pharmacological inhibitors, such as 2-deoxyglucose. By treating antitumor T cells with 2-deoxyglucose, the activity of the glycolytic pathway is inhibited, leading to increased and sustained T cell infiltration in melanoma and thus improved tumor clearance [38]. T cells treated with this regimen showed increased expression of glycolytic proteins at tumor sites and produced higher levels of inflammatory cytokines in vivo [38]. The PI3K/AKT/mTORC1 pathway is a key mechanism by which T cell induces CD28 costimulation of downstream aerobic glycolysis. Similar to 2-deoxyglucose, inhibition of AKT decreases glucose uptake and glycolysis [103]. Importantly, AKT inhibitor-treated CAR-T cells to eliminate B-cell leukemia following in vivo transfer was improved [104]. It has also been found that exogenous arginine inhibits the glycolytic metabolism of T cells while promoting oxidative pathways, thereby increasing cellular persistence and antitumor activity of T cells [43]. These results suggest that different nutrients, such as amino acids, may yield novel strategies to modify the function of antitumor T cells. Promoting antitumor activity by regulation of T cell metabolism is also accomplished by altering the mitochondrial dynamics of T cells. Pearce and colleagues observed that mitochondrial fission promotes glycolytic metabolism, while mitochondrial fusion promotes mitochondrial metabolism and fatty acid oxidation [94]. Promoting mitochondrial fusion in tumor-specific T cells in vitro limits glycolysis and promotes antitumor activity in vivo. CRISPR-based knockout screening is a powerful tool for assessing the roles of metabolic genes in T cells in different inflammatory and immune settings. REGNASE 1 is a negative regulator of BATF, regulating mitochondrial function and morphology. Regase-1-deficient CD8+ T cells exhibit a significant increase in mitochondrial function, including the production of energy and the ability to consume oxygen [105].

Another strategy to rewire mitochondria in ACT is to directly promote mitochondrial lipid metabolism specific for memory T cells. In ACT, both expression of the transcriptional coactivator PGC1α and administration of PPAR agonists, bezofibrate, or fenofibrate increase lipid oxidation, T cell function, and tumor elimination [106–108]. In summary, these studies suggest that limiting the glycolytic activity of T cells preserves their subsequent replicative capacity and lifespan, thereby mediating a successful immune response against tumors.

Glutamine metabolism is normally coupled with aerobic glycolysis to provide anaplerotic substrates to maintain the TCA cycle. However, in CD4+ Th1 and CD8+ CTL, glucose and glutamine metabolism can be uncoupled. Inhibition of glutamine metabolism leads to increased adaptation of glycolysis, which drives T cell proliferation and effector cell differentiation [102,109]. Thus, both glutaminase and inhibitors of glutamine metabolism enhance the functions of CD4+ Th1 and CD8+ CTL cells. Targeting glutamine can promote CAR-T proliferation and increase the ability of CAR-T to kill target cells. In addition, inhibition of glutamine metabolism may have the potential to activate TIL in situ, as CD8+ T cells in tumors upregulate acetate metabolism to support anabolism [34,110]. However, acetate metabolism was not changed indicating catabolic in cancer cells.

Regulation of T cell metabolism in ICB

ICB therapy is an effective method to activate antitumor immunity. ACT is to use T cells to directly immunize against tumors. ICB is most commonly used for the treatment of solid tumors, whereas the components of TME have been shown to influence the response to ICB [111]. The extent and functional orientation of the T cell infiltrated are important for therapy response.

In addition to the initial barrier that T cells must first infiltrate into tumors, ICB must overcome the exhausting states of TIL, which are accompanied by a high expression of inhibitory receptors and alteration of epigenetic markers, thus inhibiting the effector functions of TIL [112]. This state is also accoupled with various metabolic abnormalities, including impaired transfer of GLUT1 to cell surface and downregulation or inhibition of glycolytic enzymes GAPDH and Enolase [113,114]. Moreover, disruption of morphology and massive production of reactive oxygen species were observed in dysfunctional mitochondria [106,113].

One mechanism by which ICB restores T cell function is to enhance the metabolic capacity of T cells to rescue T cells from the metabolic defects caused by TME. Another mechanism is to promote the differentiation of PD1+ and TCF1+ stem-like progenitor T cells into effector T cells [115,116]. The normal function of PD-1 and CTLA4 is to inhibit PI3K/AKT signaling, thereby counteracting the CD28 family-mediated costimulation signaling [117]. PD-1 deactivates PI3K through recruiting SHP1/2 phosphatases, while CTLA4 inhibits the interaction between CD28 and ligands B7-1 and B7-2 and recruits the phosphatase PP2a. Therefore, PD-1 and CTLA4 inhibit CD28-mediated metabolic reprogramming and aerobic glycolysis [93,118].

Patsoukis et al. found that PD-1 inhibited Teff cell function by suppressing glycolysis and promoting lipolysis and fatty acid oxidation [119]. In contrast, expression of PD-L1 in tumor cells enhanced glucose uptake, further depriving the availability of Teff cells to use this key energy substrate, resulting in reduced mTOR activity, decreased glycolytic capacity, and IFN-γ production in T cells. However, blocking PD-1/PD-L1 promoted glycolysis and enhanced the antitumor activity of Teff cell [80]. Metformin, a classical regulator of glucose metabolism, has direct antitumor activity and indirect CTL-killing effects. By activating the LKB1-AMPK system, the mTOR pathway can be inhibited, interfering with glycolysis in tumors and thus inhibiting tumor growth. The regulation of PD-L1 by metformin is associated with the glycosylation of PD-L1, which further promotes the activation of the endoplasmic reticulum-associated protein degradation pathway, leading to the downregulation of PD-L1 and enhancing the killing effect of CTL [120]. Recent studies have shown that tumor cells and immune cells undergo metabolic reprogramming, with significant upregulation of cholesterol levels in cancer cells. Abnormal metabolites accumulate in the process of tumor cell development. However, the effects of traditional lipid metabolism drugs, such as statins, on T cells remain controversial. Statin was found to have antitumor functions by inhibiting lipid metabolism in tumors and lowering cholesterol levels in T cells, thereby suppressing CTL cell function [121]. On the other hand, it downregulated PD-1, 2B4, TIM-3, and LAG-3 on T cell [68]. Moreover, the ACAT-1 inhibitor avasimibe not only inhibited cholesterol esterification in tumor cells but also increased intracellular free cholesterol levels, thereby enhancing the activity of CD8+ T cells and inhibiting proliferation and metastasis of tumor cells. Avasimibe showed good antitumor effects in mouse tumor model. Combination of avasimibe and PD-1 antibodies showed better efficacy than that of monotherapy in controlling tumor progression [66]. Similarly, acetylcysteine can inhibit expression of FOXO1 by activating the PI3K/AKT signaling pathway, thereby affecting secretion of granzyme B and PD-1 expression to further enhance the antitumor capacity of in vitro-expanded T cells [122]. In contrast, CTLA-4 inhibits the expressions of SNAT1, SNAT2, and GLUT1 and glycolysis without increasing expression of CPT1A and FAO, indicating that CTLA-4 maintains immune silencing [119]. ICB promotes the metabolic reprogramming of Teff cells, thereby enhancing their antitumor effector capacity [119].

In addition to PD-1 and CTLA4, other coinhibitory molecules also regulate T-cell metabolism. Indeed, various ICB targets providing or inhibiting costimulatory signals to T cells have also been shown to affect cellular metabolism. It may be the general mechanism by which costimulation promotes or inhibits T cell function. TIM3 negatively regulates glycolysis and expression of GLUT1 and inhibits glutaminolysis [123]. Lymphocyte activation gene (LAG)-3 may prevent overproliferation of naive T cells by inhibiting IL-7-mediated STAT5 activation. LAG-3-deficient naïve T cells exhibit increased oxidative and glycolytic metabolism due to increased mitochondrial content. Therefore, targeting LAG-3 is expected to provide new insights for antitumor therapy by regulating T cell metabolism [124]. Similarly, targeting 4-1BB and OX40 enhances mitochondrial metabolism of T cells and promotes effector function and long-term survival of T cells [96,125]. Each of these metabolic regulatory functions may directly influence biosynthetic processes and further affect epigenetic markers and the consequent long-term effects on the differentiation and cell fate of T cells. However, how other members and these different T cell regulatory receptors ultimately affect the metabolism and function of antitumor T cells remains to be determined.

Conclusions and prospects

For the past few years, interest in "immunometabolism" has increased significantly. Tumor immunotherapy is now developing rapidly in both clinical and research settings.

However, the understanding of metabolic reprogramming of T cells and its impact on antitumor immune effects is still being explored. Taking advantage of the metabolic advantages of antitumor immune cells to target metabolisms of effector T cells may play an important role in triggering durable immune responses in tumors resistant to conventional therapies. Therefore, in the context of antitumor immunotherapy, therapeutic approaches targeting T cell metabolic pathways deserve further exploration.

However, striking a balance between tumor suppression and maintaining the activity of immune cells remains challenging. In addition to the effects on the efficacy of checkpoint inhibitors, further analysis of the roles of immune cell metabolism is still required. Discovery of drugs that could enhance antitumor immunity and directly kill tumors simultaneously is a priority for drug development in the future. In addition, interactions between different immune cells can affect the suppressive TME. Therefore, studies of the mechanisms of the same metabolic molecules in different cells are essential.

As immunotherapies are used in the clinic, it is important to understand the heterogeneity of the patient population. For example, different diets and obesity have different effects on patients' own systemic metabolism; there are differences in glycolysis dependence among different types of tumors, patients with the same type of tumor, different tumors in the same patient, and even different regions of the same tumor. How to integrate this information and to design immunotherapeutic strategies based on the heterogeneity of metabolism in patients is still one of the challenges we face.

References

[1] Parkin J, Cohen B. An overview of the immune system. Lancet 2001;357:1777–89. https://doi.org/10.1016/s0140-6736(00)04904-7.

[2] Madden MZ, Rathmell JC. The complex integration of T-cell metabolism and immunotherapy. Cancer Discov 2021;11:1636–43. https://doi.org/10.1158/2159-8290.Cd-20-0569.

[3] Bantug GR, Galluzzi L, Kroemer G, Hess C. The spectrum of T cell metabolism in health and disease. Nat Rev Immunol 2018;18:19–34. https://doi.org/10.1038/nri.2017.99.

[4] O'Donnell JS, Teng MWL, Smyth MJ. Cancer immunoediting and resistance to T cell-based immunotherapy. Nat Rev Clin Oncol 2019;16:151–67. https://doi.org/10.1038/s41571-018-0142-8.

[5] Chow A, Perica K, Klebanoff CA, Wolchok JD. Clinical implications of T cell exhaustion for cancer immunotherapy. Nat Rev Clin Oncol 2022;19:775–90. https://doi.org/10.1038/s41571-022-00689-z.

[6] Bagchi S, Yuan R, Engleman EG. Immune checkpoint inhibitors for the treatment of cancer: clinical impact and mechanisms of response and resistance. Annu Rev Pathol 2021;16:223–49. https://doi.org/10.1146/annurev-pathol-042020-042741.

[7] Neo SY, Xu S, Chong J, Lam KP, Wu J. Harnessing novel strategies and cell types to overcome immune tolerance during adoptive cell therapy in cancer. J Immunother Cancer 2023;11. https://doi.org/10.1136/jitc-2022-006434.

[8] Kumar BV, Connors TJ, Farber DL. Human T cell development, localization, and function throughout life. Immunity 2018;48:202–13. https://doi.org/10.1016/j.immuni.2018.01.007.

[9] Zhu J. T helper cell differentiation, heterogeneity, and plasticity. Cold Spring Harb Perspect Biol 2018;10. https://doi.org/10.1101/cshperspect.a030338.

[10] Schmidt A, Oberle N, Krammer PH. Molecular mechanisms of treg-mediated T cell suppression. Front Immunol 2012;3:51. https://doi.org/10.3389/fimmu.2012.00051.

[11] Farhood B, Najafi M, Mortezaee K. CD8(+) cytotoxic T lymphocytes in cancer immunotherapy: a review. J Cell Physiol 2019;234:8509–21. https://doi.org/10.1002/jcp.27782.

[12] Mueller SN, Gebhardt T, Carbone FR, Heath WR. Memory T cell subsets, migration patterns, and tissue residence. Annu Rev Immunol 2013;31:137–61. https://doi.org/10.1146/annurev-immunol-032712-095954.

[13] Elia I, Haigis MC. Metabolites and the tumour microenvironment: from cellular mechanisms to systemic metabolism. Nat Metab 2021;3:21–32. https://doi.org/10.1038/s42255-020-00317-z.

[14] Heintzman DR, Fisher EL, Rathmell JC. Microenvironmental influences on T cell immunity in cancer and inflammation. Cell Mol Immunol 2022;19:316–26. https://doi.org/10.1038/s41423-021-00833-2.

[15] Teijeira A, et al. Metabolic consequences of T-cell costimulation in anticancer immunity. Cancer Immunol Res 2019;7:1564–9. https://doi.org/10.1158/2326-6066.Cir-19-0115.

[16] Menk AV, et al. Early TCR signaling induces rapid aerobic glycolysis enabling distinct acute T cell effector functions. Cell Rep 2018;22:1509–21. https://doi.org/10.1016/j.celrep.2018.01.040.

[17] Chang CH, et al. Posttranscriptional control of T cell effector function by aerobic glycolysis. Cell 2013;153:1239–51. https://doi.org/10.1016/j.cell.2013.05.016.

[18] Sena LA, et al. Mitochondria are required for antigen-specific T cell activation through reactive oxygen species signaling. Immunity 2013;38:225–36. https://doi.org/10.1016/j.immuni.2012.10.020.

[19] Ho PC, Liu PS. Metabolic communication in tumors: a new layer of immunoregulation for immune evasion. J Immunother Cancer 2016;4:4. https://doi.org/10.1186/s40425-016-0109-1.

[20] Yu X, et al. Metabolic control of regulatory T cell stability and function by TRAF3IP3 at the lysosome. J Exp Med 2018;215:2463–76. https://doi.org/10.1084/jem.20180397.

[21] Chen M, Lan H, Yao S, Jin K, Chen Y. Metabolic interventions in tumor immunity: focus on dual pathway inhibitors. Cancer 2023;15. https://doi.org/10.3390/cancers15072043.

[22] Jiang S, Yan W. T-cell immunometabolism against cancer. Cancer Lett 2016;382:255–8. https://doi.org/10.1016/j.canlet.2016.09.003.

[23] Simon-Molas H, et al. PI3K-Akt signaling controls PFKFB3 expression during human T-lymphocyte activation. Mol Cell Biochem 2018;448:187–97. https://doi.org/10.1007/s11010-018-3325-9.

[24] Zhou X, et al. The deubiquitinase Otub1 controls the activation of CD8(+) T cells and NK cells by regulating IL-15-mediated priming. Nat Immunol 2019;20:879–89. https://doi.org/10.1038/s41590-019-0405-2.

[25] Wang R, et al. The transcription factor Myc controls metabolic reprogramming upon T lymphocyte activation. Immunity 2011;35:871–82. https://doi.org/10.1016/j.immuni.2011.09.021.

[26] Finlay DK, et al. PDK1 regulation of mTOR and hypoxia-inducible factor 1 integrate metabolism and migration of CD8+ T cells. J Exp Med 2012;209:2441 53. https://doi.org/10.1084/jem.20112607.

[27] MacIver NJ, Michalek RD, Rathmell JC. Metabolic regulation of T lymphocytes. Annu Rev Immunol 2013;31:259–83. https://doi.org/10.1146/annurev-immunol-032712-095956.

[28] Macintyre AN, et al. The glucose transporter Glut1 is selectively essential for CD4 T cell activation and effector function. Cell Metab 2014;20:61–72. https://doi.org/10.1016/j.cmet.2014.05.004.

[29] Angelin A, et al. Foxp3 reprograms T cell metabolism to function in low-glucose, high-lactate environments. Cell Metab 2017;25:1282–1293.e1287. https://doi.org/10.1016/j.cmet.2016.12.018.

[30] Patra KC, Hay N. The pentose phosphate pathway and cancer. Trends Biochem Sci 2014;39:347–54. https://doi.org/10.1016/j.tibs.2014.06.005.

[31] Jiang P, Du W, Wu M. Regulation of the pentose phosphate pathway in cancer. Protein Cell 2014;5:592–602. https://doi.org/10.1007/s13238-014-0082-8.

[32] Swamy M, et al. Glucose and glutamine fuel protein O-GlcNAcylation to control T cell self-renewal and malignancy. Nat Immunol 2016;17:712–20. https://doi.org/10.1038/ni.3439.

[33] Martínez-Reyes I, Chandel NS. Mitochondrial TCA cycle metabolites control physiology and disease. Nat Commun 2020;11:102. https://doi.org/10.1038/s41467-019-13668-3.

[34] Qiu J, et al. Acetate promotes T cell effector function during glucose restriction. Cell Rep 2019;27:2063–2074.e2065. https://doi.org/10.1016/j.celrep.2019.04.022.

[35] Sukumar M, Roychoudhuri R, Restifo NP. Nutrient competition: a new axis of tumor immunosuppression. Cell 2015;162:1206–8. https://doi.org/10.1016/j.cell.2015.08.064.

[36] Blagih J, et al. The energy sensor AMPK regulates T cell metabolic adaptation and effector responses in vivo. Immunity 2015;42:41–54. https://doi.org/10.1016/j.immuni.2014.12.030.

[37] Delgoffe GM, et al. The mTOR kinase differentially regulates effector and regulatory T cell lineage commitment. Immunity 2009;30:832–44. https://doi.org/10.1016/j.immuni.2009.04.014.

[38] Sukumar M, et al. Inhibiting glycolytic metabolism enhances CD8+ T cell memory and antitumor function. J Clin Invest 2013;123:4479–88. https://doi.org/10.1172/jci69589.

[39] Hosios AM, et al. Amino acids rather than glucose account for the majority of cell mass in proliferating mammalian cells. Dev Cell 2016;36:540–9. https://doi.org/10.1016/j.devcel.2016.02.012.

[40] Ren W, et al. Amino-acid transporters in T-cell activation and differentiation. Cell Death Dis 2017;8, e2655. https://doi.org/10.1038/cddis.2016.222.

[41] Sinclair LV, et al. Control of amino-acid transport by antigen receptors coordinates the metabolic reprogramming essential for T cell differentiation. Nat Immunol 2013;14:500–8. https://doi.org/10.1038/ni.2556.

[42] Martí ILAA, Reith W. Arginine-dependent immune responses. Cell Mol Life Sci 2021;78:5303–24. https://doi.org/10.1007/s00018-021-03828-4.

[43] Geiger R, et al. L-arginine modulates T cell metabolism and enhances survival and anti-tumor activity. Cell 2016;167:829–842.e813. https://doi.org/10.1016/j.cell.2016.09.031.

[44] Grohmann U, et al. Amino-acid sensing and degrading pathways in immune regulation. Cytokine Growth Factor Rev 2017;35:37–45. https://doi.org/10.1016/j.cytogfr.2017.05.004.

[45] Cronin SJF, et al. The metabolite BH4 controls T cell proliferation in autoimmunity and cancer. Nature 2018;563:564–8. https://doi.org/10.1038/s41586-018-0701-2.

[46] Nakaya M, et al. Inflammatory T cell responses rely on amino acid transporter ASCT2 facilitation of glutamine uptake and mTORC1 kinase activation. Immunity 2014;40:692–705. https://doi.org/10.1016/j.immuni.2014.04.007.

[47] Carr EL, et al. Glutamine uptake and metabolism are coordinately regulated by ERK/MAPK during T lymphocyte activation. J Immunol 2010;185:1037–44. https://doi.org/10.4049/jimmunol.0903586.

[48] Zhang L, Romero P. Metabolic control of CD8(+) T cell fate decisions and antitumor immunity. Trends Mol Med 2018;24:30–48. https://doi.org/10.1016/j.molmed.2017.11.005.

[49] Kugelberg E. T cells: nutrients guide differentiation. Nat Rev Immunol 2015;15:666. https://doi.org/10.1038/nri3923.

[50] Jin J, Byun JK, Choi YK, Park KG. Targeting glutamine metabolism as a therapeutic strategy for cancer. Exp Mol Med 2023. https://doi.org/10.1038/s12276-023-00971-9.

[51] Miska J, et al. HIF-1α is a metabolic switch between glycolytic-driven migration and oxidative phosphorylation-driven immunosuppression of Tregs in glioblastoma. Cell Rep 2019;27:226–237.e224. https://doi.org/10.1016/j.celrep.2019.03.029.

[52] O'Sullivan D, et al. Memory CD8(+) T cells use cell-intrinsic lipolysis to support the metabolic programming necessary for development. Immunity 2014;41:75–88. https://doi.org/10.1016/j.immuni.2014.06.005.

[53] Edwards-Hicks J, et al. Phosphoinositide acyl chain saturation drives CD8(+) effector T cell signaling and function. Nat Immunol 2023;24:516–30. https://doi.org/10.1038/s41590-023-01419-y.

[54] Zhang C, et al. STAT3 activation-induced fatty acid oxidation in CD8(+) T effector cells is critical for obesity-promoted breast tumor growth. Cell Metab 2020;31:148–161.e145. https://doi.org/10.1016/j.cmet.2019.10.013.

[55] Saibil SD, et al. Activation of peroxisome proliferator-activated receptors α and δ synergizes with inflammatory signals to enhance adoptive cell therapy. Cancer Res 2019;79:445–51. https://doi.org/10.1158/0008-5472.Can-17-3053.

[56] Cheng C, Geng F, Cheng X, Guo D. Lipid metabolism reprogramming and its potential targets in cancer. Cancer Commun 2018;38:27. https://doi.org/10.1186/s40880-018-0301-4.

[57] Xu T, et al. Metabolic control of T(H)17 and induced T(reg) cell balance by an epigenetic mechanism. Nature 2017;548:228–33. https://doi.org/10.1038/nature23475.

[58] van der Windt GJ, et al. Mitochondrial respiratory capacity is a critical regulator of CD8+ T cell memory development. Immunity 2012;36:68–78. https://doi.org/10.1016/j.immuni.2011.12.007.

[59] Liu C, et al. Treg cells promote the SREBP1-dependent metabolic fitness of tumor-promoting macrophages via repression of CD8(+) T cell-derived interferon-γ. Immunity 2019;51:381–397.e386. https://doi.org/10.1016/j.immuni.2019.06.017.

[60] Lee J, et al. Regulator of fatty acid metabolism, acetyl coenzyme a carboxylase 1, controls T cell immunity. J Immunol 2014;192:3190–9. https://doi.org/10.4049/jimmunol.1302985.

[61] Berod L, et al. De novo fatty acid synthesis controls the fate between regulatory T and T helper 17 cells. Nat Med 2014;20:1327–33. https://doi.org/10.1038/nm.3704.

[62] Savage PA, Klawon DEJ, Miller CH. Regulatory T cell development. Annu Rev Immunol 2020;38:421–53. https://doi.org/10.1146/annurev-immunol-100219-020937.

[63] Chakraborty P, et al. Pro-survival lipid sphingosine-1-phosphate metabolically programs T cells to limit anti-tumor activity. Cell Rep 2019;28:1879–1893.e1877. https://doi.org/10.1016/j.celrep.2019.07.044.

[64] Yanagisawa R, et al. The impacts of cholesterol, oxysterols, and cholesterol lowering dietary compounds on the immune system. Int J Mol Sci 2022;23. https://doi.org/10.3390/ijms232012236.

[65] Chen C, Wang Z, Ding Y, Qin Y. Manipulating T-cell metabolism to enhance immunotherapy in solid tumor. Front Immunol 2022;13:1090429. https://doi.org/10.3389/fimmu.2022.1090429.
[66] Yang W, et al. Potentiating the antitumour response of CD8(+) T cells by modulating cholesterol metabolism. Nature 2016;531:651–5. https://doi.org/10.1038/nature17412.
[67] Ma X, et al. Cholesterol negatively regulates IL-9-producing CD8(+) T cell differentiation and antitumor activity. J Exp Med 2018;215:1555–69. https://doi.org/10.1084/jem.20171576.
[68] Ma X, et al. Cholesterol induces CD8(+) T cell exhaustion in the tumor microenvironment. Cell Metab 2019;30:143–156.e145. https://doi.org/10.1016/j.cmet.2019.04.002.
[69] McKinney EF, Smith KGC. Metabolic exhaustion in infection, cancer and autoimmunity. Nat Immunol 2018;19:213–21. https://doi.org/10.1038/s41590-018-0045-y.
[70] Springfeld C, et al. Neoadjuvant therapy for pancreatic cancer. Nat Rev Clin Oncol 2023;20:318–37. https://doi.org/10.1038/s41571-023-00746-1.
[71] Rich TA, Evans DB, Curley SA, Ajani JA. Adjuvant radiotherapy and chemotherapy for biliary and pancreatic cancer. Ann Oncol 1994;5(Suppl 3):75–80. https://doi.org/10.1093/annonc/5.suppl_3.s75.
[72] Ethun CG, et al. Frailty and cancer: implications for oncology surgery, medical oncology, and radiation oncology. CA Cancer J Clin 2017;67:362–77. https://doi.org/10.3322/caac.21406.
[73] Kantarjian H, Rajkumar SV. Why are cancer drugs so expensive in the United States, and what are the solutions? Mayo Clin Proc 2015;90:500–4. https://doi.org/10.1016/j.mayocp.2015.01.014.
[74] Bedard PL, Hyman DM, Davids MS, Siu LL. Small molecules, big impact: 20 years of targeted therapy in oncology. Lancet 2020;395:1078–88. https://doi.org/10.1016/s0140-6736(20)30164-1.
[75] Abril-Rodriguez G, Ribas A. SnapShot: immune checkpoint inhibitors. Cancer Cell 2017;31:848–848.e841. https://doi.org/10.1016/j.ccell.2017.05.010.
[76] Ma S, et al. Current progress in CAR-T cell therapy for solid tumors. Int J Biol Sci 2019;15:2548–60. https://doi.org/10.7150/ijbs.34213.
[77] Märkl F, Huynh D, Endres S, Kobold S. Utilizing chemokines in cancer immunotherapy. Trends Cancer 2022;8:670–82. https://doi.org/10.1016/j.trecan.2022.04.001.
[78] Laskowski TJ, Biederstädt A, Rezvani K. Natural killer cells in antitumour adoptive cell immunotherapy. Nat Rev Cancer 2022;22:557–75. https://doi.org/10.1038/s41568-022-00491-0.
[79] Saxena M, van der Burg SH, Melief CJM, Bhardwaj N. Therapeutic cancer vaccines. Nat Rev Cancer 2021;21:360–78. https://doi.org/10.1038/s41568-021-00346-0.
[80] Chang CH, et al. Metabolic competition in the tumor microenvironment is a driver of cancer progression. Cell 2015;162:1229–41. https://doi.org/10.1016/j.cell.2015.08.016.
[81] Ho PC, et al. Phosphoenolpyruvate is a metabolic checkpoint of anti-tumor T cell responses. Cell 2015;162:1217–28. https://doi.org/10.1016/j.cell.2015.08.012.
[82] Wang Q, Holst J. L-type amino acid transport and cancer: targeting the mTORC1 pathway to inhibit neoplasia. Am J Cancer Res 2015;5:1281–94.
[83] Beloribi-Djefaflia S, Vasseur S, Guillaumond F. Lipid metabolic reprogramming in cancer cells. Oncogenesis 2016;5, e189. https://doi.org/10.1038/oncsis.2015.49.
[84] Dang L, et al. Cancer-associated IDH1 mutations produce 2-hydroxyglutarate. Nature 2009;462:739–44. https://doi.org/10.1038/nature08617.
[85] Tyrakis PA, et al. S-2-hydroxyglutarate regulates CD8(+) T-lymphocyte fate. Nature 2016;540:236–41. https://doi.org/10.1038/nature20165.
[86] Kohanbash G, et al. Isocitrate dehydrogenase mutations suppress STAT1 and CD8+ T cell accumulation in gliomas. J Clin Invest 2017;127:1425–37. https://doi.org/10.1172/jci90644.
[87] Fallarino F, et al. Modulation of tryptophan catabolism by regulatory T cells. Nat Immunol 2003;4:1206–12. https://doi.org/10.1038/ni1003.
[88] Munn DH, et al. Inhibition of T cell proliferation by macrophage tryptophan catabolism. J Exp Med 1999;189:1363–72. https://doi.org/10.1084/jem.189.9.1363.
[89] Borsellino G, et al. Expression of ectonucleotidase CD39 by Foxp3+ Treg cells: hydrolysis of extracellular ATP and immune suppression. Blood 2007;110:1225–32. https://doi.org/10.1182/blood-2006-12-064527.
[90] Yu J, et al. Myeloid-derived suppressor cells suppress antitumor immune responses through IDO expression and correlate with lymph node metastasis in patients with breast cancer. J Immunol 2013;190:3783–97. https://doi.org/10.4049/jimmunol.1201449.

[91] Colegio OR, et al. Functional polarization of tumour-associated macrophages by tumour-derived lactic acid. Nature 2014;513:559–63. https://doi.org/10.1038/nature13490.
[92] Maude SL, et al. Tisagenlecleucel in children and young adults with B-cell lymphoblastic leukemia. N Engl J Med 2018;378:439–48. https://doi.org/10.1056/NEJMoa1709866.
[93] Frauwirth KA, et al. The CD28 signaling pathway regulates glucose metabolism. Immunity 2002;16:769–77. https://doi.org/10.1016/s1074-7613(02)00323-0.
[94] Buck MD, et al. Mitochondrial dynamics controls T cell fate through metabolic programming. Cell 2016;166:63–76. https://doi.org/10.1016/j.cell.2016.05.035.
[95] Klein Geltink RI, et al. Mitochondrial priming by CD28. Cell 2017;171:385–397.e311. https://doi.org/10.1016/j.cell.2017.08.018.
[96] Choi BK, et al. 4-1BB signaling activates glucose and fatty acid metabolism to enhance CD8(+) T cell proliferation. Cell Mol Immunol 2017;14:748–57. https://doi.org/10.1038/cmi.2016.02.
[97] Guedan S, et al. Single residue in CD28-costimulated CAR-T cells limits long-term persistence and antitumor durability. J Clin Invest 2020;130:3087–97. https://doi.org/10.1172/jci133215.
[98] Zeng H, et al. mTORC1 and mTORC2 kinase signaling and glucose metabolism drive follicular helper T cell differentiation. Immunity 2016;45:540–54. https://doi.org/10.1016/j.immuni.2016.08.017.
[99] Sabharwal SS, et al. GITR agonism enhances cellular metabolism to support CD8(+) T-cell proliferation and effector cytokine production in a mouse tumor model. Cancer Immunol Res 2018;6:1199–211. https://doi.org/10.1158/2326-6066.Cir-17-0632.
[100] Pacella I, et al. Fatty acid metabolism complements glycolysis in the selective regulatory T cell expansion during tumor growth. Proc Natl Acad Sci USA 2018;115:E6546–e6555. https://doi.org/10.1073/pnas.1720113115.
[101] Kishton RJ, Sukumar M, Restifo NP. Metabolic regulation of T cell longevity and function in tumor immunotherapy. Cell Metab 2017;26:94–109. https://doi.org/10.1016/j.cmet.2017.06.016.
[102] Leone RD, et al. Glutamine blockade induces divergent metabolic programs to overcome tumor immune evasion. Science 2019;366:1013–21. https://doi.org/10.1126/science.aav2588.
[103] Crompton JG, et al. Akt inhibition enhances expansion of potent tumor-specific lymphocytes with memory cell characteristics. Cancer Res 2015;75:296–305. https://doi.org/10.1158/0008-5472.Can-14-2277.
[104] Klebanoff CA, et al. Inhibition of AKT signaling uncouples T cell differentiation from expansion for receptor-engineered adoptive immunotherapy. JCI Insight 2017;2. https://doi.org/10.1172/jci.insight.95103.
[105] Wei J, et al. Targeting REGNASE-1 programs long-lived effector T cells for cancer therapy. Nature 2019;576:471–6. https://doi.org/10.1038/s41586-019-1821-z.
[106] Scharping NE, et al. The tumor microenvironment represses T cell mitochondrial biogenesis to drive intratumoral T cell metabolic insufficiency and dysfunction. Immunity 2016;45:374–88. https://doi.org/10.1016/j.immuni.2016.07.009.
[107] Zhang Y, et al. Enhancing CD8(+) T cell fatty acid catabolism within a metabolically challenging tumor microenvironment increases the efficacy of melanoma immunotherapy. Cancer Cell 2017;32:377–391.e379. https://doi.org/10.1016/j.ccell.2017.08.004.
[108] Chowdhury PS, Chamoto K, Kumar A, Honjo T. PPAR-induced fatty acid oxidation in T cells increases the number of tumor-reactive CD8(+) T cells and facilitates anti-PD-1 therapy. Cancer Immunol Res 2018;6:1375–87. https://doi.org/10.1158/2326-6066.Cir-18-0095.
[109] Johnson MO, et al. Distinct regulation of Th17 and Th1 cell differentiation by glutaminase-dependent metabolism. Cell 2018;175:1780–1795.e1719. https://doi.org/10.1016/j.cell.2018.10.001.
[110] Balmer ML, et al. Memory CD8(+) T cells balance pro- and anti-inflammatory activity by reprogramming cellular acetate handling at sites of infection. Cell Metab 2020;32:457–467.e455. https://doi.org/10.1016/j.cmet.2020.07.004.
[111] Petitprez F, Meylan M, de Reyniès A, Sautès-Fridman C, Fridman WH. The tumor microenvironment in the response to immune checkpoint blockade therapies. Front Immunol 2020;11:784. https://doi.org/10.3389/fimmu.2020.00784.
[112] Paijens ST, Vledder A, de Bruyn M, Nijman HW. Tumor-infiltrating lymphocytes in the immunotherapy era. Cell Mol Immunol 2021;18:842–59. https://doi.org/10.1038/s41423-020-00565-9.
[113] Siska PJ, et al. Mitochondrial dysregulation and glycolytic insufficiency functionally impair CD8 T cells infiltrating human renal cell carcinoma. JCI Insight 2017;2. https://doi.org/10.1172/jci.insight.93411.

[114] Gemta LF, et al. Impaired enolase 1 glycolytic activity restrains effector functions of tumor-infiltrating CD8(+) T cells. Sci Immunol 2019;4. https://doi.org/10.1126/sciimmunol.aap9520.

[115] Siddiqui I, et al. Intratumoral Tcf1(+)PD-1(+)CD8(+) T cells with stem-like properties promote tumor control in response to vaccination and checkpoint blockade immunotherapy. Immunity 2019;50:195–211.e110. https://doi.org/10.1016/j.immuni.2018.12.021.

[116] Miller BC, et al. Subsets of exhausted CD8(+) T cells differentially mediate tumor control and respond to checkpoint blockade. Nat Immunol 2019;20:326–36. https://doi.org/10.1038/s41590-019-0312-6.

[117] Schildberg FA, Klein SR, Freeman GJ, Sharpe AH. Coinhibitory pathways in the B7-CD28 ligand-receptor family. Immunity 2016;44:955–72. https://doi.org/10.1016/j.immuni.2016.05.002.

[118] Parry RV, et al. CTLA-4 and PD-1 receptors inhibit T-cell activation by distinct mechanisms. Mol Cell Biol 2005;25:9543–53. https://doi.org/10.1128/mcb.25.21.9543-9553.2005.

[119] Patsoukis N, et al. PD-1 alters T-cell metabolic reprogramming by inhibiting glycolysis and promoting lipolysis and fatty acid oxidation. Nat Commun 2015;6:6692. https://doi.org/10.1038/ncomms7692.

[120] Cha JH, et al. Metformin promotes antitumor immunity via endoplasmic-reticulum-associated degradation of PD-L1. Mol Cell 2018;71:606–620.e607. https://doi.org/10.1016/j.molcel.2018.07.030.

[121] Frostegård J, Zhang Y, Sun J, Yan K, Liu A. Oxidized low-density lipoprotein (OxLDL)-treated dendritic cells promote activation of t cells in human atherosclerotic plaque and blood, which is repressed by statins: microRNA let-7c is integral to the effect. J Am Heart Assoc 2016;5. https://doi.org/10.1161/jaha.116.003976.

[122] Scheffel MJ, et al. N-acetyl cysteine protects anti-melanoma cytotoxic T cells from exhaustion induced by rapid expansion via the downmodulation of Foxo1 in an Akt-dependent manner. Cancer Immunol Immunother 2018;67:691–702. https://doi.org/10.1007/s00262-018-2120-5.

[123] Lee MJ, et al. Association of TIM-3 expression with glucose metabolism in Jurkat T cells. BMC Immunol 2020;21:48. https://doi.org/10.1186/s12865-020-00377-6.

[124] Previte DM, et al. Lymphocyte activation Gene-3 maintains mitochondrial and metabolic quiescence in naive CD4(+) T CELLS. Cell Rep 2019;27:129–141.e124. https://doi.org/10.1016/j.celrep.2019.03.004.

[125] Tsurutani N, et al. Costimulation endows immunotherapeutic CD8 T cells with IL-36 responsiveness during aerobic glycolysis. J Immunol 2016;196:124–34. https://doi.org/10.4049/jimmunol.1501217.

CHAPTER

9

Exploring the interplay of kinases and transcription factors in immunometabolism

Darby J. Ballard[a], *Liqing Wang*[a,b], *Jugal Kishore Das*[a], *Anil Kumar*[a], *Yijie Ren*[a], *and Jianxun Song*[a]

[a]Department of Microbial Pathogenesis and Immunology, Texas A&M University Health Science Center, Bryan, TX, United States [b]Department of Biochemistry and Biophysics, Texas A&M University, College Station, TX, United States

Abstract

The complex interplay between kinases and transcription factors is fundamental to the burgeoning field of immunometabolism, which investigates the intersection of cellular metabolism and immune cell function. Protein kinases, including AMP-activated protein kinase (AMPK), mechanistic target of rapamycin (mTOR), Protein Kinase C (PKC), Janus kinase (JAK), and eukaryotic elongation factor 2 kinase (eEF2K), are central regulators in this dynamic interplay. They influence key cellular signaling pathways that govern immune cell metabolism, activation, and overall function. These kinases are emerging as crucial targets for therapeutic interventions in a range of diseases, from cancer to autoimmune disorders, due to their significant impact on metabolic homeostasis and immune response modulation. In parallel, transcription factors play a pivotal role in orchestrating immune cell metabolic reprogramming, essential for effective immune responses against pathogens and tumor cells. By regulating gene expression, these factors dictate cellular phenotype and metabolic state, impacting immune cell activation, proliferation, and function. Transcription factors such as NAC1, FoxP3, Blimp1, NF-κB, STAT3, and AP-1 are particularly influential in cancer immunotherapy, affecting treatment efficacy by modulating immune responses and tumor microenvironment interactions. Their diverse effects underscore the complexity of tailoring cancer therapies to individual patients. Integrating insights into both kinases and transcription factors enhances the understanding of immunometabolism and its implications for targeted therapeutic strategies. This integrated perspective provides a foundation for developing more effective treatments and optimizing therapeutic outcomes across various disease contexts.

https://doi.org/10.1016/B978-0-443-13827-0.00015-9

Abbreviations

AICD activation-induced cell death
AMPK AMP-activated protein kinase
AP-1 activator protein-1
ATF activating transcription factor
ATP adenosine triphosphate
Blimp1 B lymphocyte-induced maturation protein-1
BTB/POZ Bric-à-brac/poxvirus and zinc finger
CAR chimeric antigen receptor
CCR4 C-C chemokine receptor type 4
CCR6 C-C chemokine receptor type 6
CCR7 C-C chemokine receptor type 7
CD4 cluster of differentiation 4
CD62L cluster of differentiation 62, L-selectin
CD8 cluster of differentiation 8
CTLA-4 cytotoxic T-lymphocyte associated protein 4
DAP diaminopimelic acid
DNA deoxyribonucleic acid
ECAR extracellular acidification rate
eEF2 eukaryotic elongation factor 2
eEF2K eukaryotic elongation factor 2 kinase
Eomes eomesodermin
eTreg effector regulatory T cell
FAO fatty acid oxidation
FoxP3 forkhead box P3
GITR glucocorticoid-induced TNFR-related protein
HIF-1α Hypoxia-inducible factor 1-alpha
IFN-γ interferon-gamma
IL-10 interleukin-10
IL-17 interleukin-17
IL-18 interleukin-18
IL-2 interleukin-2
IL-21 interleukin-21
IL-6 interleukin-6
IL-7R interleukin-7 receptor
IL-7Rα interleukin-7 receptor subunit alpha
IL-8 interleukin-8
IMR immunometabolic regulation
IRF4 interferon regulatory factor 4
JAK Janus kinase
KLRG1 killer cell lectin-like receptor subfamily G member 1
KO knockout
LA lactic acid
LDHA lactate dehydrogenase A
MAPK mitogen-activated protein kinase
MDSC myeloid-derived suppressor cell
mRNA messenger ribonucleic acid
mTOR mechanistic target of rapamycin
NAC1 nucleus accumbens-associated 1
NAD nicotinamide adenine dinucleotide
NADH reduced nicotinamide adenine dinucleotide

NF-κB	nuclear factor-kappa B
NK	natural killer
NKG2D	natural killer group 2, member D
OCR	oxygen consumption rate
Oct4	octamer-binding transcription factor 4
OXPHOS	oxidative phosphorylation
PAX2	paired box 2
PCK	protein Kinase C
PD-1	programmed death-1
PD-L1	programmed death-ligand 1
Prdm1	PR domain zinc finger protein 1
SCID	severe combined immunodeficiency
SLE	systemic lupus erythematosus
STAT3	signal transducer and activator of transcription 3
Tcm	central memory T cell
Teff	effector T cell
Tem	effector memory T cell
Tfh	follicular Th cell
TFR	follicular regulatory T
Th	helper T cell
Th1	T helper 1
Th17	T helper 17
Th2	T helper 2
TIL	tumor-infiltrating lymphocyte
TLR	toll-like receptor
TLR4	toll-like receptor 4
TLR9	toll-like receptor 9
TME	tumor microenvironment
TNFR	tumor necrosis factor receptor
Treg	regulatory T cell
Zeb1	zinc finger E-box binding homeobox 1

Conflict of interest

No potential conflicts of interest were disclosed.

Kinases in immunometabolism

Immunometabolism, the intersection of cellular metabolism and immune cell function, stands as a burgeoning field of research. Protein kinases are central to this dynamic relationship as fundamental regulators influencing cellular signaling pathways that dictate immune cell metabolism. Among these, key players such as AMP-activated protein kinase (AMPK), mechanistic target of rapamycin (mTOR), Protein Kinase C (PKC), Janus kinase (JAK), and eukaryotic elongation factor 2 kinase (eEF2K) emerge as critical regulators, exerting effects on immune cell function and metabolic homeostasis (Table 1). Understanding the mechanisms by which these kinases modulate immunometabolism provides invaluable insights into the pathogenesis of various diseases and holds promise for developing targeted therapeutic interventions. In this context, we delve into the roles of AMPK, mTOR, PKC, JAK, and eEF2K in immunometabolism and their implications for disease states, eluding their potential as therapeutic targets in combating a spectrum of diseases, from cancer to autoimmune disorders.

TABLE 1 The role of key kinases in immune cell metabolism.

Kinase	Metabolic impact	Role in immunometabolism
AMPK	• Key regulator of cellular energy homeostasis • Promotes ATP production by enhancing glucose uptake and FAO • Inhibits energy-consuming processes (e.g., protein synthesis)	• Activation in immune cells supports metabolic adaptation during energy stress • Promotes survival and effector functions • Regulates T cell differentiation and cytokine production
mTOR	• Integrates nutrient and energy signals to regulate cell growth, metabolism, and proliferation • Promotes glycolysis, protein synthesis, and lipid metabolism in activated immune cells	• Signaling controls metabolic pathways in immune cells • Modulates effector functions and differentiation • Dysregulated activity can lead to immune dysfunction and inflammation
eEF2K	• Regulates protein synthesis by phosphorylating eEF2, impacting energy utilization • Modulates translation in response to cellular stress and nutrient availability	• Influences immune cell function • Regulates protein synthesis and metabolic responses • Impacts T cell activation and differentiation through translational control
JAK	• Intracellular tyrosine kinases that transduce signals from cytokine receptors • Impact gene expression and metabolic pathways • Regulate immune cell activation and proliferation	• Modulates immune cell metabolism in response to cytokines • Regulates metabolic pathways essential for immune responses • Dysregulated activity is associated with immune-related disorders
PKC	• Regulates various cellular processes, including metabolism • Isoforms influence glucose uptake, lipid metabolism, and mitochondrial function	• Signaling impacts immune cell activation, differentiation, and metabolic adaptation • Regulates cellular responses to environmental stimuli and cytokine signals

AMP-activated protein kinase (AMPK)

AMPK, a metabolic regulator, is a central controller of cellular energy homeostasis and metabolic adaptation. As a sentinel of cellular stress response, AMPK finely senses fluctuations in cellular energy status, notably triggered by environmental cues such as glucose deprivation or hypoxia. Upon activation, AMPK begins a cascade downstream to rectify energy imbalances by augmenting adenosine triphosphate (ATP) production while concurrently restraining ATP-consuming pathways [1].

A nuanced understanding of AMPK's role in immunometabolism highlights its impact on the modulation of metabolic pathways within immune cells, encompassing glycolysis, fatty acid oxidation (FAO), and mitochondrial biogenesis. This metabolic rewriting impacts the bioenergetic profile of immune cells, influencing their functional phenotype and regulating their bioenergetic profile. Notably, AMPK-mediated metabolic rewiring in immune cells improves the effector functions of antitumor T cells, enhancing their cytotoxic potential against malignant cells [2,3]. Preclinical investigations have illuminated the therapeutic potential of targeting AMPK signaling pathways to potentiate antitumor immune responses. Activation

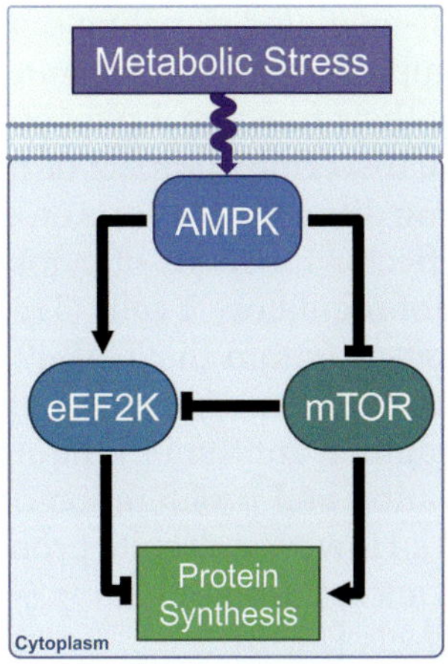

FIG. 1 AMPK becomes activated during metabolic stress events, resulting in protein synthesis.

of AMPK enhances the efficacy of cancer immunotherapy by fostering the activation and proliferation of cytotoxic T lymphocytes, increasing antitumor immunity [3]. These findings underscore the substantial role of AMPK in influencing immunometabolic dynamics within the tumor microenvironment (TME), offering a compelling avenue for therapeutic intervention [4,5].

Furthermore, expanding upon the complex molecular mechanisms underlying AMPK-mediated immunometabolic regulation (IMR) holds promise for developing innovative therapeutic strategies to reprogram the metabolic landscape of immune cells to potentiate antitumor immune responses. By harnessing the multifaceted capabilities of AMPK in regulating immune cell metabolism, novel therapeutic modalities can be envisioned to exploit the synergistic interplay between metabolic reprogramming and immune cell function in combating cancer [5,6]. Dysregulation of AMPK signaling is associated with various disease states, including metabolic disorders such as obesity and type 2 diabetes mellitus, where AMPK activity contributes to metabolic dysfunction and disease progression [4,7]. Moving forward, further contributing to the intricacies of AMPK-mediated IMR holds immense promise for developing innovative therapeutic strategies to combat cancer and mitigate metabolic disorders (Fig. 1).

Mechanistic target of rapamycin (mTOR)

mTOR, an evolutionarily conserved serine/threonine protein kinase, serves a central role in modulating the web of cellular metabolism and growth in response to diverse environmental cues, including nutrient availability, growth factors, and stress signals. At the nexus of metabolic regulation, mTOR modulates many metabolic pathways essential for cellular homeostasis and function. Within immune cell biology, mTOR is a master regulator of metabolic reprogramming, dictating immune cells' bioenergetic and biosynthetic demands in response to various stimuli. Notably, mTOR-mediated signaling cascades regulate key metabolic processes within immune cells, encompassing glycolysis, mitochondrial respiration, and amino acid metabolism, thereby shaping their functional phenotype and effector responses [8].

Of particular significance is the discernible impact of mTOR signaling on the functional plasticity of T lymphocytes, exerting differential effects on distinct T cell subsets. Specifically, mTOR activation augments the effector functions of cytotoxic T cells. Concurrently, mTOR dampens the suppressive activity of regulatory T cells (Tregs), culminating in a net enhancement of antitumor immune responses within the TME [9]. In cancer immunotherapy, evidence from preclinical investigations underscores the therapeutic potential of targeting mTOR signaling pathways to strengthen antitumor immune responses. Activation of mTOR is associated with enhanced activation and proliferation of cytotoxic T cells, thereby potentiating antitumor immunity [10,11]. However, despite promising preclinical data, the clinical translation of mTOR inhibitors in cancer immunotherapy is hampered by concerns surrounding their toxicity and off-target effects [12,13].

Consequently, there is a pressing need to develop alternative strategies to target mTOR signaling in a more selective and efficacious manner. Innovative approaches, including exploring selective mTOR inhibitors or rational combination therapies, promise to overcome the limitations associated with conventional mTOR inhibitors and enhance their therapeutic utility in cancer immunotherapy [14–16]. Dysregulation of mTOR signaling is associated with various disease states, including autoimmune disorders such as systemic lupus erythematosus (SLE) and metabolic disorders such as obesity and type 2 diabetes mellitus [17–20]. In summary, understanding the mechanistic foundations of mTOR-mediated IMR and exploring therapeutic modalities aimed at targeting mTOR signaling represent promising avenues for advancing the field of cancer immunotherapy and addressing diverse pathological conditions (Fig. 2).

Protein kinase C (PKC)

PKC is a prominent serine/threonine kinase family member that plays a central role in finely tuning immune cell function and signaling cascades. Operating with remarkable precision, the diverse array of PKC isoforms modulates metabolic pathways within immune

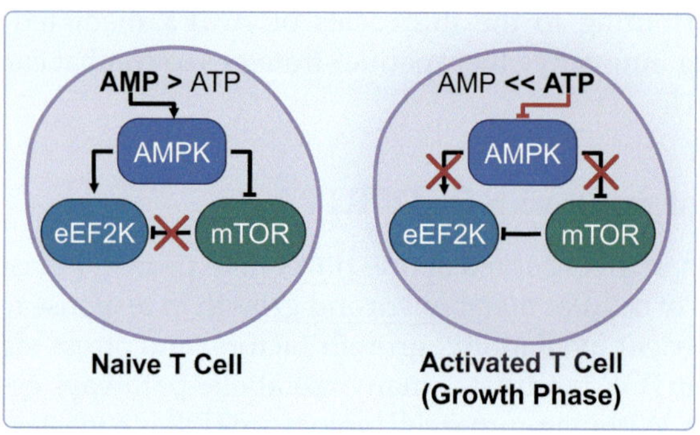

FIG. 2 The presence of AMP and ATP impacts the AMPK signaling pathway.

cells, impacting cellular metabolism and function. Notably, PKC isoforms have been linked to critical metabolic processes, including glycolysis, FAO, and mitochondrial biogenesis, thereby wielding influence over the bioenergetic landscape of immune cells [21–23].

PKC signaling shapes the functional phenotype of T lymphocytes, with consequential implications for antitumor immune responses. Activation of PKC signaling pathways is associated with heightened activity and enhanced effector functions of cytotoxic T cells. Preclinical investigations advocate targeting PKC signaling to amplify the efficacy of cancer immunotherapy. Strategies to modulate PKC have demonstrated the controlled activation and proliferation of antitumor T cells, potentiating antitumor immune responses within the TME [24,25]. Much like mTOR, the clinical translation of PKC modulators for immunotherapy has encountered hurdles due to concerns regarding toxicity and off-target effects [26,27].

Furthermore, dysregulation of PKC signaling is associated with the pathogenesis of various diseases, including autoimmune disorders such as rheumatoid arthritis and inflammatory bowel disease. Unchecked PKC activity contributes to immune dysregulation and chronic inflammation in these conditions, driving disease progression [28,29]. Conversely, deficient PKC signaling is associated with primary immunodeficiencies, impairing immune cell function, and increased susceptibility to infections [30].

Janus kinase (JAK)

JAK, a prominent family of nonreceptor protein tyrosine kinases, is central in signaling cascades necessary for immune cell function and regulation. At the crossroads of cytokine and growth factor signaling pathways, JAKs are crucial mediators in transducing signals from an array of immunomodulatory molecules. The array of molecules that JAKs impact includes interferons, interleukins, and growth hormone, demonstrating its effects on immune cell activation and effector responses. Beyond their classical role in immune cell signaling, previous findings indicate JAKs as crucial regulators of metabolic reprogramming within immune cells, establishing an intimate link between immune cell function and metabolic homeostasis. Notably, JAK signaling pathways modulate metabolic pathways essential for immune cell function, such as glycolysis and FAO, thereby sculpting the metabolic landscape of immune cells and shaping their effector functions [31–33].

The activation of JAK signaling pathways increases the effector functions of cytotoxic T cells while modulating their metabolic pathways. Specifically, JAKs influence cytotoxic T cell activation and proliferation, producing heightened antitumor immune responses. Preclinical studies have underscored the therapeutic promise of targeting JAK signaling pathways to elevate antitumor immune responses in cancer immunotherapy. By enhancing the functionality and growth of antitumor T cells, targeting JAK signaling holds the potential for amplifying the efficacy of cancer immunotherapy [34–36]. However, the clinical translation of JAK inhibitors in cancer immunotherapy is hampered by concerns regarding their toxicity and off-target effects, necessitating exploring alternative strategies to modulate JAK signaling more precisely and efficaciously [37–39].

Furthermore, dysregulation of JAK signaling is associated with the pathogenesis of diverse diseases. For instance, excess activation of JAK signaling is associated with autoimmune disorders such as rheumatoid arthritis and inflammatory bowel disease, in which excessive JAK-mediated cytokine signaling contributes to chronic inflammation and tissue damage

[36,40,41]. Conversely, deficient JAK signaling is linked to primary immunodeficiencies, exemplified by severe combined immunodeficiency (SCID), where mutations in JAK genes impair immune cell development and function [42–44]. Innovative strategies, including developing selective JAK inhibitors or rational combination therapies, offer promising avenues for overcoming the limitations associated with conventional JAK inhibitors [32,45,46]. Advancing our understanding of the mechanism of JAK-mediated IMR holds excellent potential for devising targeted therapeutic interventions to mitigate disease pathology and advance precision medicine.

eEF2K

eEF2K is a pivotal regulator at the interface of cellular metabolism and protein synthesis, exerting substantial effects on immune cell function and the pathophysiology of various disease states. Central to its mode of action is the phosphorylation-mediated regulation of eukaryotic elongation factor 2 (eEF2), a key component of the translational machinery involved in peptide chain elongation during protein synthesis [47].

Mechanistically, eEF2K phosphorylates and inhibits eEF2, attenuating the peptide chain elongation rate and protein synthesis. This regulatory mechanism represents a crucial checkpoint in cellular energy conservation under stress or metabolic perturbation conditions, in which the reduction in global protein synthesis serves to conserve energy and promote cell survival [48]. Significantly, when eEF2K blocks protein synthesis, it facilitates the reallocation of cellular resources toward energy-generating metabolic pathways, such as glycolysis and FAO, to meet the heightened metabolic demands of activated immune cells or proliferating cancer cells [49].

The regulatory role of eEF2K in protein synthesis is a critical determinant of immune cell function and effector responses. Notably, eEF2K-mediated inhibition of protein synthesis is linked to the modulation of T cell activation and differentiation, where the attenuation of protein synthesis serves to fine-tune the balance between effector and regulatory T cell subsets [50]. For instance, the reduction of protein synthesis by eEF2K promotes the differentiation of Tregs, dampening excessive immune responses and maintaining immune homeostasis [18]. Conversely, in the context of antitumor immunity, eEF2K-mediated protein synthesis inhibition within effector T cells may restrain their cytotoxic potential, thereby facilitating immune evasion and tumor progression [49,51].

Furthermore, dysregulation of eEF2K signaling is tied to the pathogenesis of various disease states, including cancer, neurodegenerative disorders, and autoimmune diseases [52–54]. In cancer, excessive activation of eEF2K-mediated signaling pathways is associated with tumor cell survival and resistance to therapy [49,55]. Contrarily, the inhibition of protein synthesis promotes tumor cell adaptation to nutrient-deprived microenvironments and confers resistance to apoptosis-inducing stimuli [56,57]. In neurodegenerative disorders such as Alzheimer's disease, dysregulation of eEF2K signaling is attributed to synaptic dysfunction and neuronal degeneration, where the inhibition of protein synthesis exacerbates neuronal vulnerability to cellular stressors and contributes to disease progression [52].

These signaling molecules—AMPK, mTOR, eEF2K, JAK, and PKC—play critical roles in immune cell metabolism by sensing and responding to changes in nutrient availability, energy status, and cellular stress. They integrate signaling pathways to modulate metabolic programs that support immune cell functions, including proliferation, differentiation, and effector responses. Dysregulated signaling in these pathways can impact immune cell metabolism and contribute to immune-related diseases, highlighting their importance as potential therapeutic targets for immunomodulation and metabolic regulation in health and disease. Understanding the interplay between signaling pathways and immunometabolism is essential for developing targeted interventions to modulate immune responses and treat immune-related disorders.

Transcription factors in immunometabolism

In immune responses directed against pathogens or tumor cells, immune cells undergo substantial metabolic reprogramming to meet their heightened energy demands. Specifically, cells must increase their metabolism to produce the necessary cytokine, chemokine, and inflammatory mediators, leading to immune cell activation and expansion. Transcription factors, essential proteins regulating gene expression and dictating cell phenotype in specific microenvironments, are pivotal in orchestrating these metabolic alterations and shaping immune cell function. Manipulating cellular energy metabolism influences the nature and outcome of immune cell responses. For instance, promoting glycolytic metabolism in naive CD4 T cells postactivation can bias them toward an inflammatory phenotype. However, directing these cells to utilize FAO instead may skew them toward a regulatory phenotype [58]. Furthermore, promoting glycolytic metabolism in differentiated Tregs can destabilize their regulatory phenotype and hasten the loss of the lineage-defining transcription factor FoxP3 [59,60].

In cancer immunotherapy, transcription factors wield considerable influence over treatment efficacy by modulating immune responses. This section delves into various transcription factors associated with cancer immunometabolism, delineating their roles in defining pertinent cells' stability and metabolic signature. Notable transcription factors in this context include NAC1, FoxP3, Blimp1, NF-κB, STAT3, and AP-1. Despite their significance in cancer immunotherapy, understanding the impact of transcription factors on treatment outcomes is complex (Table 2). The influence of specific transcription factors on cancer therapy efficacy can vary depending on the intricacies of the TME. Consequently, implementing diverse strategies may be imperative to enhance the effectiveness of cancer treatment regimens (Fig. 3).

Comprehending the roles of transcription factors in cancer immunotherapy is indispensable for devising efficacious treatment strategies. While certain transcription factors may be valuable biomarkers for monitoring treatment response, their nuanced effects within the tumor environment necessitate careful consideration. Ultimately, advancing the field's understanding of how transcription factors modulate cancer immunotherapy can empower clinicians to tailor treatment approaches to individual patient needs, optimizing therapeutic outcomes.

NAC1

Nucleus accumbens-associated protein 1 (NAC1) is a BTB/POZ domain-containing nuclear protein and transcriptional corepressor with widespread roles in stem cell maintenance and cancer progression. Initially identified as a cocaine-inducible transcript in the nucleus

TABLE 2 The effect of transcription factors in immunometabolism.

Transcription factor	Metabolic impact	Role in immunometabolism
NAC1	• Regulates cancer cell metabolism • Enhances glycolysis and stabilizes HIF-1α • Contributes to altered metabolic profiles in tumors	• Promotes tumor growth and progression • Regulates immune responses in the tumor microenvironment • Impacts T cell function and metabolism
FoxP3	• Critical for regulating the metabolic profile of Tregs • Suppresses glycolysis and promotes OXPHOS in Tregs • Supporting Tregs' suppressive function	• Influences immune tolerance • Alters Treg function in the tumor microenvironment • Impacts antitumor immunity
Blimp1	• Regulates cellular metabolism • Modulates glycolysis and mitochondrial function in immune cells • Influences Teff differentiation and function	• Affects immune responses and tolerance • Dysregulated expression is implicated in autoimmune disorders
NF-κB	• Modulates glucose metabolism, lipid synthesis, and mitochondrial function in immune cells • Supports inflammatory responses	• Activation influences immune cell metabolism • Impacts cytokine production and immune cell survival • Dysregulated activity is associated with inflammatory diseases
STAT3	• Modulates gene expression • Regulates cellular metabolism by promoting glycolysis, lipid synthesis, and mitochondrial function in immune cells • Supports immune cell survival and effector functions	• Activation influences immune cell metabolism • Modulates T cell differentiation and cytokine production • Dysregulated signaling is linked to immune disorders and cancer
AP-1	• Regulates cellular metabolism by modulating glycolysis, lipid metabolism, and antioxidant responses in immune cells • Influences immune cell activation and differentiation	• Activation impacts immune cell metabolism • Modulates immune responses and inflammation • Dysregulated activity is implicated in autoimmune diseases and cancer

accumbens, a neuroanatomical structure integral to motivational processes, NAC1's involvement extends beyond addictive behaviors to fundamental cellular processes [61,62]. As a crucial partner of NANOG, a key pluripotency-associated factor, NAC1 is pivotal in repressing Zeb1 and ensuring the proper expression of E-cadherin during cellular reprogramming and the reactivation of the Oct4 distal enhancer [63,64].

The elevated expression of NAC1 is a hallmark feature across various cancer types, including ovarian, cervical, and uterine cancers [65,66]. NAC1 is particularly evident in recurrent ovarian serous carcinoma, where its upregulation correlates with the development of drug resistance [67]. Mechanistically, NAC1 functions as a positive regulator of glycolysis in

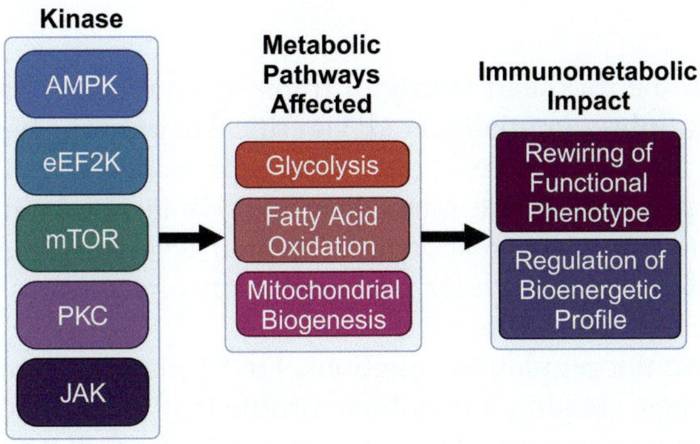

FIG. 3 Multiple kinases impact immunometabolism in cells, leading to various downstream effects.

ovarian cancer by stabilizing HIF-1α, thereby modulating cellular metabolism within tumor cells [68]. Moreover, NAC1 is pivotal in orchestrating immune homeostasis within the TME [69–71].

Recent research has unveiled the pivotal role of tumorous NAC1 overexpression in shaping the dynamics of antitumor immunity and immune evasion mechanisms, primarily mediated through LDHA-driven lactic acid (LA) production. Within the nutrient-deficient TME, LDHA-mediated accumulation of lactic acid fosters an inhospitable atmosphere for tumor-infiltrating lymphocytes (TILs), facilitating immune evasion. This hostile environment is characterized by diminished expressions of tumor antigens and significant histocompatibility complexes alongside heightened levels of inhibitory checkpoint molecules, such as PD-L1, culminating in immune suppression and tumor progression.

NAC1 inhibition, exemplified by NIC3, effectively disrupts NAC1 homodimerization, leading to notable antitumor effects [72]. Investigations involving NAC1 knockout (KO) mice have illuminated the relationship between NAC1 and immune regulation. Notably, NAC1 KO mice exhibit resistance to autoimmunity and elevated metabolic and immune-suppressive activities, underscored by the expansion of CD4+ Tregs in peripheral tissues [71]. Mechanistically, NAC1 KO Tregs demonstrate elevated expression of FoxP3 and IL-10, attributed to enhanced acetylation and stabilization of FoxP3, consequently modulating the metabolic profile and suppressive activity of TILs within the TME.

Functional analyses reveal that Tregs derived from NAC1 KO mice exhibit significantly heightened metabolic activity, exemplified by elevated oxygen consumption rate (OCR) and extracellular acidification rate (ECAR), compared to their wild-type counterparts. NAC1 also influences the survival and metabolic status of CD8+ T cells, as evidenced by lower ECAR and oxidative phosphorylation (OXPHOS) levels observed in NAC1 KO CD8+ T cells. These findings are further corroborated by the heightened expression of Interferon Regulatory Factor 4 (IRF4), a critical transcription factor linked with T cell development and receptor signaling pathways, in NAC1 KO CD8+ T cells [69,73].

NAC1 is a pivotal regulator influencing metabolic activities and immune responses within the TME. By modulating metabolic and immune-suppressive pathways, NAC1 plays a central role in shaping the functional phenotype of effector immune cells, offering novel therapeutic avenues for cancer immunotherapy.

FoxP3

FoxP3, a member of the forkhead family of transcription factors, is crucial in generating and maintaining Tregs. Subsequentially, FoxP3 is critical to immune homeostasis and peripheral tolerance [74–77]. Tregs, characterized by the expression of FoxP3, exert immunosuppressive functions to prevent autoimmune disorders while also being exploited by tumor cells to evade antitumor immune responses. Elevated levels of FoxP3+ Tregs within the TME have been correlated with adverse prognostic outcomes across various cancer types, highlighting their role in tumor immune evasion [78].

In addition to its immunoregulatory functions, FoxP3 plays a pivotal role in shaping the metabolic profile of Tregs, creating a metabolic profile that favors OXPHOS over glycolysis [79]. By transcriptionally repressing MYC and promoting OXPHOS, FoxP3 enhances the NAD:NADH ratio within Tregs. This increased ratio leads to increased metabolic resilience as it enables the utilization of lactate as an energy source in the glucose-deprived, lactate-rich TME [80–82]. This metabolic adaptation gives Tregs a survival advantage over Teffs, facilitating their persistence and immunosuppressive functions within the tumor environment.

Genetic ablation of FoxP3 causes immune dysregulation and lymphoproliferative disorders, such as fatal autoimmune disorders. FoxP3 additionally regulates the expression of CCR4, leading to Treg infiltration into the TME. This enhanced infiltration leads to a suppression of effector cell functions, ultimately contributing to tumor progression and immune evasion [83,84]. Conversely, genetic ablation of FoxP3 has been shown to downregulate CCR4+ Treg infiltration, restoring antitumor immunity and inhibiting tumor growth.

Until recently, FoxP3 expression was known to be specific for T cell lineage, specifically Tregs. Intriguingly, several recent studies have detected FoxP3 expression in tumor cells of human melanoma tissue as well as cell lines derived from other solid tumors [85–87]. Several reports confirmed that FoxP3 is widely expressed in tumor cells apart from the Tregs, suggesting a potential role in carcinogenesis beyond its canonical function. Distinct from Tregs, tumor cell lines express several isoforms of FoxP3 lacking one or more exons. Specifically, the FoxP3 mRNA variant lacking exons 3 and 4 was identified in tumor cell lines [79,87,88]. Remarkably, predominant FoxP3 localization was observed in the cytoplasm of tumor cells, whereas nuclear localization was observed in Tregs.

FoxP3 expression in Tregs suppresses effector immune cells, causing this transcription factor to play a tumor-enhancing role. However, when present in tumor cells, FoxP3 functions as a potent suppressor of several oncogenes, thereby suppressing carcinogenesis. FoxP3-mediated metabolic programming of infiltrating CD8+ cytotoxic T cells enhances their antitumor efficacy, proliferation, and cytotoxic activity within the TME [89,90]. Transcriptomic analysis of tumor-infiltrating FoxP3+ cytotoxic T cells reveals metabolic reprogramming driven by enhanced FAO and OXPHOS. This altered metabolic profile enables cytotoxic T cells to adapt to the nutrient-deficient TME. Notably, FoxP3+ cytotoxic T cells exhibit metabolic plasticity, engaging both glycolysis and FAO to meet their energy demands and sustain proliferation within the TME. Interestingly, FoxP3+ cytotoxic T cells could activate glycolysis and engage FAO to further drive OXPHOS to compensate for their energy demands.

FoxP3 is a central regulator of immune function and metabolism in cancer immunotherapy, orchestrating a complex interplay between immune regulation, metabolic adaptation, and tumor progression. Further elucidation of FoxP3's role in tumor immunity and metabolism holds promise for developing novel therapeutic strategies targeting immune-metabolic pathways in cancer treatment.

Blimp1

B lymphocyte-induced maturation protein-1 (Blimp1) is a transcriptional repressor encoded by the PRDM1 gene [91,92]. Blimp1 is well known for its role in promoting the terminal differentiation of B cells into plasma cell phenotype and restraining memory cell differentiation. Recent studies have indicated that Blimp1 is expressed in immune cells other than B cells, including T cells, granulocytes, macrophages, and epithelial cells [93–95]. Additionally, Blimp1 expression is attributed to the exhaustion of CD8+ T cells during viral infection [96].

Despite the functional and transcriptional differences, the expression of Blimp1 plays a crucial role in the differentiation of late B cells and Teffs, including effector regulatory T cells (eTregs) [97,98]. Notably, Blimp1 acts as a key regulator of T cell homeostasis. Blimp1 regulates Th cell lineage by promoting Th2 lineage and suppressing the differentiation of IFN-γ-secreting Th1 cells. Blimp1 expression is primarily restricted to activated T cells and coordinates with the transcription factors to regulate expression of IL-2, IL-21, and IL-10 in Teffs [98]. Transgenic expression of Blimp1 in T cells regulates and suppresses IL-17-secreting Th17 cells, while Blimp1-deficiency leads to an increase in Th17 cells [99]. Interestingly, autoimmune diabetes is mediated by Th17 cells, indicating that Blimp1 impacts the development of autoimmune diseases [99].

Blimp1 deletion in T cells results in the downregulation of IL-10 and upregulation of IL-2 and IFN-γ [98,100]. Furthermore, Blimp1 deficiency results in enhanced proliferation and reduced activation-induced cell death (AICD) within inactivated T cells. Clinically, Blimp-1 deficiency leads to a higher frequency of activated CD4+ T cells with dysregulated Teffs cells or memory T cell phenotype, contributing to severe colitis. Blimp1 and IRF4 are indispensable for cooperative regulation and expression of IL-10 and CCR6 by eTreg cells [101,102]. By collaborating with IRF4, Blimp1 can maintain Treg function by antagonizing follicular Th (Tfh) cells [97].

Blimp1 controls the function of the effector CD8+ T cells, including migration, cytotoxicity, survival, proliferation, and cytokine production. Akin to its role in B cells, Blimp1 plays a similar function in the terminal differentiation of CD8+ T cells. Expression of Blimp1 influences CD8+ into effector and "effector memory" (Tem) CD8+ T cells, restraining their differentiation into long-lived central memory (Tcm) T cells [96]. Conversely, Blimp1 deletion promoted the development of memory cell phenotype. Blimp1 additionally regulates the generation of effector cells (KLRG1^{hi} IL-7R^{lo}), effector memory (KLRG1^{hi} IL-7Rα^{hi} CD62L^{lo} CCR7^{lo}) cells, and central memory (KLRG1^{lo} IL-7Rα^{hi} CD62L^{hi} CCR7^{hi}) cells.

Infiltration of FoxP3+ Treg cells in tumors often characterizes a significant mechanism for immune evasion and a critical barrier to immunotherapy and antitumor immunity. Recent studies confer the elevated expression of Blimp1 in activated tumor-infiltrating Tregs. Blimp1+ Treg and follicular regulatory T (TFR) cells were enriched in tumors, and higher tumoral TFR indicated an increased risk of melanoma metastasis. Genetic ablation of Blimp1 in tumor-infiltrating Tregs reduced their suppressive function and metabolic reprogramming into a Teff phenotype.

When compared to Tregs present in the spleen and lymph nodes, Blimp1 + tumor-infiltrating Tregs showed enhanced expression of Blimp1 and regulatory signatures like FoxP3, GITR, Helios, and CTLA-4 [103,104]. Blimp1 skews intratumoral Tregs toward a suppressive signature phenotype through the suppression of the Eomesodermin (Eomes) expression. Deletion of Eomes in Blimp1-deficient Tregs restored tumor growth and attenuated antitumor immunity [103,104]. These reports establish the role of Blimp1 as a new critical regulator of tumor-infiltrating Treg cells and a crucial target for antitumor immunotherapy.

NF-κB

NF-κB plays a critical role in regulating the immune response [105,106]. Once activated, NF-κB translocates to the nucleus and binds to specific DNA sequences, leading to the transcriptional activation of target genes [105]. NF-κB is activated in response to various stimuli, including cytokines, microbial products, and cellular stress [107]. Notably, NF-κB is critical for T cell activation and proliferation [108,109].

The mechanism by which NF-κB influences the B cell is still unknown and a controversial topic within the field. From B cell germinal centers, NF-κB manipulates B cell maturation and differentiation [110]. However, human melanoma secretomes will induce NF-κB-related genes in tumor-associated B cells [111]. In innate immune cells (macrophages, MDSCs, dendritic cells, or NK cells), myeloid cell deletion of NF-κB signaling boosts glioblastoma growth [112]. Contrarily, inhibiting NF-κB in tumor-associated macrophages increases their cytotoxicity to tumor cells [113].

IL-18 is important for NF-κB and improves NK cell antitumor immunity [114]. Many TLR agonist therapies have been proven to enhance NF-κB activities in DCs during cancer treatment [115]. In the pancreatic cancer model, polysaccharide boosts NK cell antitumor cytotoxicity through TLR4/MAPKs/NF-κB pathway [116].

NF-κB modulates the expression of cytokines and chemokines, which are essential for recruiting immune cells to tumors [117,118]. In cancer, NF-κB also regulates the expression of genes involved in cell survival and proliferation, contributing to tumorigenesis [106,118]. Specifically, breast cancer invasion and metastasis are associated with NF-κB activation [119]. Inhibition of NF-κB enhances the effectiveness of immunotherapy in different cancer disease models [106]. For example, inhibition of cancer NF-κB enhances the antitumor immune response in a mouse model of melanoma treated with anti-PD-L1 immunotherapy [120]. Furthermore, elevated NF-κB expression is linked to increased success of adoptive T cell therapy when treating MC57-SIY tumors [121].

The role of NF-κB in cancer immunotherapy is quite complicated. While NF-κB could support cancer cell progression, the antitumor role of immune cells depends on NF-κB expression. Consequently, directly using NF-κB inhibitors for cancer treatment is a complex, delicate puzzle. The multifaceted roles of NF-κB in different cells need various strategies to improve cancer immunotherapy. Despite learning of its correlation with cancer more than 30 years ago, there is still a long road to deciphering the role of NF-κB.

STAT3

Signal transducer and activator of transcription 3 (STAT3) is an oncogene related to cellular transformation and apoptosis blockage [122]. STAT3 promotes multiple types of tumor progression, and, in response to cytokine stimulation, plays an important role in the immune and hematopoietic systems [123,124]. Notably, STAT3 deletion leads to disorders and interrupts innate and adaptive immunity [124].

In T cells, STAT3 inhibits Th1 polarization and promotes Th17 and Treg development, indicating that STAT3 seems to act in an immunosuppressive role during tumor progression [125]. In innate cells, NK cell proliferation and function predominantly largely occur through the STAT3 signaling pathway when activated with IL-21, increasing the NKG2D expression [126]. However, other researchers have discovered that tumor-secreted IL-6 and IL-8 damage NK cell function through the STAT3 pathway [127]. This has led to the assumption that the level at which STAT3 interrupts NK cells is dependent on various upstream ligands.

In tumor cells, STAT3 is frequently intertwined with NF-κB overexpression during tumor progression, metastasis, and antiapoptosis [128]. Blocking tumor STAT3 increases the level of proinflammatory chemokines and cytokines, thus improving the killing activity of immune cells [129]. Specifically, the IL-6/STAT3 signaling pathway hijacks estrogen receptors to boost breast cancer metastasis [130].

There have been numerous progressions in the field of STAT3-based immunotherapies. In oral carcinoma and gastric cancer cells, highly expressed cyclin D1 is linked to the activation of STAT3 [131,132]. This discovery has led to the development of a STAT3 inhibitor/anti-PD-1 combination immunotherapy that has proven to be an effective melanoma treatment [133]. Using TLR9-targeted oligonucleotide STAT3 inhibitors to treat B cell lymphoma has also been successful [134]. Additionally, a STAT3 inhibitor, WP1066, has been used to improve T cell cytotoxicity by inhibiting Tregs during melanoma treatment [135]. Harnessing the adaptive immune system, STAT3-depleted dendritic cell therapy has recently been shown to induce the ability of Th1 cells and restrain MC38 tumor progression [136].

Despite these promising advancements, targeting STAT3 for cancer treatment is not straightforward. In cancers, inhibiting STAT3 is promising for tumor treatment; however, other immune cells do not always benefit from STAT3 inhibition. The numerous upstream regulators of STAT3 must be further analyzed to create effective therapeutics with minimal off-target effects.

AP-1

Activator Protein 1 (AP-1) is a group of transcriptional factors that contain multiple complexes of ATF, c-Jun, and c-Fos [137]. AP-1 family proteins regulate numerous cell activities, including cell proliferation, apoptosis, and programmed cell death [138,139].

AP-1 proteins are related to multiple types of cancer progression, including colon, breast, and melanoma cancers. For instance, when activated by PAX2, AP-1 proteins promote colon cancer cell proliferation and the expression of cyclin D1 [140]. Moreover, overexpression of c-Jun/AP-1 proteins increases tumorigenic signaling in breast cancer [141]. C-Jun/AP-1 additionally drives melanoma progression and supports its malignancy, indicating that AP-1 will be a critical target in combating cancer [142].

In T cells, the AP-1 family also interrupts cancer immunotherapy efficacy. AP-1 protein-transduced CAR-T cells enhanced tumor rejection and prevented themselves from exhaustion through the coexpression of BATF and IRF4 [143]. Coexpression with c-Jun in DAP-CAR-T cells does not affect T cell expansion but increases the Tcm ratio in solid tumors [144]. The TNF-α+ macrophages determine types and prognosis for innate immune cells by reprogramming AP-1 enhancers in pancreatic cancer [144]. Deletion of c-Maf in myeloid cells causes reduced tumor burden with boosted antitumor T cells [145].

Currently, the mechanism by which AP-1 proteins regulate B cells, DCs, or NK cells during tumor progression is unknown. The current strategy for tumor immunotherapies involving AP-1 includes blocking AP-1 in tumors or tumor-associated macrophages while boosting the AP-1 expression in CAR-T cells.

Understanding the roles of NAC1, FoxP3, Blimp1, NF-κB, STAT3, and AP-1 in immune cell metabolism provides insights into their impact on immune responses, inflammation, and disease pathogenesis. Dysregulated expression and activity of these transcription factors and signaling molecules can influence immune cell function and metabolism, contributing to immune-related disorders and cancer progression. Targeting these pathways may offer therapeutic opportunities to modulate immunometabolism and improve immune responses to health and disease.

Acknowledgments

This work was supported by National Institutes of Health Grants R01CA221867, R01CA273002, and R21AI167793 to J. S., and Department of Defense Grant LC210150 to J. S.

References

[1] González A, Hall MN, Lin SC, Hardie DG. AMPK and TOR: the Yin and Yang of cellular nutrient sensing and growth control. Cell Metab 2020;31(3):472–92.

[2] Gao Y, Päivinen P, Tripathi S, Domènech-Moreno E, Wong IPL, Vaahtomeri K, et al. Inactivation of AMPK leads to attenuation of antigen presentation and immune evasion in lung adenocarcinoma. Clin Cancer Res 2022;28 (1):227–37.

[3] Pokhrel RH, Acharya S, Ahn JH, Gu Y, Pandit M, Kim JO, et al. AMPK promotes anti-tumor immunity by downregulating PD-1 in regulatory T cells via the HMGCR/p38 signaling pathway. Mol Cancer 2021;20(1):133.

[4] Sadria M, Seo D, Layton AT. The mixed blessing of AMPK signaling in cancer treatments. BMC Cancer 2022;22 (1):105.

[5] Finisguerra V, Dvorakova T, Formenti M, Van Meerbeeck P, Mignion L, Gallez B, et al. Metformin improves cancer immunotherapy by directly rescuing tumor-infiltrating CD8 T lymphocytes from hypoxia-induced immunosuppression. J Immunother Cancer 2023;11(5), e005719.

[6] Zhang Z, Li F, Tian Y, Cao L, Gao Q, Zhang C, et al. Metformin enhances the anti-tumor activity of CD8+ T lymphocytes via the AMPK–miR-107–Eomes–PD-1 pathway. J Immunol 2020;204(9):2575–88.

[7] Rodríguez C, Muñoz M, Contreras C, Prieto D. AMPK, metabolism, and vascular function. FEBS J 2021;288 (12):3746–71.

[8] Gabriel SS, Tsui C, Chisanga D, Weber F, Llano-León M, Gubser PM, et al. Transforming growth factor-β-regulated mTOR activity preserves cellular metabolism to maintain long-term T cell responses in chronic infection. Immunity 2021;54(8):1698–1714.e5.

[9] Wang S, Raybuck A, Shiuan E, Cho SH, Wang Q, Brantley-Sieders DM, et al. Selective inhibition of mTORC1 in tumor vessels increases anti-tumor immunity. JCI Insight 2020;5(15), e139237.

[10] Yang Q, Liu X, Liu Q, Guan Z, Luo J, Cao G, et al. Roles of mTORC1 and mTORC2 in controlling γδ T1 and γδ T17 differentiation and function. Cell Death Differ 2020;27(7):2248–62.

[11] Silic-Benussi M, Sharova E, Ciccarese F, Cavallari I, Raimondi V, Urso L, et al. mTOR inhibition downregulates glucose-6-phosphate dehydrogenase and induces ROS-dependent death in T-cell acute lymphoblastic leukemia cells. Redox Biol 2022;51, 102268.

[12] Ding Z, Shen H, Xu K, Wu Y, Wang S, Yi F, et al. Comprehensive analysis of mTORC1 signaling pathway-related genes in the prognosis of HNSCC and the response to chemotherapy and immunotherapy. Front Mol Biosci 2022;9. [Internet]. [cited 2024 Apr 4]. Available from: https://www.frontiersin.org/articles/10.3389/fmolb.2022.792482.

[13] Gupta H, Kari S, Salinas E, Bai H, Osta E, Kornepati A, et al. 298 tumor cell-intrinsic mTORC1 signaling through raptor makes melanoma and ovarian cancer immunotherapy resistant by regulating interferon-gamma responsiveness and promoting tumor-initiating cells. J Immunother Cancer 2021;9(Suppl 2). [Internet]. [cited 2024 Apr 4]. Available from: https://jitc.bmj.com/content/9/Suppl_2/A321.

[14] Sagers JE, Beauchamp RL, Zhang Y, Vasilijic S, Wu L, DeSouza P, et al. Combination therapy with mTOR kinase inhibitor and dasatinib as a novel therapeutic strategy for vestibular schwannoma. Sci Rep 2020;10(1):4211.

[15] Laukkanen S, Veloso A, Yan C, Oksa L, Alpert EJ, Do D, et al. Therapeutic targeting of LCK tyrosine kinase and mTOR signaling in T-cell acute lymphoblastic leukemia. Blood 2022;140(17):1891–906.

[16] Qiu J, Feng M, Yang G, Cao Z, Liu Y, You L, et al. mTOR inhibitor, gemcitabine and PD-L1 antibody blockade combination therapy suppresses pancreatic cancer progression via metabolic reprogramming and immune microenvironment remodeling in Trp53flox/+LSL-KrasG12D/+Pdx-1-Cre murine models. Cancer Lett 2023;554, 216020.

[17] Murayama G, Chiba A, Kuga T, Makiyama A, Yamaji K, Tamura N, et al. Inhibition of mTOR suppresses IFNα production and the STING pathway in monocytes from systemic lupus erythematosus patients. Rheumatology 2020;59(10):2992–3002.

[18] Peng T, Pei H. The role of autophagy-related gene mTOR signaling pathway in the pathogenesis of systemic lupus erythematosus. J Med Postgrad 2020;(12):62–6.

[19] Tsai K, Tullis B, Jensen T, Graff T, Reynolds P, Arroyo J. Differential expression of mTOR related molecules in the placenta from gestational diabetes mellitus (GDM), intrauterine growth restriction (IUGR) and preeclampsia patients. Reprod Biol 2021;21(2), 100503.

[20] Wang Y, Rijal B, Xu M, Li Z, An Y, Zhang F, et al. Renal denervation improves vascular endothelial dysfunction by inducing autophagy via AMPK/mTOR signaling activation in a rat model of type 2 diabetes mellitus with insulin resistance. Acta Diabetol 2020;57(10):1227–43.

[21] Lepique AP, Boccardo E, de Miranda FS. Metabolic reprogramming and cancer. In: Camara NOS, Alves-Filho JC, de Moraes-Vieira PMM, Andrade-Oliveira V, editors. Essential aspects of immunometabolism in health and disease. Cham: Springer International Publishing; 2022. p. 177–201. [Internet]. [cited 2024 Apr 4] Available from: https://doi.org/10.1007/978-3-030-86684-6_9.

[22] Takuwa Y, Matsui T, Abe Y, Nagafuji T, Yamashita K, Asano T. Alterations in protein kinase C activity and membrane lipid metabolism in cerebral vasospasm after subarachnoid hemorrhage. J Cereb Blood Flow Metab 1993;13(3):409–15.

[23] Domenicotti C, Paola D, Lamedica A, Ricciarelli R, Chiarpotto E, Marinari UM, et al. Effects of ethanol metabolism on PKC activity in isolated rat hepatocytes. Chem Biol Interact 1996;100(2):155–63.

[24] Lei G, Horbath A, Li Z, Gan B. PKCβII–ACSL4 pathway mediating ferroptosis execution and anti-tumor immunity. Cancer Commun 2022;42(7):583–6.

[25] Satow R, Suzuki Y, Asada S, Ota S, Idogawa M, Kubota S, et al. Downregulation of protein kinase C gamma reduces epithelial property and enhances malignant phenotypes in colorectal cancer cells. iScience 2022;25(12):105501.

[26] Lautala S. Molecular dynamics simulations of the lipid environment in protein kinase C targeting, 2024. [Internet]. [cited 2024 Apr 4] Available from: https://helda.helsinki.fi/items/81cfe47f-803b-4588-84fd-cebc2fc8111a.

[27] Park E, Chen J, Moore A, Leitges M, Frietze SE, Ringshausen I. Overcoming multidrug resistance in B cell malignancies by antagonism of stromal mediated TGF-β signalling. Blood 2020;136:17.

[28] Papa R, Schena F, Corcione A, Bocca P, Drago E, Malattia C, et al. Seropositive polyarthritis and diffuse lymphadenopathy associated with PRKCD mutation. Rheumatology 2023;63, kead639.

[29] Oikawa T, Yamada K, Tsubota A, Saeki C, Tago N, Nakagawa C, et al. Protein kinase C delta is a novel biomarker for hepatocellular carcinoma. Gastro Hep Adv 2023;2(1):83–95.

[30] Leitges M, Schmedt C, Guinamard R, Davoust J, Schaal S, Stabel S, et al. Immunodeficiency in protein kinase Cβ-deficient mice. Science 1996;273(5276):788–91.

[31] Fujita Y, Matsuoka N, Temmoku J, Furuya-Yashiro M, Asano T, Sato S, et al. JAK inhibitors impair GM-CSF-mediated signaling in innate immune cells. BMC Immunol 2020;21(1):35.

[32] Nocturne G, Pascaud J, Ly B, Tahmasebi F, Mariette X. JAK inhibitors alter NK cell functions and may impair immunosurveillance against lymphomagenesis. Cell Mol Immunol 2020;17(5):552–3.

[33] Bouwman W, Verhaegh W, Holtzer L, van de Stolpe A. Quantitative measurement of activity of JAK-STAT signaling pathways in blood samples and immune cells to predict innate and adaptive cellular immune response to viral infection and accelerate vaccine development. bioRxiv; 2020 2020.05.13.092759. [Internet]. [cited 2024 Apr 5]. Available from: https://www.biorxiv.org/content/10.1101/2020.05.13.092759v1.

[34] Courtois L, Cabannes-Hamy A, Kim R, Delecourt M, Pinton A, Charbonnier G, et al. IL-7 receptor expression is frequent in T-cell acute lymphoblastic leukemia and predicts sensitivity to JAK inhibition. Blood 2023;142 (2):158–71.

[35] Zhang H, Zhao H, He X, Xi F, Liu J. JAK-STAT domain enhanced MUC1-CAR-T cells induced esophageal cancer elimination. Cancer Manag Res 2020;(12):9813–24.

[36] Gerlach K, Lechner K, Popp V, Offensperger L, Zundler S, Wiendl M, et al. The JAK1/3 inhibitor to tofacitinib suppresses T cell homing and activation in chronic intestinal inflammation. J Crohn's Colitis 2021;15(2):244–57.

[37] Müller D. Immunocytokines with activity-on-demand by combination with small molecule inhibitors. EMBO Mol Med 2024;16:1–2.

[38] Rai S, Stetka J, Usart M, Hao-Shen H, Park Y, Houston R, et al. The second generation type II JAK2 inhibitor, AJ1-10502, demonstrates enhanced selectivity, improved therapeutic efficacy and reduced mutant cell fraction compared to type I JAK2 inhibitors in models of myeloproliferative neoplasms (MPNs). Blood 2022;140(Suppl 1):6722–3.

[39] Kim DM, Kim MJ, Moon JH, Lee EY, Hong JK, Lee S, et al. Inhibition of JAK1/2 can overcome EGFR-TKI resistance in human NSCLC. Biochem Biophys Res Commun 2020;527(1):305–10.

[40] Parlato M, Nian Q, Charbit-Henrion F, Ruemmele FM, Rodrigues-Lima F, Cerf-Bensussan N, et al. Loss-of-function mutation in PTPN2 causes aberrant activation of JAK signaling via STAT and very early onset intestinal inflammation. Gastroenterology 2020;159(5):1968–1971.e4.

[41] Chakrabarti S, Visweswariah SS. Intramacrophage ROS primes the innate immune system via JAK/STAT and toll activation. Cell Rep 2020;33(6), 108368.

[42] Wang X, Tian R, Zhang H, Abdalla M, Bai L, Lv Y, et al. Combination of synonymous and missense mutations in *JAK3* gene contributes to severe combined immunodeficiency in one child. Hum Mutat 2023;2023, e6633251.

[43] Sinha S, Nandi S, Das MK, Basu S. Severe combined immunodeficiency. Med J DY Patil Univ 2021;14(5):562.

[44] Silberger D, Pham D. Immunodeficiency: consequences of mutations in JAK-STAT signaling. In: JAK-STAT signaling in diseases. CRC Press; 2020.

[45] Lescoat A, Lelong M, Jeljeli M, Piquet-Pellorce C, Morzadec C, Ballerie A, et al. Combined anti-fibrotic and anti-inflammatory properties of JAK-inhibitors on macrophages *in vitro* and *in vivo*: perspectives for scleroderma-associated interstitial lung disease. Biochem Pharmacol 2020;178, 114103.

[46] Daniele S, Bunick C. JAK inhibitor safety compared to traditional systemic immunosuppressive therapies. J Drugs Dermatol 2022;21(12):1298–303.

[47] Temme L, Asquith CRM. eEF2K: an atypical kinase target for cancer. Nat Rev Drug Discov 2021;20(8):577.

[48] Ju Y, Ben-David Y, Rotin D, Zacksenhaus E. Inhibition of eEF2K synergizes with glutaminase inhibitors or 4EBP1 depletion to suppress growth of triple-negative breast cancer cells. Sci Rep 2021;11(1):9181.

[49] Xiao M, Xie J, Wu Y, Wang G, Qi X, Liu Z, et al. The eEF2 kinase-induced STAT3 inactivation inhibits lung cancer cell proliferation by phosphorylation of PKM2. Cell Commun Signal 2020;18(1):25.

[50] Cen L, Gu Q, Zhou X, Lu H, Yang Q. Eukaryotic extension factor 2 kinase may affect the occurrence and development of glioblastoma through immune cell infiltration. Neurochem Res 2022;47(12):3670–81.

[51] Gurbuz N, Kahraman N, Sonmez HE, Mokhlis HA, Kosar PA, Ozpolat B. miRNA-193b-5p suppresses pancreatic cancer cell proliferation, invasion, epithelial mesenchymal transition, and tumor growth by inhibiting eEF2K. Anti Cancer Agents Med Chem 2022;22:2607–18. Ingenta Connect. [Internet]. [cited 2024 Apr 5]. Available from: https://www.ingentaconnect.com/content/ben/acamc/2022/00000022/00000014/art00012.

[52] Wang Y, Zhan D, Wang L. Ribosomal proteins are blood biomarkers and associated with CD4+ T cell activation in Alzheimer's disease: a study based on machine learning strategies and scRNA-Seq data validation. Am J Transl Res 2023;15(4):2498–514.

[53] Fu J, Yang Y, Zhu L, Chen Y, Liu B. Unraveling the roles of protein kinases in autophagy: an update on small-molecule compounds for targeted therapy. J Med Chem 2022;65(8):5870–85.
[54] Onder FC, Siyah P, Durdagi S, Ay M, Ozpolat B. Novel etodolac derivatives as eukaryotic elongation factor 2 kinase (eEF2K) inhibitors for targeted cancer therapy. RSC Med Chem 2022;13(7):840–9.
[55] De Gassart A, Demaria O, Panes R, Zaffalon L, Ryazanov AG, Gilliet M, et al. Pharmacological eEF2K activation promotes cell death and inhibits cancer progression. EMBO Rep 2016;17(10):1471–84.
[56] Cheng Y, Ren X, Zhang Y, Shan Y, Huber-Keener KJ, Zhang L, et al. Integrated regulation of autophagy and apoptosis by EEF2K controls cellular fate and modulates the efficacy of curcumin and velcade against tumor cells. Autophagy 2013;9(2):208–19.
[57] Cheng Y, Ren X, Zhang Y, Patel R, Sharma A, Wu H, et al. eEF-2 kinase dictates cross-talk between autophagy and apoptosis induced by Akt inhibition, thereby modulating cytotoxicity of novel Akt inhibitor MK-2206. Cancer Res 2011;71(7):2654–63.
[58] Michalek RD, Gerriets VA, Jacobs SR, Macintyre AN, MacIver NJ, Mason EF, et al. Cutting edge: distinct glycolytic and lipid oxidative metabolic programs are essential for effector and regulatory CD4+ T cell subsets. J Immunol 2011;186(6):3299–303.
[59] Huynh A, DuPage M, Priyadharshini B, Sage PT, Quiros J, Borges CM, et al. Control of PI(3) kinase in Treg cells maintains homeostasis and lineage stability. Nat Immunol 2015;16:188–96.
[60] Shrestha S, Yang K, Guy C, Vogel P, Neale G, Chi H. Treg cells require the phosphatase PTEN to restrain TH1 and TFH cell responses. Nat Immunol 2015;16:178–87.
[61] Cha XY, Pierce RC, Kalivas PW, Mackler SA. NAC-1, a rat brain mRNA, is increased in the nucleus accumbens three weeks after chronic cocaine self-administration. J Neurosci 1997;17:6864–71.
[62] Korutla L, Wang P, Jackson TG, Mackler SA. NAC1, a POZ/BTB protein that functions as a corepressor. Neurochem Int 2009;54:245–52.
[63] Faiola F, Yin N, Fidalgo M, Huang X, Saunders A, Ding J, et al. NAC1 regulates somatic cell reprogramming by controlling Zeb1 and E-cadherin expression. Stem Cell Rep 2017;9:913–26.
[64] Ruan Y, He J, Wu W, He P, Tian Y, Xiao L, et al. Nac1 promotes self-renewal of embryonic stem cells through direct transcriptional regulation of c-Myc. Oncotarget 2017;8:47607–18.
[65] Rahman MT, Nakayama K, Ishikawa M, Rahman M, Katagiri H, Katagiri A, et al. NAC1, a BTB/POZ protein overexpressed in uterine sarcomas. Anticancer Res 2012;32:3841–5.
[66] Chen F, Yin Y, Yan Z, Cao K, Zhong K. NAC1 promotes the migration of prostate cancer cells and participates in osteoclastogenesis by negatively regulating IFNbeta. Oncol Lett 2018;15:2921–8.
[67] Ueda SM, Yap KL, Davidson B, Tian Y, Murthy V, Wang TL, et al. Expression of fatty acid synthase depends on NAC1 and is associated with recurrent ovarian serous carcinomas. J Oncol 2010; 285191. https://doi.org/10.1155/2010/285191.
[68] Zhang Y, Ren YJ, Guo LC, Ji C, Hu J, Zhang HH, et al. Nucleus accumbens-associated protein-1 promotes glycolysis and survival of hypoxic tumor cells via the HDAC4-HIF-1alpha axis. Oncogene 2017;36:4171–81.
[69] Wang L, Kumar A, Das JK, Ren Y, Peng HY, Ballard DJ, et al. Expression of NAC1 restrains the memory formation of CD8+ T cells during viral infection. Viruses 2022;14:1713.
[70] Ren Y, Kumar A, Das JK, Peng HY, Wang L, Ballard D, et al. Tumorous expression of NAC1 restrains anti-tumor immunity through the LDHA-mediated immune evasion. J Immunother Cancer 2022;10(9), e004856.
[71] Yang JM, Ren Y, Kumar A, Xiong X, Das JK, Peng HY, et al. NAC1 modulates autoimmunity by suppressing regulatory T cell-mediated tolerance. Sci Adv 2022;8:0183.
[72] Wang X, Ji C, Zhang H, Shan Y, Ren Y, Hu Y, et al. Identification of a small-molecule compound that inhibits homodimerization of oncogenic NAC1 protein and sensitizes cancer cells to anticancer agents. J Biol Chem 2019;294:10006–17.
[73] Huber M, Lohoff M. IRF4 at the crossroads of effector T-cell fate decision. Eur J Immunol 2014;44:1886–95.
[74] Sakaguchi S. Naturally arising Foxp3-expressing CD25+CD4+ regulatory T cells in immunological tolerance to self and non-self. Nat Immunol 2005;6:345–52.
[75] Fontenot JD, Gavin MA, Rudensky AY. Foxp3 programs the development and function of CD4+CD25+ regulatory T cells. Nat Immunol 2003;4:330–6.
[76] Yuan X, Cheng G, Malek TR. The importance of regulatory T-cell heterogeneity in maintaining self-tolerance. Immunol Rev 2014;259:103–14.
[77] Sakaguchi S, Yamaguchi T, Nomura T, Ono M. Regulatory T cells and immune tolerance. Cell 2008;133:775–87.

[78] Saleh R, Elkord E. FoxP3(+) T regulatory cells in cancer: prognostic biomarkers and therapeutic targets. Cancer Lett 2020;490:174–85.
[79] Gong Z, Jia H, Yu J, Liu Y, Ren J, Yang S, et al. Nuclear FOXP3 inhibits tumor growth and induced apoptosis in hepatocellular carcinoma by targeting c-Myc. Oncogenesis 2020;9:97.
[80] Foxp3 reprograms T cell metabolism to function in low-glucose, high-lactate environments. Cell Metab 2017;25:1282–1293.e7.
[81] Watson MJ, Vignali PDA, Mullett SJ, Overacre-Delgoffe AE, Peralta RM, Grebinoski S, et al. Metabolic support of tumour-infiltrating regulatory T cells by lactic acid. Nature 2021;591:645–51.
[82] Wang H, Franco F, Tsui YC, Xie X, Trefny MP, Zappasodi R, et al. CD36-mediated metabolic adaptation supports regulatory T cell survival and function in tumors. Nat Immunol 2020;21:298–308.
[83] Matsuo K, Itoh T, Koyama A, Imamura R, Kawai S, Nishiwaki K, et al. CCR4 is critically involved in effective anti-tumor immunity in mice bearing intradermal B16 melanoma. Cancer Lett 2016;378:16–22.
[84] Marshall LA, Marubayashi S, Jorapur A, Jacobson S, Zibinsky M, Robles O, et al. Tumors establish resistance to immunotherapy by regulating Treg recruitment via CCR4. J Immunother Cancer 2020;8(2):e000764.
[85] Ebert LM, Tan BS, Browning J, Svobodova S, Russell SE, Kirkpatrick N, et al. The regulatory T cell-associated transcription factor FoxP3 is expressed by tumor cells. Cancer Res 2008;68:3001–9.
[86] Zhang H, Chen Y, Liao W, Wang L, Xie X, Fei R, et al. FOXP3 expression in FOXP3(+) tumor cells promotes hepatocellular cells metastasis. Transl Cancer Res 2020;9:5868–81.
[87] Takenaka M, Seki N, Toh U, Hattori S, Kawahara A, Yamaguchi T, et al. FOXP3 expression in tumor cells and tumor-infiltrating lymphocytes is associated with breast cancer prognosis. Mol Clin Oncol 2013;1:625–32.
[88] Du J, Wang Q, Yang S, Chen S, Fu Y, Spath S, et al. FOXP3 exon 2 controls Treg stability and autoimmunity. Sci Immunol 2022;7(72):eabo5407.
[89] Conde E, Casares N, Mancheno U, Elizalde E, Vercher E, Capozzi R, et al. FOXP3 expression diversifies the metabolic capacity and enhances the efficacy of CD8 T cells in adoptive immunotherapy of melanoma. Mol Ther 2023;31:48–65.
[90] Heeren AM, Rotman J, Stam AGM, Pocorni N, Gassama AA, Samuels S, et al. Efficacy of PD-1 blockade in cervical cancer is related to a CD8(+)FoxP3(+)CD25(+) T-cell subset with operational effector functions despite high immune checkpoint levels. J Immunother Cancer 2019;7:43.
[91] Keller AD, Maniatis T. Identification and characterization of a novel repressor of beta-interferon gene expression. Genes Dev 1991;5:868–79.
[92] Nadeau S, Martins GA. Conserved and unique functions of Blimp1 in immune cells. Front Immunol 2021;12, 805260.
[93] Chiang MF, Yang SY, Lin IY, Hong JB, Lin SJ, Ying HY, et al. Inducible deletion of the Blimp-1 gene in adult epidermis causes granulocyte-dominated chronic skin inflammation in mice. Proc Natl Acad Sci USA 2013;110:6476–81.
[94] Martins G, Calame K. Regulation and functions of Blimp-1 in T and B lymphocytes. Annu Rev Immunol 2008;26:133–69.
[95] Lee H, Huang DY, Chang HC, Lin CY, Ren WY, Dai YS, et al. Blimp-1 upregulation by multiple ligands via EGFR transactivation inhibits cell migration in keratinocytes and squamous cell carcinoma. Front Pharmacol 2022;13, 763678.
[96] Rutishauser RL, Martins GA, Kalachikov S, Chandele A, Parish IA, Meffre E, et al. Transcriptional repressor Blimp-1 promotes CD8(+) T cell terminal differentiation and represses the acquisition of central memory T cell properties. Immunity 2009;31:296–308.
[97] Cretney E, Kallies A, Nutt SL. Differentiation and function of Foxp3(+) effector regulatory T cells. Trends Immunol 2013;34:74–80.
[98] Martins GA, Cimmino L, Shapiro-Shelef M, Szabolcs M, Herron A, Magnusdottir E, et al. Transcriptional repressor Blimp-1 regulates T cell homeostasis and function. Nat Immunol 2006;7:457–65.
[99] Lin MH, Chou FC, Yeh LT, Fu SH, Chiou HY, Lin KI, et al. B lymphocyte-induced maturation protein 1 (BLIMP-1) attenuates autoimmune diabetes in NOD mice by suppressing Th1 and Th17 cells. Diabetologia 2013;56:136–46.
[100] Kallies A, Hawkins ED, Belz GT, Metcalf D, Hommel M, Corcoran LM, et al. Transcriptional repressor Blimp-1 is essential for T cell homeostasis and self-tolerance. Nat Immunol 2006;7:466–74.

[101] Fu SH, Yeh LT, Chu CC, Yen BL, Sytwu HK. New insights into Blimp-1 in T lymphocytes: a divergent regulator of cell destiny and effector function. J Biomed Sci 2017;24:49.
[102] Cretney E, Xin A, Shi W, Minnich M, Masson F, Miasari M, et al. The transcription factors Blimp-1 and IRF4 jointly control the differentiation and function of effector regulatory T cells. Nat Immunol 2011;12:304–11.
[103] Yoshikawa T, Wu Z, Inoue S, Kasuya H, Matsushita H, Takahashi Y, et al. Genetic ablation of PRDM1 in antitumor T cells enhances therapeutic efficacy of adoptive immunotherapy. Blood 2022;139:2156–72.
[104] Dixon ML, Luo L, Ghosh S, Grimes JM, Leavenworth JD, Leavenworth JW. Remodeling of the tumor microenvironment via disrupting Blimp1(+) effector Treg activity augments response to anti-PD-1 blockade. Mol Cancer 2021;20:150.
[105] Liang Y, Zhou Y, Shen P. NF-kappaB and its regulation on the immune system. Cell Mol Immunol 2004;1: 343–50.
[106] Zhang T, Ma C, Zhang Z, Zhang H, Hu H. NF-κB signaling in inflammation and cancer. MedComm 2021;2: 618–53.
[107] Liu T, Zhang L, Joo D, Sun SC. NF-κB signaling in inflammation. Signal Transduct Target Ther 2017;2:17023.
[108] Busuttil V, Droin N, McCormick L, Bernassola F, Candi E, Melino G, et al. NF-kappaB inhibits T-cell activation-induced, p73-dependent cell death by induction of MDM2. Proc Natl Acad Sci USA 2010;107:18061–6.
[109] Evaristo C, Spranger S, Barnes SE, Miller ML, Molinero LL, Locke FL, et al. Cutting edge: engineering active IKKβ in T cells drives tumor rejection. J Immunol 2016;196:2933–8.
[110] Heise N, Silva NS, Silva K, Carette A, Simonetti G, Pasparakis M, et al. Germinal center B cell maintenance and differentiation are controlled by distinct NF-κB transcription factor subunits. J Exp Med 2014;211:2103–18.
[111] Griss J, Bauer W, Wagner C, Simon M, Chen M, Grabmeier-Pfistershammer K, et al. B cells sustain inflammation and predict response to immune checkpoint blockade in human melanoma. Nat Commun 2019;10:4186.
[112] Achyut BR, Angara K, Jain M, Borin TF, Rashid MH, Iskander ASM, et al. Canonical NFκB signaling in myeloid cells is required for the glioblastoma growth. Sci Rep 2017;7:13754.
[113] Hagemann T, Lawrence T, McNeish I, Charles KA, Kulbe H, Thompson RG, et al. "Re-educating" tumor-associated macrophages by targeting NF-kappaB. J Exp Med 2008;205:1261–8.
[114] Ni J, Wang X, Stojanovic A, Zhang Q, Wincher M, Bühler L, et al. Single-cell RNA sequencing of tumor-infiltrating NK cells reveals that inhibition of transcription factor HIF-1α unleashes NK cell activity. Immunity 2020;52:1075–1087.e8.
[115] Javaid N, Choi S. Toll-like receptors from the perspective of cancer treatment. Cancers (Basel) 2020;12(2):297.
[116] Xie X, Ma L, Zhou Y, Shen W, Xu D, Dou J, et al. Polysaccharide enhanced NK cell cytotoxicity against pancreatic cancer via TLR4/MAPKs/NF-κB pathway in vitro/vivo. Carbohydr Polym 2019;225, 115223.
[117] Richmond A. Nf-kappa B, chemokine gene transcription and tumour growth. Nat Rev Immunol 2002;2:664–74.
[118] Xia Y, Shen S, Verma IM. NF-κB, an active player in human cancers. Cancer Immunol Res 2014;2:823–30.
[119] Wang W, Nag SA, Zhang R. Targeting the NFκB signaling pathways for breast cancer prevention and therapy. Curr Med Chem 2015;22:264–89.
[120] Du F, Qi X, Zhang A, Sui F, Wang X, Proud CG, et al. MRTF-A-NF-κB/p65 axis-mediated PDL1 transcription and expression contributes to immune evasion of non-small-cell lung cancer via TGF-β. Exp Mol Med 2021;53:1366–78.
[121] Barnes SE, Wang Y, Chen L, Molinero LL, Gajewski TF, Evaristo C, et al. T cell-NF-κB activation is required for tumor control in vivo. J Immunother Cancer 2015;3:1.
[122] Bromberg JF, Wrzeszczynska MH, Devgan G, Zhao Y, Pestell RG, Albanese C, et al. Stat3 as an oncogene. Cell 1999;98:295–303.
[123] Tolomeo M, Cascio A. The multifaced role of STAT3 in cancer and its implication for anticancer therapy. Int J Mol Sci 2021;22:603.
[124] Hillmer EJ, Zhang H, Li HS, Watowich SS. STAT3 signaling in immunity. Cytokine Growth Factor Rev 2016;31:1–15.
[125] Wang Y, Shen Y, Wang S, Shen Q, Zhou X. The role of STAT3 in leading the crosstalk between human cancers and the immune system. Cancer Lett 2018;415:117–28.
[126] Zhu S, Phatarpekar PV, Denman CJ, Senyukov VV, Somanchi SS, Nguyen-Jackson HT, et al. Transcription of the activating receptor NKG2D in natural killer cells is regulated by STAT3 tyrosine phosphorylation. Blood 2014;124:403–11.

[127] Wu J, Gao FX, Wang C, Qin M, Han F, Xu T, et al. IL-6 and IL-8 secreted by tumour cells impair the function of NK cells via the STAT3 pathway in oesophageal squamous cell carcinoma. J Exp Clin Cancer Res 2019;38:321.
[128] Fan Y, Mao R, Yang J. NF-κB and STAT3 signaling pathways collaboratively link inflammation to cancer. Protein Cell 2013;4:176–85.
[129] Wang T, Niu G, Kortylewski M, Burdelya L, Shain K, Zhang S, et al. Regulation of the innate and adaptive immune responses by Stat-3 signaling in tumor cells. Nat Med 2004;10:48–54.
[130] Siersbæk R, Scabia V, Nagarajan S, Chernukhin I, Papachristou EK, Broome R, et al. IL6/STAT3 signaling hijacks estrogen receptor α enhancers to drive breast cancer metastasis. Cancer Cell 2020;38:412–423.e9.
[131] Deepak Roshan VG, Sinto MS, Thomas S, Kannan S. Cyclin D1 overexpression associated with activation of STAT3 in oral carcinoma patients from South India. J Cancer Res Ther 2018;14:403–8.
[132] Luo J, Yan R, He X, He J. Constitutive activation of STAT3 and cyclin D1 overexpression contribute to proliferation, migration and invasion in gastric cancer cells. Am J Transl Res 2017;9:5671–7.
[133] Kim TW, Kim Y, Keum H, Jung W, Kang M, Jon S. Combination of a STAT3 inhibitor with anti-PD-1 immunotherapy is an effective treatment regimen for a vemurafenib-resistant melanoma. Mol Ther Oncolytics 2022;26:1–14.
[134] Zhao X, Zhang Z, Moreira D, Su YL, Won H, Adamus T, et al. B cell lymphoma immunotherapy using TLR9-targeted oligonucleotide STAT3 inhibitors. Mol Ther 2018;26:695–707.
[135] Kong LY, Wei J, Sharma AK, Barr J, Abou-Ghazal MK, Fokt I, et al. A novel phosphorylated STAT3 inhibitor enhances T cell cytotoxicity against melanoma through inhibition of regulatory T cells. Cancer Immunol Immunother 2009;58:1023–32.
[136] Iwata-Kajihara T, Sumimoto H, Kawamura N, Ueda R, Takahashi T, Mizuguchi H, et al. Enhanced cancer immunotherapy using STAT3-depleted dendritic cells with high Th1-inducing ability and resistance to cancer cell-derived inhibitory factors. J Immunol 2011;187:27–36.
[137] Eferl R, Wagner EF. AP-1: a double-edged sword in tumorigenesis. Nat Rev Cancer 2003;3:859–68.
[138] Shaulian E, Karin M. AP-1 in cell proliferation and survival. Oncogene 2001;20:2390–400.
[139] Shaulian E, Karin M. AP-1 as a regulator of cell life and death. Nat Cell Biol 2002;4:131–6.
[140] Zhang HS, Yan B, Li XB, Fan L, Zhang YF, Wu GH, et al. PAX2 protein induces expression of cyclin D1 through activating AP-1 protein and promotes proliferation of colon cancer cells. J Biol Chem 2012;287:44164–72.
[141] He H, Sinha I, Fan R, Haldosen LA, Yan F, Zhao C, et al. c-Jun/AP-1 overexpression reprograms ERα signaling related to tamoxifen response in ERα-positive breast cancer. Oncogene 2018;37:2586–600.
[142] Kappelmann-Fenzl M, Gebhard C, Matthies AO, Kuphal S, Rehli M, Bosserhoff AK. C-Jun drives melanoma progression in PTEN wild type melanoma cells. Cell Death Dis 2019;10:584.
[143] Seo H, González-Avalos E, Zhang W, Ramchandani P, Yang C, Lio CJ, et al. BATF and IRF4 cooperate to counter exhaustion in tumor-infiltrating CAR T cells. Nat Immunol 2021;22:983–95.
[144] Xu T, Wang C, Chen X, Bai J, Wang E, Sun M. Coexpression of c-Jun in multiple-chain DAP-CAR-engineered T-cells for solid tumor therapy. Immunotherapy 2022;14:1457–66.
[145] Liu M, Tong Z, Ding C, Luo F, Wu S, Wu C, Albeituni S, He L, Hu X, Tieri D, Rouchka EC, Hamada M, Takahashi S, Gibb AA, Kloecker G, Zhang H-G, Bousamra 2nd M, Hill BG, Zhang X, Yan J. Transcription factor c-Maf is a checkpoint that programs macrophages in lung cancer. J Clin Invest 2020;130:2081–96.

CHAPTER

10

Targeting engineered bacteria to improve cancer immunotherapy

Aiswarya Dash[a], *Sonakshi Bahuguna*[b], *and Jugal Kishore Das*[b]

[a]Department of Medicinal Chemistry and Biophysics Kemihuset (K), Campus, Umeå University, Umeå, Sweden [b]Department of Life Science, School of Natural Sciences, Shiv Nadar University, Uttar Pradesh, India

Abstract

Advances in anticancer therapy have transformed the management of disease in patients, raising hopes of finding an effective cure for the disease. Nevertheless, many challenges in modern cancer therapy arise due to the nonspecificity of newly developed anticancer drugs, severe off-target toxicity of established cancer therapeutic regimens, as well as the emerging drug resistance of tumor cells. The limited efficacy of the advanced anticancer treatment strategies, such as cancer immunotherapy, can be attributed to the highly complex and immunosuppressive tumor microenvironment. Designing a cancer immunotherapy regimen with the potential to overcome immunosuppression will remove the major drawbacks of resistance of cancer to immunotherapy.

Targeting microbes in the tumor microenvironment is an excellent therapeutic alternative to counter the limitations of presently accessible anticancer therapeutic strategies. Moreover, recent advances in genetic engineering technologies have given rise to engineered bacteria with versatile capability to restrain tumor-associated malignancies. These microbes accumulate precisely in the immunosuppressive tumor microenvironment and proliferate in favorable conditions in this unique niche, enhancing therapeutic outcomes by augmenting the immune response. Microbe-based anticancer therapy and their combination with other conventional therapeutic methods have also greatly advanced disease management in patients. This chapter summarizes the current status and future developments in tumor-targeted microbial cancer immunotherapy. The chapter also discusses the advances in engineering microbes for improved targeting and efficiency in the tumor microenvironment.

Abbreviations

BCG	*Bacillus Calmette-Guerin*
CAR-T cells	chimeric antigen receptor T cells
CFU	colony forming units
E. coli	*Escherichia coli*
EcN	*E. coli NISSLE 1917*

T Cell Metabolism and Cancer Immunotherapy
https://doi.org/10.1016/B978-0-443-13827-0.00004-4

GALT	gut-associated lymphoid tissue
GRAS	generally regarded as safe
IBD	inflammatory bowel disease
ICB	immune checkpoint blockade
L. plantarum	*Lactobacillus plantarum*
L. reuteri	*Lactobacillus reuteri*
PAMPs	pathogen-associated molecular patterns
S. Typhimurium *ΔppGpp*	*Salmonella* Typhimurium ΔppGpp-
S. Typhimurium	*Salmonella enterica* serovar Typhimurium
TLRs	toll-like receptors
TME	tumor microenvironment
T_{reg} Cells	regulatory T cells

No conflict statement

The authors declare no conflict of interest.

Introduction

The human microbiome consists of a diverse set of bacteria inhabiting not only the local tissues but also the tumor in cancer patients. The composition of the microbiome modulates the progression of several types of cancer [1,2]. Intriguingly, although the association of microbiome to tumor development has been known since ancient times, employing microbes as an instrument for cancer therapy was only attempted in the early nineteenth century by a physician named William Coley.

Coley injected a combination of heat-killed Gram-positive streptococcus bacteria along with heat-killed Gram-negative *Serratia marcescens* (*S. marcescens*) bacteria into several cancer patients with inoperable cancers. This combination of microbes and their products used for cancer therapy came to be known as Coley's toxins. The methodology adopted by Coley was the first known instance of anticancer immunotherapy using microbes. Coley hypothesized that the malignant tumor would shrink as a byproduct of the infection he created, due to stimulation of the immune responses, and to his surprise, this hypothesis turned out to be true [3]. However, despite his excellent reported success in curing cancer, especially in soft tissue and bone sarcomas, Coley's toxins came under criticism from his colleagues and therefore had limited global clinical impact. Although Coley's strategy was initially incapable of imparting maximal therapeutic benefit to the patients, it has laid the foundations for the development of highly effective modern immunotherapeutic approaches such as immune checkpoint blockade (ICB) therapy and chimeric antigen receptor T cells (CAR-T) anticancer treatment approaches. However, clinical studies in recent years have revealed the restraints of these immunotherapeutic approaches arising from cancer heterogenicity, adverse side effects, and lack of accessibility of patients to these expensive treatment modalities [4,5]. To increase the applicability of immunotherapeutic approaches and obtain robust clinical therapeutic results, more precise treatment has become the need of the hour.

Recent advances in the field of bacterial genetic engineering and understanding of tumor immunology have therefore made microbe-based anticancer immunotherapy an attractive and precise anticancer treatment strategy. Obligate intracellular bacteria such as *Bifidobacterium* and Clostridium spp. and facultative anaerobic bacteria like *Salmonella* spp., *Listeria* spp., and *E. coli* have intrinsic tumor-targeting and killing activity [6–8]. These microbes target the tumor microenvironment by employing a variety of passive as well as active penetration, survival, and proliferation strategies within the anaerobic tumor core regions [9]. The bacteria may initially enter the tumor microenvironment (TME) from the blood circulation by passive entrapment in the tumor vasculature followed by active chemotaxis and multiplication in a highly favorable necrotic, immunosuppressive, and hypoxic TME [10,11]. In addition to their capability to preferentially colonize in the TME, these bacteria also possess the ability to stimulate the immune system effectively. Interestingly, it has also been recently shown that an attenuated *Brucella melitensis* bacterial strain lacking the type-IV secretion system preferentially colonizes in the tumor and potently enhances the efficacy of targeted chimeric antigen receptor T (CAR-T) cell cancer immunotherapy in a murine model of colorectal cancer [10].

The microbes penetrating and multiplying within the tumor modulate the TME by decreasing its immunosuppressive characteristic via driving inflammation in these regions [10,11]. Moreover, the change in the composition of the tumor microbiome, driven by the introduction of these microbes into the TME, also increases the immunosurveillance of the cancer cells. These remarkable attributes of the microbes in the TME greatly limit the growth and spread of cancer cells in the host. The lack of success of other conventional cancer therapy methods, combined with accumulating evidence of highly efficacious outcomes of microbe-based cancer immunotherapy, has paved the way for engineering attenuated microbes to increase the efficacy of anticancer immunotherapy [12].

Microbes such as *Salmonella, E. coli, Brucella* spp., and *Lactobacillus* spp. have been engineered to improve their tumor-targeting mechanisms [9–12]. The construction of auxotrophic mutant bacterial strains that cannot replicate in normal tissues, due to the lack of specific essential nutrients outside the TME, improves the safety of the bacteria while also ensuring targeted delivery into TME. For instance, *Salmonella* A1-R strain utilizes leucine and arginine that are enriched in the tumor but not in normal tissues and therefore replicate only inside the tumor [13,14]. Similarly, rationally designed bacterial strains have also been used to deliver inflammatory cytokines and other targeted payloads into the TME [13–15].

The promising preclinical as well as clinical outcomes of genetic engineering of microbes to alter disease outcomes have created exciting avenues for future research. One such area of research is engineering commensal microbes at their original ecological niche to enhance their antitumor activity [15]. This development will lead to safe, targeted, precise, and sustained anticancer treatment options. Moreover, engineered probiotics could be developed with tumor targeting and immune stimulation characteristics to serve as local mediators in modulation of the TME [16]. The integration of commensal and probiotic microbes for treatment of cancer will be a safer alternative to using attenuated pathogens with potential for reversal to virulent types, in anticancer immunotherapy. However, noncommensal bacteria have the natural ability to evade the immune system while localizing in the TME. On reaching the TME, the bacteria may present their pathogen-associated molecular patterns (PAMPs) to stimulate the immune responses in an otherwise immunosuppressive microenvironment. This unique ability of noncommensal microbes makes them ideal candidates for use in anticancer immunotherapy with integration of safety switches. This chapter discusses the potential of using both generally regarded as safe (GRAS) *Lactobacillus* strains as well as attenuated pathogens like *Salmonella, E. coli,* and *Brucella* spp. for anticancer immunotherapy. Future research will focus on improving the safety and usability of these microbial strains by integrating gene circuits to deploy the bacteria to tune to the external cues triggering bacterial lysis when required.

Genetically modified microbes in cancer immunotherapy

Bacterial-mediated anticancer therapy offers a novel approach for the treatment of cancers that are either insensitive to medicines or gain resistance over time. Cytotoxic substances can successfully kill cancer cells but often destroy them at the expense of harming normal tissues, resulting in harmful patient toxicities. Small-molecule inhibitors may not be sufficiently specific, and patients treated with targeted therapy frequently experience resistance or relapse due to intrinsically resistant genetic changes or selection for a fraction of cancer cells with those changes. Tumor-targeting bacteria provide an alternative method to overcome the limitations of cytotoxic or conventional cancer drugs. Bacteria can thrive physiologically in hypoxic TME by acquiring nutrients in tumor necrotic zones, thus competing with cancer cells for nutrition. Bacteria are rich in PAMPs that can elicit immune responses in the tumor immunosuppressive microenvironment; they can enhance the specific immune recognition and elimination of tumor cells [9,17].

These criteria can be used to the advantage of cancer therapy by engineering bacteria to produce antitumor or immunotherapeutic agents. Bacillus Calmette-Guérin (BCG), a live attenuated *Mycobacterium tuberculosis* vaccine, was used for high-risk, nonmuscle-invasive bladder cancer. However, further development of this biological therapy was withheld due to the infection-associated toxicity and the insufficient comprehension of tumor immunity at the time [17].

Different bacteria, including *Salmonella, E. coli, Brucella* spp., *and Lactobacillus* spp., have been studied in several clinical trials targeting various cancers [17]. Bacteria can be genetically modified to create very effective therapeutic payloads such as cytotoxic drugs, cytokines, prodrug-converting enzymes, small interfering RNAs, and nanobodies. These immunomodulators or therapeutic payloads corroborate with microorganisms to modify the TME through interactions with tumor cells, immune cells, and other TME components. This reprogramming is performed by infiltration and activation of immune cells and induction of appropriate immunomodulatory responses [9,17].

However, drug development using live bacteria is currently difficult. Pathogenic infection continues to be a risk factor, and while attenuated bacteria have low virulence and may be removed with antibiotics, it can harm immunocompromised people. To reduce the hazards of horizontal gene transfer, bacterial strains expressing antibiotic resistance genes must be excluded. Some of the usages of unedited or edited microbes for cancer immunotherapy are listed below:

Genetically modified *Salmonella* in cancer immunotherapy

The goal of bacterial-mediated cancer therapy is to maximize tumor cell death while limiting infection-related damage to normal cells; hence, targeted therapy is required. *Salmonella* Typhimurium meets numerous criteria for cancer therapy, including tumor-homing ability, intrinsic tumor-killing action, the ability to induce proinflammatory responses in the host, and the ease of genetic engineering [9]. *Salmonella* infects a wide range of hosts, including humans and livestock. Moreover, it can be delivered orally and induce both local and systemic immune responses, indicating its potential as a model vector for anticancer therapy. For *Salmonella*-mediated cancer therapy to be successful, it is critical to discover how to utilize *Salmonella* as an ideal anticancer agent or delivery vector with high tumor-targeting ability [9].

Attenuated *Salmonella enterica* serovar Typhimurium (*S.* Typhimurium) is a facultative intracellular bacterium that mediates antitumor effects by activating tumor-specific humoral and cellular immune responses. One of the attenuated *Salmonella* strains (ΔppGpp/ *relA* and *spoT* double mutant), which lacks the ability to synthesize ppGpp (guanosine pentaphosphate), was observed to be avirulent in BALB/c rodents. Immunizing mice with the ΔppGpp strain resulted in systemic and mucosal antibody responses, as well as cell-mediated immunity, as demonstrated by various assays [18].

Salmonella treatment has been shown to increase the number of monocytes, macrophages, and CD8+ T cells in tumors and the expression of proinflammatory cytokines-IFN-γ and IL-1β. In tumor tissues, *Salmonella* increases the expression of calreticulin, a protein that is part of the immunogenic cell death pathway [19,20]. Studies have reported enhanced macrophages, neutrophils, and CD8+ T lymphocyte infiltration in mouse tumors following *Salmonella* infection. Upon administration into animals, *Salmonella* was reported to be more enriched in the tumors than normal tissues, leading to a strong antitumor impact in animal models. Although the initial *Salmonella* levels in tumors and other tissues were not significantly different, the circulatory system and other tissues cleared the bacteria within hours to days, while the *Salmonella* that entered tumors took hold, colonized, and thrived well. The depleted oxygen concentration in the tumor microenvironment is beneficial to *Salmonella* tumor-homing ability. The uneven and highly vascularized tumor microenvironment promotes hypoxia, allowing *Salmonella* to survive as a facultative anaerobe [19,21].

Salmonella flourishes in the hypoxic zone of tumor tissues, where oxygen, pH, and metabolite levels change throughout tumor growth and colonization. Such stressors can activate some bacterial genes, which serve as internal triggers for gene expression. Likewise, drugs or small molecules can be used to enhance expression of therapeutic genes bioengineered in *Salmonella* [9]. A recent study developed a doxycycline-inducible gene switch system capable of producing drug payloads for therapeutic purposes with a single induction of the promoter [22]. The researchers created an attenuated strain of *S.* Typhimurium that secretes *Vibrio vulnificus* flagellin B (FlaB) conjugated to human and mouse interleukin-15 proteins in the presence of L-arabinose [22]. The fusion proteins possessed both IL15 and FlaB functional activity, inhibiting tumor formation and preventing recurrence more effectively than bacteria secreting IL15 or FlaB alone. Strong vaccination adjuvant FlaB stimulates anticancer immunity by attracting innate immune cells to tumor tissue. However, FlaB did not attract a sufficient number of CD4+ T, CD8+ T, natural killer (NK), and NKT cells in tumor tissue, implying that IL15 is more effective. The latter is critical to the development, maturation, and activation of CD8+ T cells, NK, and NKT cells, as well as the survival and proliferation of CD8+ T memory cells [22].

A study by Liang et al. showed that *Salmonella* can be genetically modified to enhance tumor targeting [23]. This modification involves truncating the lipopolysaccharide (LPS) layer and displaying targeting peptides on the bacterial membrane surface [23]. This approach leads to strong antitumor effects through targeted medication delivery. The RGD4C peptide remains obscured in *Salmonella* strains that express full-length LPS. This peptide is known to mediate the binding of drug molecules or nanocarriers to αvβ3 integrin with high affinity and selectivity, thereby increasing their efficiency in targeting tumor blood vessels or certain tumor cells. Therefore, it is hypothesized that surface display of RGD peptides on *Salmonella* may also confer high tumor targeting to the bacteria. Strains with only lipid A and partial core oligosaccharides, lacking O-units (particularly strain D2 ΔrfaJ-RGD4C), can efficiently display RGD4C peptide. This improved targeting ability allows *Salmonella* to adhere to HUVECs expressing αvβ3 integrin and colonize the tumor tissue. αvβ3 integrin is highly expressed on the surface of proliferating endothelial cells and other types of tumor cells, but not in normal tissue blood vessels. Animal experiments further indicated that this tumor-targeting *Salmonella* strain could simultaneously deliver endostatin and TRAIL, two drugs with different antitumor properties [23].

Genetically modified *E. coli* in cancer immunotherapy

E. coli NISSLE 1917 (EcN) is a Gram-negative strain that has probiotic characteristics and is used to treat intestinal disorders such as diarrhea and inflammatory bowel disease (IBD), specifically ulcerative colitis. EcN has been shown to selectively colonize and proliferate within solid tumors. The strain has been utilized successfully to deliver vaccines, cytokines, and other therapeutics into the TME [24]. A recent study established the feasibility of using altered EcN in adenoma identification by noninvasive stool and urine testing. EcN was bioengineered to produce chemokine-GM-CSF and immunotherapeutics for delivery within adenomas, leading to a significant decrease in adenoma load in colorectal cancer mouse models [7].

Yang S et al. reported a novel bacterial therapeutic platform by engineering nonpathogenic *E. coli* K-12 DH5α to produce potent cytokines for improved immunotherapy [25]. Bacteria that expressed the murine decoy-resistant IL18 mutein (DR18) triggered immunological responses in mice with colorectal cancer and melanoma. This led to considerable tumor reduction and increased survival of mice with colorectal carcinoma and melanoma. The modified bacteria exhibit tumor tropism and induce immunologic memory. *E. coli* K-12 DH5a was modified to express human DR18-induced mesothelin-targeting CAR NK cells. This resulted in increased control and survival in xenograft mice with mesothelioma tumor cells, which were previously resistant to natural killer cells. This study expanded the translational potential of this method and shed light on the mechanisms behind CAR-natural killer (NK) cell activation [25].

Genetically modified *B. melitensis* strain in cancer immunotherapy

The successful use of *Salmonella* and *E. coli* in cancer immunotherapy has reinvigorated research in using other microbes for effective anticancer outcomes. An engineered *B. melitensis* (*Bm*Δ*vjbR*) bacterial strain lacking the Type-IV secretion system was generated as a live attenuated safe *Brucella* vaccine candidate [26]. *Bm*Δ*vjbR* was found to be safe in all tested animal models and homed specifically to tumor tissue while improving the TME in a murine model of solid cancer [10]. This live attenuated bacterial treatment could overcome the resistance of cancer cells to CAR-T-cell therapy by remodeling the TME toward promoting proinflammatory macrophage and T cell-mediated antitumor immunity. This preclinical study is an excellent example of using previously unexplored microbes in combination with directed CAR-T cell therapy to modulate the TME beneficially and inhibit cancer. Current research focuses on engineering of the *Bm*Δ*vjbR* bacterial strain to further enhance its safety and antitumor activities in a variety of different tumor models [10,26].

GRAS *Lactobacillus* strains in cancer immunotherapy

Commensal bacteria such as rod-shaped Gram-positive *Lactobacillus* bacteria reside in the intestinal microbiome of humans [27,28]. These microbes are one of the major probiotic bacteria in the intestine and play a critical role in lactic acid fermentation to strengthen the intestinal barrier [29]. *Lactobacillus* species play a critical role in maintaining systemic immunity by regulating gut homeostasis. By modulating the composition and function of the gut microbiota, *Lactobacillus* influences the development and function of immune cells systemically and within the gut-associated lymphoid tissue (GALT). Dysbiosis, characterized by alterations in the gut microbiota composition, has been implicated in various diseases, including cancer [30,31]. *Lactobacillus* supplementation restores microbial balance and enhances antitumor immune responses in preclinical models [30,31].

L. plantarum (*L. plantarum*) is one of the most well-studied *Lactobacillus* strains in prevention of infection as well as in cancer [30]. The L. *plantarum* L-14 extract was shown to restrain the growth, relocation, and gene expression profile of human A375 carcinoma cells [30]. *Lactobacillus* strains have also been shown to exert immunomodulatory effects to inhibit cancer. For instance, it has been shown that *Lactobacillus casei* down-regulates IL-22 and upregulates caspases, thereby inducing apoptosis in cancer cells [32]. Similarly, a study carried out in a melanoma mouse model showed that *L. reuteri* (*L. reuteri*) FLRE5K strain induces elevated levels of the inflammatory cytokines such as TNF-α and IFN-γ, thereby inhibiting the proliferation of melanoma cells [32].

Mechanism of action of bacteria in targeting and inhibiting tumors

Bacteria, with their inherent ability to selectively migrate toward hypoxic and necrotic regions within solid tumors, present a promising avenue for immune-activated tumor killing and targeted drug delivery into the TME. Bacteria may penetrate the TME by employing a diverse set of active and passive strategies depending upon the microbial strain. The TME is characterized by a distinctive set of conditions such as hypoxic microenvironment, acidic pH, and an altered extracellular matrix composition. These factors provide a suitable homing niche for several microbes possessing tropisms toward such environments such as obligate intracellular bacteria. Directed chemotaxis of bacteria is an effective mechanism employed by bacteria to inhabit favorable niches in the TME. Bacteria traverse through complex environments by responding to gradients of various chemical signals. Once inside the tumor, these microbes replicate in the immunosuppressive and hypoxic conditions within the TME more effectively than outside the TME. For instance, it was reported that the intratumoral levels of *S*. Typhimurium reached very high levels (>100 million colony forming units (CFUs) after intravenous (*i.v*) administration, and the intratumoral to extratumoral bacterial ratio exceeded 10,000:1 [33]. Similarly, Kasinskas et al. have also shown that chemical receptors of *Salmonella* have an important effect on the tumor tropism of bacteria by silencing different chemical receptors on the bacteria. The study also showed that metabolites such as galactose, serine, and ribose helped in the intratumoral mobility of *Salmonella* [34]. Similarly, in another study, the authors detected *Bifidobacterium* 1 week post systemic administration, although it was notably absent from other organs at this time-point [35]. Remarkably, modern genetic engineering tools have further improved the tumor homing and intratumoral survival of these bacteria. The incorporation of tumor-specific promoters into their genomes has generated bacteria that can express virulence factors or therapeutic payloads selectively within the TME. This approach ensures precise targeting while minimizing off-target effects. Auxotrophic bacteria have been engineered analogous to exact nutrient utilization exclusively within the tumor by the help of gene editing technologies. For instance, a leucine-arginine *Salmonella* auxotroph was designed to specifically target the TME [36] and survive in this niche. Moreover, bacteria have been engineered to express tumor antibodies or tumor-homing peptides on their surface improving their tumor-targeting ability. An example of this approach was demonstrated in a study by Park et al., which showed that altering the arginine–glycine–aspartate peptide residues on the outer membrane surface protein of *S.* Typhimurium significantly enhanced the tumor tropism of the bacteria [23]. Similarly, Massa et al. have also demonstrated that the expression of surface antibodies targeted against tumor-associated antigens improves the specificity of *Salmonella* to home into the TME [37]. Hypoxia, nutrient deprivation, acidic pH, and metabolic products generated in the TME may also serve as additional chemical cues for bacterial migration. Anaerobic bacteria thrive in these conditions, making them ideal candidates for tumor-targeted therapies.

After penetration into the TME, the gene-edited bacteria may independently prime the immune system to enhance inflammation and inhibit the growth and proliferation of cancer cells. Alternatively, the bacteria may colonize the TME and alter the nature of the resident microbiome, modifying its composition and function. The engineered microbes may downregulate the activity of antiinflammatory regulatory (T_{reg}) T cells, which results in enhanced cytotoxicity and functionality of CD8+ T cells and tumor cell killing. Similarly, the genetically modified microbes may act by enhancing the activity of natural killer (NK) cells and macrophages to initiate the regression of tumors (Fig. 1).

Perspectives on clinical trials

One of the most successful anticancer microbial therapies has been carried out with the BCG vaccine. This live vaccine made up of attenuated bovine tuberculosis suspension has been widely used in clinical practice due to its safety and efficacy [38]. The demonstrated success, safety, and immunogenicity of the BCG vaccine for treatment of tuberculosis, combined with its promising preclinical efficacy in cancer therapy regimens, resulted in the repurposing of this classic microbial strain as an antidote for several cancers, like bladder cancer. Furthermore, the notable successes of anticancer BCG therapy also inspired several other microbial-based clinical trials, as listed in Table 1. It is interesting to note that some of these bacteria have shown excellent therapeutic efficacy when used in combination with other drugs or treatment strategies. Recent studies have also reported several clinical trials using combinatorial therapeutic approaches to overcome cancer.

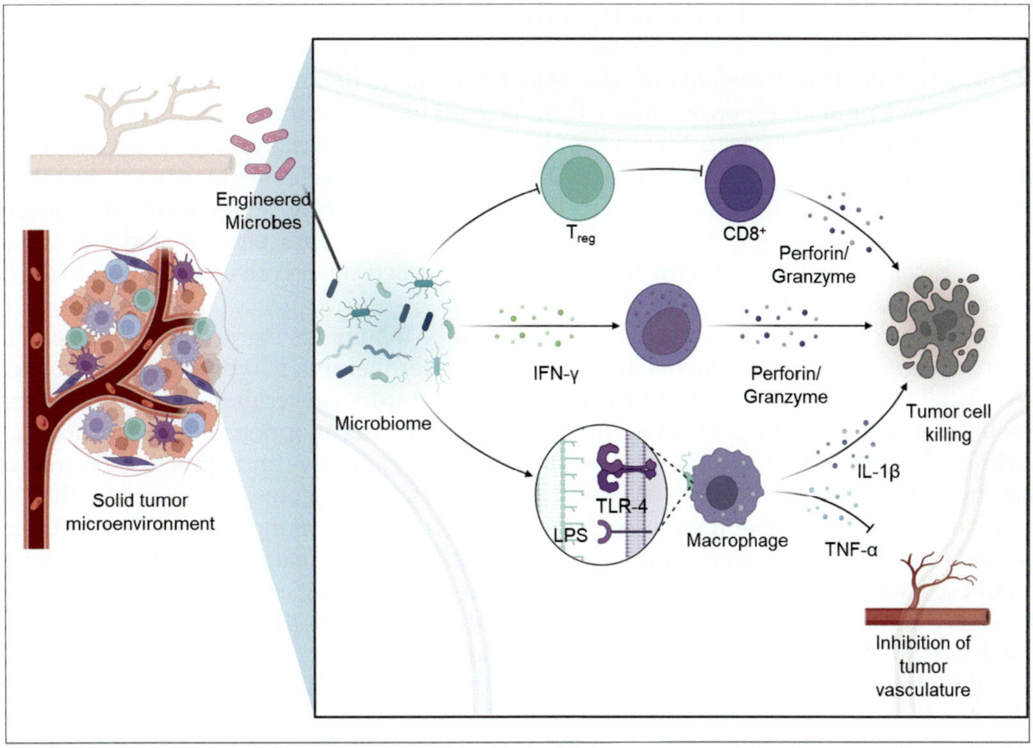

FIG. 1 **Mechanism of tumor inhibition by targeted bacterial therapy.** Tumor-penetrating engineered bacteria downregulate the functions of T_{reg} cells, thereby activating cytotoxic CD8+ T cells. The activated and functional cytotoxic CD8+ T cells produce perforin and granzyme, which results in inhibition of tumor cells. The altered microbiome may also produce IFN-γ-activating NK cells, which produce perforin and granzyme, resulting in tumor cell killing and inhibition of cancer. The microbiome also produces several unique molecular patterns, like the LPS which binds to toll-like receptors (TLRs) on the surface of microbes to enhance the activity of macrophages, facilitating tumor cell killing.

TABLE 1 Important clinical trials in bacterial immunotherapy.

Bacterial strain	Clinical trial phase	Cancer type	References
S. Typhimurium VNP200009	Phase-1	Melanoma	[39]
S. Typhimurium VNP200009	Phase-1	Head, neck, and esophagus cancer	[39]
Pseudomonas	Phase 1	Glioblastoma	[40]
S. Typhimurium expressing human IL-2	Phase-1	Liver cancer	https://www.clinicaltrials.gov/ct2/show/NCT01099631
Listeria monocytogenes	Phase 2	Pancreatic tumors	[41]
Clostridium novyi-NT	Phase-1	Solid metastatic tumors	https://www.clinicaltrials.gov/ct2/show/NCT01118819

Future perspectives and conclusion

The ability of microbes to not only precisely target the TME but also accelerate a sequence of antitumor immune responses makes them elegant microbots for cancer therapy. The advent of gene technology has further improved the ability of bacterial vectors to have accurate tumor targeting and enhanced safety in the TME. Gene editing technologies have also enabled the creation of syngeneic bacterial strains that can produce anticancer elements for constant effectiveness in the host-bearing cancer. These biological characteristics make bacterial carriers excellent delivery systems that have outperformed other conventional and modern anticancer treatment strategies.

The advancement of anticancer immunotherapeutic treatments has arguably been one of the most significant developments in the field of cancer therapy. ICB therapy and directed CAR-T cell therapy have shown great promise as the future of cancer therapy. However, several deficiencies and limitations have emerged with increasing adoption of these technologies in clinics. Microbe-based immunotherapy provides an inexpensive and attractive alternative that overcomes the limitations of other immunotherapy regimes. The combination of microbe-based cancer therapeutic regimens with current immunotherapeutic treatment methods has led to the development of increasingly refined treatment modalities aiming to impose maximal damage to the cancer cells while also avoiding collateral tissue and organ damage. Therefore, the future of bacterial-based cancer immunotherapy will be the design and advancement of combinatorial treatment strategies.

The continued adoption and improvement of these microbe-based immunotherapy treatment regimens of safe and local immune stimulation within the TME will lead to the design of improved clinical anticancer therapies with decreased toxicity and better survival outcomes in the future.

Acknowledgments

J.K.D. is supported by the Ramalingaswami Fellowship and would like to acknowledge the Department of Biotechnology, Government of India for the funding support.

References

[1] Doocey CM, Finn K, Murphy C, Guinane CM. The impact of the human microbiome in tumorigenesis, cancer progression, and biotherapeutic development. BMC Microbiol 2022;22.

[2] Bourzac K. Microbiome: the bacterial tightrope. Nature 2014;516:S14–6.

[3] Starnes CO. Coley's toxins in perspective; 1992.

[4] Guo Y, Chen Y, Liu X, Min JJ, Tan W, Zheng JH. Targeted cancer immunotherapy with genetically engineered oncolytic Salmonella typhimurium. Cancer Lett 2020;469:102–10.

[5] Zhao Z, Zheng L, Chen W, Weng W, Song J, Ji J. Delivery strategies of cancer immunotherapy: recent advances and future perspectives. J Hematol Oncol 2019;12.

[6] Swofford CA, Van Dessel N, Forbes NS. Quorum-sensing Salmonella selectively trigger protein expression within tumors. Proc Natl Acad Sci U S A 2015;112(11):3457–62.

[7] Gurbatri CR, Radford GA, Vrbanac L, Im J, Thomas EM, Coker C, et al. Engineering tumor-colonizing E. Coli Nissle 1917 for detection and treatment of colorectal neoplasia. Nat Commun 2024;15(1).

[8] Arunima A, van Schaik EJ, Samuel JE. The emerging roles of long non-coding RNA in host immune response and intracellular bacterial infections. vol. 13. Frontiers Media S.A.; 2023. Frontiers in Cellular and Infection Microbiology.

[9] Wang D, Wei X, Kalvakolanu DV, Guo B, Zhang L. Perspectives on oncolytic Salmonella in Cancer immunotherapy—a promising strategy. Front Immunol 2021;12. Frontiers Media S.A.
[10] Guo F, Das JK, Kobayashi KS, Qin QM, Ficht TA, Alaniz RC, et al. Live attenuated bacterium limits cancer resistance to CAR-T therapy by remodeling the tumor microenvironment. J Immunother Cancer 2022;10(1).
[11] Zhou S, Gravekamp C, Bermudes D, Liu K. Tumour-targeting bacteria engineered to fight cancer. Nat Rev Cancer 2018;18:727–43. Nature Publishing Group.
[12] Plocica J, Guo F, Das JK, Kobayashi KS, Ficht TA, Alaniz RC, et al. Engineering live attenuated vaccines: Old dogs learning new tricks. J Transl Autoimmun 2023;6. Elsevier B.V.
[13] Hoffman RM. Tumor-seeking Salmonella amino acid auxotrophs. Curr Opin Biotechnol 2011;22(6):917–23.
[14] Zhao M, Geller J, Ma H, Yang M, Penman S, Hoffman RM. Monotherapy with a tumor-targeting mutant of *Salmonella typhimurium* cures orthotopic metastatic mouse models of human prostate cancer. Proc Natl Acad Sci USA 2007;104(24):10170–4.
[15] Hwang IY, Chang MW. Engineering commensal bacteria to rewire host–microbiome interactions. Curr Opin Biotechnol 2020;62:116–22.
[16] Gurbatri CR, Lia I, Vincent R, Coker C, Castro S, Treuting PM, et al. Engineered probiotics for local tumor delivery of checkpoint blockade nanobodies. Sci Transl Med 2020;12.
[17] Nguyen DH, Chong A, Hong Y, Min JJ. Bioengineering of bacteria for cancer immunotherapy. Nat Commun 2023;14:3553.
[18] Na HS, Kim HJ, Lee HC, Hong Y, Rhee JH, Choy HE. Immune response induced by Salmonella typhimurium defective in ppGpp synthesis. Vaccine 2006;24(12):2027–34.
[19] Lee CH, Wu CL, Shiau AL. Toll-like receptor 4 mediates an antitumor host response induced by Salmonella choleraesuis. Clin Cancer Res 2008;14(6):1905–12.
[20] Lee CH, Hsieh JL, Wu CL, Hsu HC, Shiau AL. B cells are required for tumor-targeting Salmonella in host. Appl Microbiol Biotechnol 2011;92(6):1251–60.
[21] Lee CH, Hsieh JL, Wu CL, Hsu PY, Shiau AL. T cell augments the antitumor activity of tumor-targeting Salmonella. Appl Microbiol Biotechnol 2011;90(4):1381–8.
[22] Ngo HTT, Nguyen DH, You SH, Van Nguyen K, Kim SY, Hong Y, et al. Reprogramming a doxycycline-inducible gene switch system for Bacteria-mediated Cancer therapy. Mol Imaging Biol 2024;26(1):148–61.
[23] Park SH, Zheng JH, Nguyen VH, Jiang SN, Kim DY, Szardenings M, et al. RGD peptide cell-surface display enhances the targeting and therapeutic efficacy of attenuated salmonella-mediated cancer therapy. Theranostics 2016;6(10):1672–82.
[24] Li R, Helbig L, Fu J, Bian X, Herrmann J, Baumann M, et al. Expressing cytotoxic compounds in Escherichia coli Nissle 1917 for tumor-targeting therapy. Res Microbiol 2019;170(2):74–9.
[25] Romee R, Yang S, Sheffer M, Kaplan I, Wang Z, Tarannum M, et al. Nonpathogenic *E. coli* engineered to surface display cytokines as a new platform for immunotherapy. Available from: https://doi.org/10.21203/rs.3.rs-4031911/v1.
[26] Hensel ME, Garcia-Gonzalez DG, Chaki SP, Hartwig A, Gordy PW, Bowen R, et al. Vaccine candidate *Brucella melitensis* 16M ΔvjbR is safe in a pregnant sheep model and confers protection. mSphere 2020;5(3).
[27] Das JK, Mishra D, Ray P, Tripathy P, Beuria TK, Singh N, et al. In vitro evaluation of anti-infective activity of a Lactobacillus plantarum strain against Salmonella enterica serovar Enteritidis. Gut Pathog 2013;5(1).
[28] Das JK, Mahapatra RK, Patro S, Goswami C, Suar M. Lactobacillus acidophilus binds to MUC3 component of cultured intestinal epithelial cells with highest affinity. FEMS Microbiol Lett 2016;363(8).
[29] Hemarajata P, Versalovic J. Effects of probiotics on gut microbiota: mechanisms of intestinal immunomodulation and neuromodulation. Therap Adv Gastroenterol 2013;6:39–51.
[30] Park J, Kwon M, Lee J, Park S, Seo J, Roh S. Anti-cancer effects of lactobacillus plantarum l-14 cell-free extract on human malignant melanoma A375 cells. Molecules 2020;25(17).
[31] Jeong S, Kim Y, Park S, Lee D, Lee J, Hlaing SP, et al. Lactobacillus plantarum metabolites elicit anticancer effects by inhibiting autophagy-related responses. Molecules 2023;28(4).
[32] Tiptiri-Kourpeti A, Spyridopoulou K, Santarmaki V, Aindelis G, Tompoulidou E, Lamprianidou EE, et al. Lactobacillus casei exerts anti-proliferative effects accompanied by apoptotic cell death and up-regulation of TRAIL in colon carcinoma cells. PloS One 2016;11(2).
[33] Zheng JH, Nguyen VH, Jiang SN, Park SH, Tan W, Hong SH, Shin MG, Chung IJ, Hong Y, Bom HS, Choy HE, Lee SE, Rhee JH, Min JJ. Two-step enhanced cancer immunotherapy with engineered *Salmonella typhimurium* secreting heterologous flagellin. Sci Transl Med 2017;9(376):eaak9537. https://doi.org/10.1126/scitranslmed (aak9537, PMID: 28179508).

[34] Kasinskas RW, Forbes NS. Salmonella typhimurium lacking ribose chemoreceptors localize in tumor quiescence and induce apoptosis. Cancer Res 2007;67(7):3201–9.
[35] Shi Y, Zheng W, Yang K, Harris KG, Ni K, Xue L, et al. Intratumoral accumulation of gut microbiota facilitates CD47-based immunotherapy via STING signaling. J Exp Med 2020;217(5).
[36] Zhao M, Yang M, Ma H, Li X, Tan X, Li S, et al. Targeted therapy with a Salmonella typhimurium leucine-arginine auxotroph cures orthotopic human breast tumors in nude mice. Cancer Res 2006;66(15):7647–52.
[37] Massa PE, Paniccia A, Monegal A, De Marco A, Rescigno M. Salmonella engineered to express CD20-targeting antibodies and a drug-converting enzyme can eradicate human lymphomas. Blood 2013;122(5):705–14.
[38] Tran V, Liu J, Behr MA. BCG Vaccines, 2014. Available from: https://journals.asm.org/journal/spectrum.
[39] King I, Itterson M, Bermudes D. Tumor-targeted salmonella typhimurium overexpressing cytosine deaminase: a novel, tumor-selective therapy. Methods Mol Biol 2009;542.
[40] Puri RK. Development of a recombinant interleukin-4-pseudomonas exotoxin for therapy of glioblastoma. Toxicol Pathol 1999;27(1):53–7.
[41] Le DT, Wang-Gillam A, Picozzi V, Greten TF, Crocenzi T, Springett G, et al. Safety and survival with GVAX pancreas prime and Listeria monocytogenes-expressing mesothelin (CRS-207) boost vaccines for metastatic pancreatic cancer. J Clin Oncol 2015;33(12).

CHAPTER

11

Salmonella: A bacterial warrior in cancer immunotherapy

Wei Kong[a], *Lingchen Fu*[a], *and Yixin Shi*[b]

[a]Biodesign Center for Applied Structural Discovery, Arizona State University, Tempe, AZ, United States [b]School of Life Sciences, Arizona State University, Tempe, AZ, United States

Abstract

Salmonella has emerged as a promising agent in the field of cancer immunotherapy due to its unique ability to selectively target and colonize tumor tissues due to their hypoxic and nutrient-rich microenvironment, which is less prevalent in normal tissues, thereby stimulating robust antitumor immune responses. The natural propensity of *Salmonella* to invade hypoxic tumor environments and its ability to stimulate both innate and adaptive immune responses make it an attractive candidate for cancer treatment. Genetic engineering further enhances the therapeutic potential of *Salmonella* by enabling it to deliver a range of antitumor agents, including tumor-associated antigens, cytotoxic compounds, immunomodulators, pro-apoptotic proteins, and nucleic acids, which directly within the tumor microenvironment (TME), thereby minimizing systemic toxicity and enhancing local immune responses to improve efficacy. Preclinical and early-phase clinical trials have demonstrated promising results, with significant tumor regression and prolonged survival in treated subjects. The integration of *Salmonella* into cancer immunotherapy addresses the limitations of conventional treatments and offers a novel approach to overcoming tumor resistance and enhancing the overall immune response against cancer. Thus, *Salmonella* stands as a potent "bacterial warrior" in the armamentarium of cancer immunotherapy, offering new hope for more effective and targeted cancer treatments.

Abbreviations

ALDH	aldehyde dehydrogenase
APCs	antigen-presenting cells
CAR	constitutive androstane receptor
CDP	caspase-1 dependent pyroptosis
CdtB	cytolethal-distending toxin B
ClyA	cytolysin A
CMV E/P	cytomegalovirus immediate-early gene enhancer/promoter
CPG2	carboxypeptidase G2
CSC	cancer stem cell
CTLs	cytotoxic T lymphocytes

T Cell Metabolism and Cancer Immunotherapy
https://doi.org/10.1016/B978-0-443-13827-0.00012-3

DCs	dendritic cells
DTS	DNA nuclear-targeting sequences
DUB	deubiquitinase
EMT	epithelial-to-mesenchymal transition
FZD6	frizzled 6
GMS	genetically modified *Salmonella*
IFN-γ	interferon-gamma
IL-2	interleukin-2
LGR5	leucine-rich-repeat-containing G-protein-coupled receptor 5
LPS	lipid A
LPS	lipopolysaccharide
M cells	microfold cells
MDSCs	myeloid-derived suppressor cells
MHC	major histocompatibility complex
MPLA	monophosphoryl lipid A
mTOR	mammalian target of rapamycin
NF-κB	nuclear factor kappa-light-chain-enhancer of activated B cells
NK cells	natural killer cells
PAMPs	pathogen-associated molecular patterns
PD-1	programmed cell death protein 1
PD-L1	programmed cell death-ligand 1
PE38	*Pseudomonas* exotoxin A
PKB	protein kinase B
PMNs	polymorphonuclear neutrophils
PPARD	peroxisome proliferator-activated receptor delta
RGD4C	arginine-glycine-aspartic acid peptide
ROS	reactive oxygen species
S. enterica	*Salmonella enterica*
SCV	*Salmonella*-containing vacuole
SD-GMS	self-destructing genetically modified *Salmonella*
shRNA	small hairpin RNA
shRNA	small hairpin RNA
SIRS	systemic inflammatory response syndrome
SOX2	SRY-box transcription factor 2
SOX9	SRY-box transcription factor 9
SPI	*Salmonella* pathogenicity island
SRF	Science Research Fund
T3SS	Type 3 secretion system
TAAs	tumor-associated antigens
Tar	aspartate chemoreceptor
TGFα	transforming growth factor alpha
THB	tumor homing bacteria
TLR	toll-like receptors
TME	tumor microenvironment
TME	tumor microenvironment
TNF	tumor necrosis factor
TNF-α	tumor necrosis factor-alpha
TRAIL	tumor necrosis factor-related apoptosis-inducing ligand
Tregs	regulatory T cells
Trg	ribose chemoreceptor
Tsr	serine chemoreceptor

Conflict statement

The authors declare no conflict of interest.

Introduction

In recent years, the utilization of *Salmonella enterica* in cancer immunotherapy has emerged as a captivating frontier. The ability of *Salmonella* to selectively target tumor tissues and activate immune responses has positioned it as a promising candidate for novel therapeutic interventions.

The concept of using bacteria to treat cancer dates to the late 19th century, when William Coley observed tumor regression in patients infected with *Streptococcus pyogenes* [1]. Since then, advancements in genetic engineering have enabled the development of attenuated *Salmonella* strains tailored for cancer therapy [2–9]. These engineered bacteria possess an intrinsic tropism for hypoxic and necrotic regions within tumors, exploiting the unique microenvironment of cancerous tissues for selective colonization and propagation.

Beyond its tumor-targeting capabilities, the interaction of *Salmonella* with the host immune system plays a pivotal role in its therapeutic efficacy. By eliciting both innate and adaptive immune responses, *Salmonella* stimulates the production of pro-inflammatory cytokines, activation of antigen-presenting cells, and recruitment of cytotoxic T lymphocytes, culminating in tumor regression and enhanced antitumor immunity [10–20].

However, the translation of *Salmonella*-based cancer immunotherapy from bench to tumor bed is not devoid of challenges. Safety concerns regarding bacterial dissemination and systemic toxicity necessitate meticulous optimization of bacterial strains and treatment protocols. Additionally, the intricate interplay between *Salmonella,* the TME, and host immune responses underscores the importance of deciphering the underlying mechanisms to maximize therapeutic outcomes.

In this chapter, we provide an overview of the current landscape of *S. enterica* and the continuous development and innovation in *Salmonella*-based cancer therapies, aiming to shed light on the potential of *Salmonella*-based strategies in revolutionizing cancer treatment paradigms.

The oncolytic potential of *Salmonella*

Salmonella-based cancer immunotherapy is an innovative approach that leverages the unique properties of bacteria to target and eradicate tumors. *Salmonella,* a facultative anaerobe, preferentially colonizes acidic, hypoxic, and necrotic TME, potentially by sensing specific chemicals released by tumor cells, which guide *Salmonella* toward tumor sites [11,21,22]. The accumulation of *S. enterica* within tumors has also been associated with the bacterium forming biofilms as a protective measure against phagocytosis by tumor or immune cells and with dampened innate and adaptive antitumor immune responses in tumor beds [23,24].

After colonization in the TME, *Salmonella* triggers mechanisms aimed at eliminating the cancer cells, including decrease of angiogenesis [25], activation of autophagy [26], and release of antitumor bacterial toxins [27,28]. The ability of *Salmonella* to reprogram the TME and compete with tumor cells for nutrients further contributes to its anticancer efficacy [11,29]. The metabolic competition between *Salmonella* and tumor cells for nutrients disrupts the tumor's metabolic advantage, potentially reprogramming immune cells to regain their antitumor capacity [30]. Also, *Salmonella*-based therapies have been effective in reducing metastasis by inhibiting the epithelial-to-mesenchymal transition and suppressing key signaling pathways like the protein kinase B/mammalian target of rapamycin, which are crucial for cancer cell migration and invasion [31–33].

The antitumor activity of *Salmonella* has been demonstrated in various preclinical studies. Administration of *S. enterica* reduced tumor growth and increased survival in various cancer models, including colorectal cancer [13,34,35], breast cancer [36], prostate cancer [37,38], pancreatic cancer [10], melanoma [39,40], lymphoma [41,42], and others. The antitumor activity of *S. enterica* VNP20009 strain has also been evaluated in clinical trials of solid tumors with no side effects, although only modest colonization with no significant antineoplastic activity [6].

Immunomodulatory effects of *Salmonella*

Salmonella can activate both innate and adaptive immune responses, which is crucial for its antitumor activities, making it a potent immunomodulatory agent [32]. Internalized *Salmonella* can infect nonphagocytic cells through the type 3 secretion system (T3SS), which may promote the elimination of tumor cells by presenting bacterial antigens on their surface, recognizable by *Salmonella*-specific T lymphocytes [43].

Salmonella activates anticancer innate immune responses through several mechanisms that involve both direct and indirect stimulation of the immune system. One key mechanism is the activation of natural killer (NK) cells, which produce interferon-gamma (IFN-γ) during early *Salmonella* infection. IFN-γ promotes the accumulation, activation, and cytotoxicity of NK cells, which directly eliminate cancer cells, thereby blocking metastasis [44,45]. Also, the ability of *Salmonella* to activate Toll-like receptors (TLR) and other immune pathways further amplifies its antitumor effects, even in immunosuppressive tumor microenvironments [46]. Toll-like receptors recognize bacterial components, including pathogen-associated molecular pattern molecules, lipopolysaccharide, flagellin, and CpG sequences, to activate signaling pathways involved in the innate and adaptive immune responses. Flagellin, recognized through TLR5, suppresses tumor cell proliferation via TLR5 and induces an antitumor response mediated by NK cells [47,48]. The binding of flagellin to TLR5 activates the NF-κ B signaling pathway [49], increasing levels of TNF-α-mediated apoptosis [50]. Furthermore, LPS binds to TLR4 and induces TNF-α expression allowing the transformation of immunosuppressive myeloid-derived suppressor cells into TNF-α-secreting cells with neutrophil characteristics, reducing regulatory T cells and enhancing tumor-specific cytotoxic T lymphocytes [7,51–54], lineages to infiltrate and eliminate tumor cells.

Salmonella can also promote changes in the tumor microenvironment and activate the antitumor adaptive immune responses that turn an immunosuppressed into an immunogenic phenotype [55]. This involves breaking immune tolerance by reducing T-regulatory lymphocyte levels via LPS and lipoprotein of *Salmonella* [56]. Consequently, tumor cells infected by *Salmonella* can present bacterial antigens to T lymphocytes, inducing the elimination of transformed cells in various solid tumors [57,58]. Also, in the TME, *Salmonella* activates cells involved in the adaptive immune response, such as B and T lymphocytes and dendritic cells, facilitating cross-presentation of tumor antigens [59,60]. In addition, *Salmonella* infection upregulates connexin 43 [61], a protein involved in forming gap junctions between tumor cells and DCs, allowing transfer of tumor cell preprocessed peptides to DCs, inducing cross-presentation and cross-priming of specific tumor antigens [62].

Using *Salmonella* as a delivery system in cancer therapy

The ability of *Salmonella* to colonize tumors and its natural tropism for hypoxic environments make it an effective vector for delivering therapeutic molecules. The unique characters of the TME, hypoxia, necrosis, acidity, and immune response suppression, allow the migration, infection, and prolonged presence of *Salmonella* within the TME to activate its intrinsic oncolytic properties, break immune tolerance, and induce tumor cell elimination. Once *Salmonella* reaches the tumor, it acts as a factory producing antitumor heterologous molecules [59,63,64]. Advances in synthetic biology and genetic engineering have enabled the modification of *Salmonella* to minimize toxicity and improve tumor-targeting properties, thereby enhancing safety and efficacy [11,65,66]. The safety and efficacy of these approaches are supported by the well-understood physiology of *Salmonella,* which allows for precise genetic modifications to balance attenuation and immunogenicity [67]. For instance, engineered strains of *Salmonella* have been used to deliver prodrug-converting enzymes, cytokines, and tumor antigens, which can trigger a robust antitumor immune response and inhibit tumor growth and metastasis [68–71].

Salmonella has been genetically engineered to express and deliver tumor-associated antigens (TAAs) effectively and directly to antigen-presenting cells (APCs) such as dendritic cells and macrophages, thereby stimulating robust cellular, humoral, and mucosal immune responses [11,72]. The use of *Salmonella* T3SS allows for the efficient translocation of TAAs into the cytosol of professional APCs, enhancing antigen presentation and subsequent T cell activation [14,73–77]. Additionally, modifications such as the truncation of lipopolysaccharides and the display of targeting peptides like arginine-glycine-aspartic acid peptide (RGD4C) have been shown to improve the tumor-targeting capabilities of *Salmonella,* leading to more effective delivery of therapeutic agents like endostatin and tumor necrosis factor (TNF)-related apoptosis-inducing ligand (TRAIL) [44,78]. Moreover, the engineered *Salmonella* strain Saltikva, which carries the gene for human IL-2, has shown promising results in preclinical and early-phase clinical trials, demonstrating significant tumor colonization and immune activation without pathogenicity [79–81]. The versatility of *Salmonella* is also highlighted by its capacity to deliver antigens via both major histocompatibility complex (MHC) MHC-I and MHC-II pathways, potentially overcoming immune evasion strategies employed by tumors [53].

Attenuated *Salmonella* strains have been used to deliver antitumor cytotoxic agents selectively and effectively in tumor sites. *S. enterica* serovar Gallinarum, for instance, has been engineered to express transforming growth factor α (TGFα) fused to *Pseudomonas* exotoxin A (PE38), showing significant antitumor effects in mice without notable systemic toxicity [82]. The use of *Salmonella choleraesuis* to deliver cytolethal-distending toxin B (CdtB) has also been shown to induce apoptosis in tumor cells and reduce tumor growth in mice [83]. Engineered strains of *Salmonella* Typhimurium have also been shown to express inducible cytotoxins like cytolysin A (ClyA), significantly suppressing tumor growth in vivo [84]. Moreover, studies have used it to express and release carboxypeptidase G2 (CPG2), potentiating oncolytic effects in mouse cancer models [85]. The ability of *Salmonella* to target tumors, combined with genetic engineering to enhance its therapeutic efficacy, positions it as a viable vector for delivering antitumor cytotoxic agents.

Salmonella can be modified to express and deliver various antigens, cytokines, and other immunomodulatory molecules, enhancing both innate and adaptive immune responses against tumors. For instance, *Salmonella* has been engineered to deliver small hairpin RNA (shRNA) against the immunosuppressive molecule indoleamine 2,3-dioxygenase 1 [18], leading to significant tumor cell death and enhanced immune responses through the recruitment of reactive oxygen species (ROS)-producing polymorphonuclear neutrophils (PMNs) [32,86,87]. The ability of *Salmonella* to reduce the expression of immune checkpoint molecules like PD-L1 further supports its role in modulating the immune environment to favor antitumor activity [88–90]. Overall, the continuous development and innovation in *Salmonella*-based therapies hold significant promise for their use in cancer treatment and immunotherapy.

Recognizing the complementary nature of cancer therapies, researchers are exploring synergistic approaches that combine *Salmonella*-based immunotherapy with conventional treatments such as chemotherapy, radiotherapy, and immune checkpoint inhibitors. Furthermore, ongoing efforts to engineer *Salmonella* strains with further enhanced safety profiles, tumor-targeting capabilities, deliver efficiency, and immunomodulatory properties hold promise for expanding the therapeutic repertoire against cancer.

Genetically modified *Salmonella* (GMS) engineered for safe and efficient delivery of antitumor agents

The transition from preclinical success to clinical application has presented formidable challenges. While early-phase clinical trials have demonstrated the safety and feasibility of *Salmonella*-based immunotherapy, issues such as bacterial clearance, systemic toxicity, and delivery efficiency remain significant hurdles to overcome. Addressing these challenges will be paramount to realizing the full potential of *Salmonella* in the clinic.

The Kong lab at the Biodesign Institute, Arizona State University, created self-destructing GMS that are designed to undergo programmed lysis after fulfilling their intended function, such as delivering vaccines or therapeutic agents [91–93]. These self-destructing GMS strains are developed to efficiently colonize host tissues and then self-destruct to release their contents, thereby stimulating mucosal, systemic, and cellular immunities against various pathogens [91–93], while minimizing the risk of persistent infection or environmental contamination. This innovative approach represents a significant advancement in the field of vaccine delivery and therapeutic interventions, offering a versatile and safe platform for combating a wide range of diseases.

This system utilizes *Salmonella* strains with a deletion of the *asdA* gene and arabinose-regulated expression of the *murA* gene, both essential for the synthesis of the peptidoglycan layer of bacterial cell wall. Additional mutations are incorporated to enhance bacterial cell lysis and antigen/therapeutic delivery. The lysis vector pYA3681 (Fig. 1) collaborates with its host self-destructing GMS strain to enable arabinose-dependent bacterial cell wall synthesis required for bacterial reproduction [91]. Once the bacteria invade host tissues, an arabinose-free environment, cell wall synthesis eventually stops. This cessation leads to bacterial cell lysis and the release of bacterial cell contents (such as therapeutic proteins or vaccines) after the bacteria accumulate in host tissues (Fig. 2). Ultimately, this process ensures the self-destructing of *Salmonella*, thereby facilitating bacterial clearance and increasing delivery efficiency [91–93]. To enhance safety by dismissing systemic toxicity, the toxic Lipid A component of lipopolysaccharide (LPS) was modified [94,95]. The inner membrane phosphatase

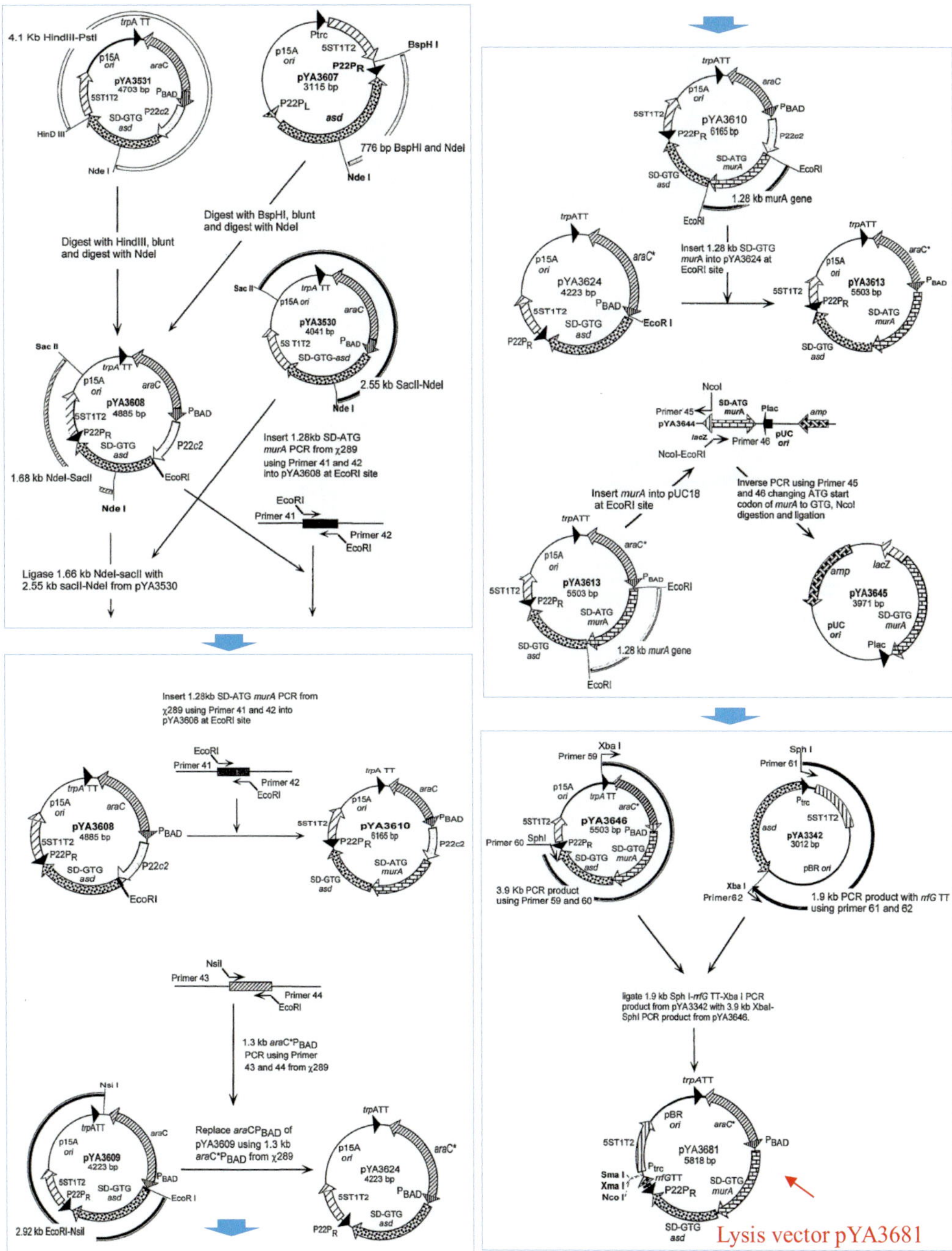

FIG. 1 Diagram of a long journey for construction of lysis vector pYA3681. Maps of plasmids used to construct pYA3681, a critical element in developing self-destructing GMS strains. Plasmid sequences include the *trpA, rrfG,* and 5S ribosomal RNA transcriptional terminators; the P_{BAD}, P_{trc}, and P22 P_R promoters; the *araC* gene; and the start codon-modified *murA* and *asdA* genes.

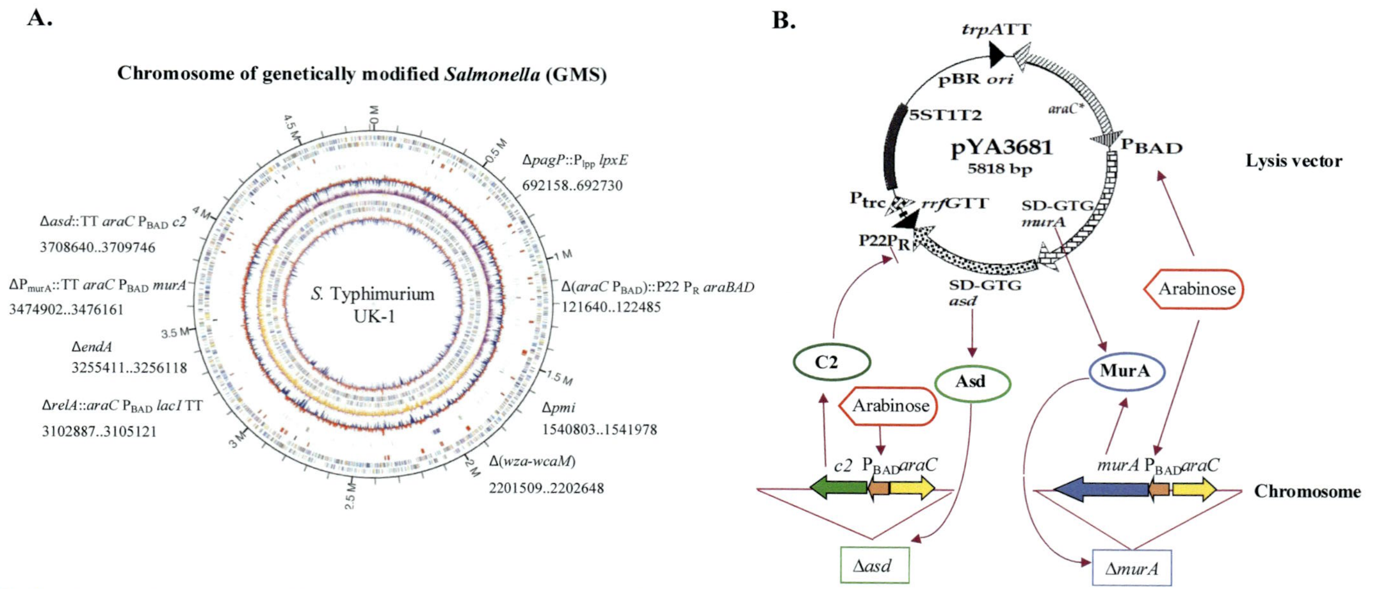

FIG. 2 Genetically modified self-destructing *Salmonella* (GMS). (A) Genetic attributes, deletions, and deletion-insertions, included in self-destructing GMS chromosome. The parent strain of self-destructing GMS is the *S*. Typhimurium UK-1 strain. (B) Diagram of the model depicting the regulatory interactions within the self-destructing GMS.

LpxE from *Francisella tularensis* can selectively remove the 1-phosphate group of lipid A in living *Salmonella* cells, producing an analog of monophosphoryl lipid A (MPLA) [96]. MPLA, a TLR4 agonist and adjuvant, enhances vaccine efficacy, induces dendritic cell maturation, primarily stimulates a Th1 response, lowers the activation threshold for Th1 cells, and upregulates MHC class II molecules [97–99]. Consequently, an LpxE-expressing cassette was introduced into the self-destructing GMS strain, retaining or enhancing immunostimulatory benefits while reducing toxicity.

Additionally, these self-destructing GMS strains could be engineered to deliver DNA vaccines, enhancing the delivery and expression of protective antigens or therapeutics within host cells [93]. This system includes a DNA vaccine vector with the same regulatory domain as pYA3681 but uses a eukaryotic expression promoter. Nuclear import of plasmid DNA relies on DNA nuclear-targeting sequences (DTS), which often contain transcription factor binding sites. The SV40 enhancer is an excellent DTS, requiring a single 72-bp element combined with the CMV immediate-early gene enhancer/promoter (CMV E/P) for functionality. *Salmonella* infection stimulates the synthesis of eukaryotic transcription factors like NF-κB and AP-2, which can bind to nonviral DNA vaccine vectors and facilitate their nuclear import. To enhance the efficiency of DNA vaccine vector delivery, the 72-bp DTS of the SV40 enhancer and the BGH polyadenylation signal were included. Additionally, artificial DTS for NF-κB and AP-2 are inserted upstream of CMV E/P, creating the improved DNA vaccine vector pYA4545. This vector allows rapid nuclear import and high-level synthesis of encoded genes. Moreover, pYA4545 contains 24 immune-enhancing CpG motifs, which may enhance the innate inflammatory response.

When *Salmonella* invades host cells, it typically resides in an endosomal compartment known as the *Salmonella*-containing vacuole (SCV). To facilitate the escape of *Salmonella* into the cytosol, the *sifA* gene, which encodes a Type III secreted effector protein responsible for converting SCVs into filaments, is deleted. This inactivation allows *Salmonella* to exit the endosome and rapidly multiply in the cytoplasm. Also, *S.* Typhimurium invades host cells by traversing the epithelial monolayer through microfold (M) cells. The *hilA* regulator, part of the OmpR/ToxR family, activates invasion genes in response to environmental and genetic factors. To enhance M cell-mediated invasion, the *hilA* promoter is replaced with the P_{trc} promoter, enabling constitutive synthesis of HilA. Moreover, invasive *Salmonella* induces pyroptosis/apoptosis in macrophages, which can hinder DNA vaccine transcription after nuclear trafficking. *Salmonella* strains expressing the SPI-1 T3SS and flagellin trigger caspase-1-dependent pyroptosis, while mutations in SPI-1 T3SS genes prevent cytotoxicity. Additionally, the SPI-2 T3SS is involved in delayed macrophage death, and SseL, a *Salmonella* deubiquitinase, is crucial for this cytotoxicity. Strains with Δ*sseL* mutations exhibit reduced cytotoxic effects. TlpA, a temperature-sensing protein with a long α-helical coiled coil motif, also contributes to host cell apoptosis, and Δ*tlpA* mutations reduce this effect. With these additional genetic attributes, these self-destructing GMS strains enable immediate escape from the endosome compartment into the cytosol and display hyper-invasive features with reduced pyroptosis/apoptosis-inducing phenotype for efficient DNA vaccine delivery.

The technology also extends to cancer treatment [100–102]. Tumors often release small molecules that attract *Salmonella* [103,104]. The aspartate chemoreceptor (Tar) triggers chemotaxis toward tumor cylindroids, the serine chemoreceptor (Tsr) enables cylindroid penetration, and the ribose chemoreceptor (Trg) directs *Salmonella* to the necrotic areas [104]. The

Kong lab reprogrammed the chemotaxis system to create tumor-targeting and self-destructing GMS strains for therapeutic and DNA vaccine delivery, transforming self-destructing *Salmonella* into universal tumor-targeting delivery vehicles for cancer therapy. The reprogrammed chemotaxis system endows self-destructing GMS strains with superior tumor homing, cancer cell targeting, cellular attachment, and invasion capabilities compared to their parent strain, which lacks the genetically engineered chemotaxis system. Advancements in understanding the mechanisms of programmed cell death have led to new agents capable of reactivating apoptosis in malignant cells. The extrinsic apoptosis pathway is triggered by TRAIL [105–108]. The Kong lab has successfully delivered TRAIL using their tumor-targeting and self-destructing GMS strains. These strains actively seek out tumors (both primary and metastatic), colonize them, and release cancer-killing materials, resulting in tumor cell death, antigen release, innate immune cell attraction and activation, and a potent antitumor immune response [100].

Exploring new anticancer targets associated with PD-L1 for tumor-targeting and self-destructing GMS-mediated cancer immunotherapy

Among the notable advancements of cancer therapy is the targeting of programmed cell death-ligand 1 (PD-L1), a key player in immune regulation and a prominent feature in various cancer types. PD-L1, often overexpressed in tumor cells, serves as a shield against immune attack by binding to its receptor PD-1 on T cells, thereby suppressing their activity. This immune evasion mechanism has spurred intense interest in developing therapies that disrupt the PD-L1/PD-1 axis, unleashing the immune system to mount an effective antitumor response. By harnessing the power of the immune system to combat malignancies, PD-L1-targeted therapies offer renewed hope for patients battling cancer and pave the way toward personalized and precision medicine approaches [109–112]. Combining *Salmonella* with PD-L1 blockade has shown promising results in preclinical models, significantly inhibiting tumor growth and increasing effector T cell infiltration [69,113–115]. Targeting PD-L1 with *Salmonella*-mediated approaches offers a novel and potentially effective strategy for enhancing anticancer immune responses and overcoming tumor immune evasion.

The Kong lab aims to explore novel anticancer targets associated with PD-L1 for tumor-targeting and self-destructing GMS-mediated cancer immunotherapy. Cancer stem cells (CSCs) are pivotal in tumor initiation, progression, metastasis, and resistance to treatment [31,116–119]. Various proteins, including leucine-rich-repeat-containing G-protein-coupled receptor 5 (LgR5), CD44, ALDH, SOX2, SOX9, and PPARD, are recognized as stem cell markers due to their heightened expression in CSCs [120–125]. The β-catenin signaling pathway is crucial for maintaining CSC stemness and proliferation by regulating the expression of several stem cell markers [126–131]. While LgR5 is a well-known stem cell marker influenced by the Wnt/β-catenin pathway, its precise regulatory mechanisms remain largely unexplored. The activation of Wnt/β-catenin signaling via LgR5 has been observed in both normal and cancerous stem cells [116,132–134]. Moreover, receptor Frizzled 6 (FZD6) acts as a receptor for Wnt proteins, initiating β-catenin signaling activation, although its other ligands are not extensively studied. Elevated levels of FZD6 have been noted in various cancer types, including colorectal and breast cancers [135–137]. Additionally, PD-L1 was found to regulate lung cancer progression by impacting WIP and β-catenin signaling [138]. While some studies suggest the involvement of PD-L1 in CSC maintenance, its precise role and underlying mechanisms remain elusive [139–144].

The Kong lab discovered that PD-L1 interacts with FZD6 to activate β-catenin signaling and increase β-catenin targeted gene expression, such as a putative stem cell marker LgR5 [145]. Blockage of PD-L1 function, using a specific small hairpin RNA or a specific antibody, inhibits disease progression by reducing the CSC population in both colorectal and breast tumors. Moreover, β-catenin conversely regulates PD-L1 expression through a β-catenin complex binding site in the PD-L1 promoter [145]. Their discoveries reveal that besides assistant tumor cell immune escaping, PD-L1 and β-catenin signaling form a positive feedback loop to promote cancer progression through CSC maintenance and expansion (Fig. 3) [145]. These findings shed light on the intricate mechanisms underlying the novel function of PD-L1 in CSC regulation in colorectal and breast cancers. These newly uncovered anticancer targets associated with PD-L1 will be perfect therapeutic candidates for tumor-targeting and self-destructing GMS-mediated cancer immunotherapy. The Kong lab is currently evaluating the selected PD-L1-associated anticancer targets in preclinical animal models.

Clinical trials and translational challenges

Salmonella-based cancer therapy has shown promising potential in preclinical studies, but its clinical application remains in the early stages. Clinical trials have faced challenges. For example, the genetically modified strain VNP20009 entered phase I clinical trials but showed higher toxicity and limited efficacy compared to preclinical expectations [6]. Nonetheless, the ΔppGpp strain has shown reduced toxicity and enhanced immune responses, suggesting a safer and more effective alternative for future clinical applications [6]. Furthermore, the promising results in clinical trials were reported with VNP20009 TAPET-CD expressing cytosine deaminase [146–148]. A phase I clinical trial result utilizing an *S.* Typhimurium strain Saltikva, engineered to express human IL-2, an immune stimulant with anticancer properties, showed no adverse effects at the designed dose, with an observed increase in circulating NK cell response, suggesting a potential immune-mediated therapeutic effect [80]. The ongoing Phase II clinical trial will be orally administered in combination with standard treatment to patients with metastatic stage IV pancreatic cancer [149].

Although *S.* Typhimurium has shown potential in selectively colonizing tumors and inducing immune responses, its clinical translation is hindered by the need for more rigorous preclinical investigations to ensure safety and efficacy. One significant hurdle is the incomplete understanding of the mechanisms and side effects associated with microbial cancer therapy, which complicates the development of safe and effective therapeutic strains. Issues such as bacterial clearance, systemic toxicity, and patient heterogeneity remain significant obstacles to overcome. The ability of *Salmonella* to target metastasis, which accounts for most cancer-related deaths, is promising but requires further validation in clinical settings. The combination of conventional therapies like chemotherapy and radiotherapy has shown synergistic effects, but the optimization of such combinatorial approaches for clinical use remains complex. Moreover, the immunomodulatory effects of *Salmonella,* which involve both innate and adaptive immune responses, need to be better understood to harness their full

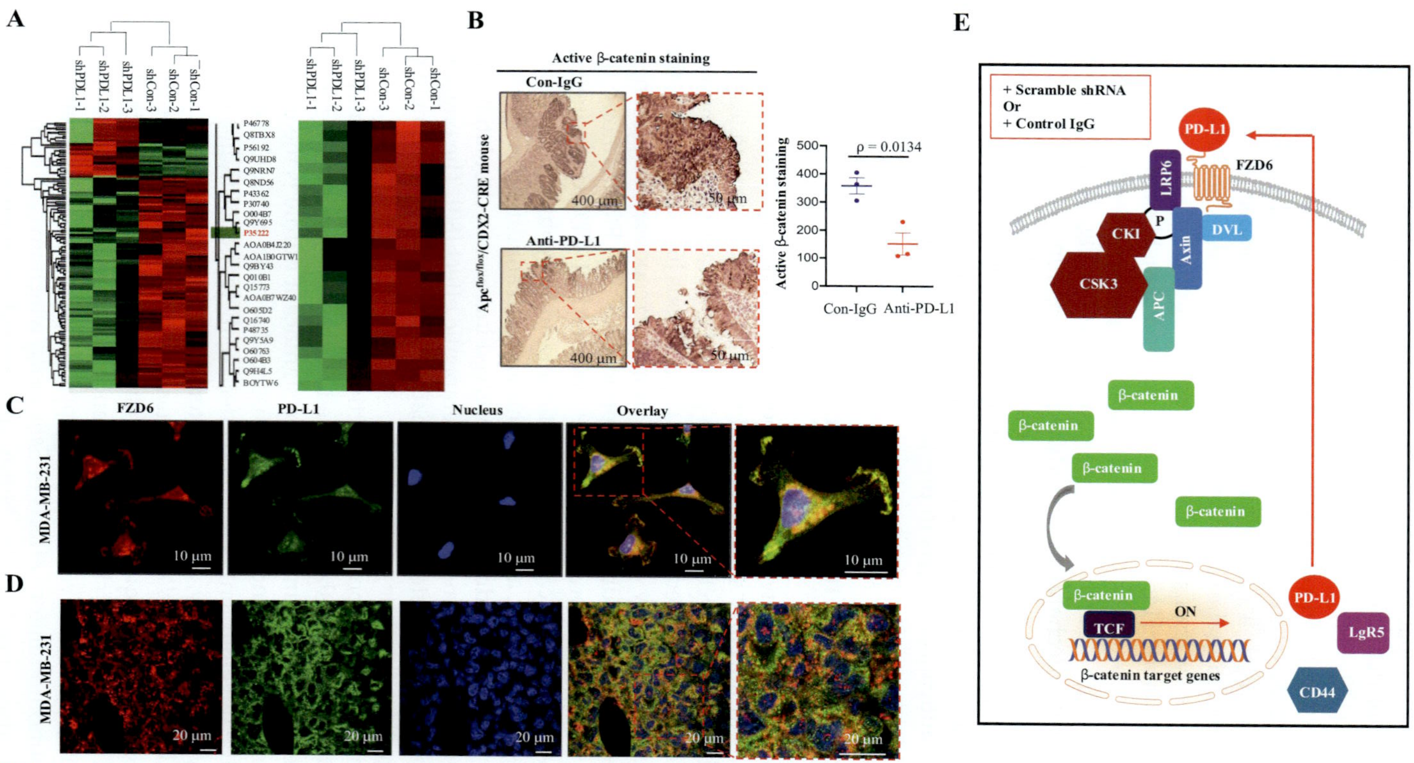

FIG. 3 PD-L1 interacts with Frizzled 6 to activate β-catenin and form a positive feedback loop to promote cancer stem cell expansion. (A) Hierarchical clustering for significant proteins. The abundances were z-score transformed and were displayed as colors ranging from *green* to *red* as shown in the key. The distribution of the z scores lies in between −2 to+1.5 (left panel). The heatmap is zoomed in to show the expression of β-catenin (P35222) in different samples (right panel). (B) Immunohistochemistry staining of activated β-catenin in the colon from $Apc^{flox/flox}$/CDX2-CRE mice treated with control IgG or anti-PD-L1 antibody (left panel). Activated β-catenin staining was quantified (right panel, $N = 3$). (C) Representative immunofluorescent images of MDA-MB-231 cells stained with anti-FDZ6 and anti-PD-L1, and then counterstained with Alexa Fluor 488 and Alexa Fluor 569 conjugated secondary antibodies, respectably. Images were created using confocal microscopy. Scale bars represent 10 mm. (D) Representative images of primary tumor tissue from MDA-MB-231/shCon NSG™ mice were stained with anti-FZD6 and anti-PD-L1 antibodies, and then were counterstained with Alexa Fluor 488 and Alexa Fluor 569 conjugated secondary antibodies respectably. Images were obtained using confocal microscopy. Scale bars represent 20 mm. (E) A graphical overview of the discovery that PD-L1 interacts with Frizzled 6 to activate β-catenin and form a positive feedback loop to promote cancer stem cell expansion.

therapeutic potential. The results from early-phase clinical trials have been unsatisfactory, highlighting the need for more comprehensive studies to address these translational challenges. Finally, the potential of *Salmonella* to act as a delivery vector for anticancer molecules is promising, but the precise control over its therapeutic actions and the prevention of unintended effects are critical areas that require further research. Addressing these challenges will be paramount to realizing the full potential of *Salmonella* in the clinic.

Future directions

The future directions of *Salmonella*-based cancer therapy are promising and multifaceted, focusing on enhancing its efficacy, safety, and applicability. One key area is the genetic engineering of *Salmonella* strains to improve tumor targeting and reduce virulence, thereby increasing safety and specificity for cancer cells. Additionally, combining *Salmonella* with other immunostimulatory agents has shown superior tumor control and enhanced T cell-mediated antitumor immunity, suggesting that combinatorial approaches could be highly effective. The role of innate immune responses is crucial for metastasis suppression, indicating that future therapies could focus on harnessing these immune pathways. Moreover, the potential of *Salmonella* to deliver a variety of therapeutic agents, including TAAs, cytotoxic molecules, and immunomodulatory proteins, opens avenues for personalized and targeted cancer treatments. Despite the promising preclinical results, more research is needed to address the metabolic defects observed in T cells during bacterial cancer therapy and to optimize the balance between immune activation and tumor suppression. Clinical trials and further preclinical investigations are essential to validate these approaches and ensure their safety and efficacy in humans. Overall, the continuous development and innovation in *Salmonella*-based cancer therapy hold significant potential for it to become a mainstream therapeutic intervention, addressing the limitations of current cancer treatments and improving patient outcomes.

Conclusion

Salmonella-based cancer therapy represents a promising and innovative approach in the field of cancer therapy, leveraging the unique properties of *S.* Typhimurium to target and treat tumors. This therapy exploits the bacterium's ability to selectively colonize tumor tissues, where it can exert direct oncolytic effects and modulate the immune system to enhance antitumor responses. Additionally, *Salmonella* can be genetically modified to deliver therapeutic agents, such as tumor-associated antigens, antitumor cytotoxic agents, and other immunomodulatory molecules, directly to the tumor microenvironment, thereby enhancing the specificity and efficacy of the treatment. Despite the promising preclinical results, clinical trials have faced challenges, including low tumor regression and dose-dependent side effects, highlighting the need for further optimization and combination with other therapeutic modalities like chemotherapy, radiotherapy, and immunotherapy to improve outcomes. The continuous development and innovation in this field, including the use of attenuated and genetically engineered strains, hold the potential to overcome current limitations and establish *Salmonella*-based cancer therapy as a mainstream treatment option in the future.

Overall, while more research and clinical validation are required, the unique advantages of *Salmonella*, such as high tumor specificity, deep tissue penetration, and engineering plasticity, make it a compelling candidate for advancing cancer treatment strategies.

References

[1] Coley II WB. Contribution to the knowledge of sarcoma. Ann Surg 1891;14(3):199–220.
[2] Eisenstein TK, et al. Immunotherapy of a plasmacytoma with attenuated Salmonella. Med Oncol 1995;12(2):103–8.
[3] Saltzman DA, et al. Attenuated *Salmonella typhimurium* containing interleukin-2 decreases MC-38 hepatic metastases: a novel anti-tumor agent. Cancer Biother Radiopharm 1996;11(2):145–53.
[4] Medina E, et al. *Salmonella* vaccine carrier strains: effective delivery system to trigger anti-tumor immunity by oral route. Eur J Immunol 1999;29(2):693–9.
[5] Clairmont C, et al. Biodistribution and genetic stability of the novel antitumor agent VNP20009, a genetically modified strain of *Salmonella typhimurium*. J Infect Dis 2000;181(6):1996–2002.
[6] Toso JF, et al. Phase I study of the intravenous administration of attenuated *Salmonella typhimurium* to patients with metastatic melanoma. J Clin Oncol 2002;20(1):142–52.
[7] Zhang L, et al. Intratumoral delivery and suppression of prostate tumor growth by attenuated *Salmonella enterica serovar typhimurium* carrying plasmid-based small interfering RNAs. Cancer Res 2007;67(12):5859–64.
[8] Lewen S, et al. A Legumain-based minigene vaccine targets the tumor stroma and suppresses breast cancer growth and angiogenesis. Cancer Immunol Immunother 2008;57(4):507–15.
[9] al-Ramadi BK, et al. Potent anti-tumor activity of systemically-administered IL2-expressing *Salmonella* correlates with decreased angiogenesis and enhanced tumor apoptosis. Clin Immunol 2009;130(1):89–97.
[10] Nagakura C, et al. Efficacy of a genetically-modified *Salmonella typhimurium* in an orthotopic human pancreatic cancer in nude mice. Anticancer Res 2009;29(6):1873–8.
[11] Forbes NS. Engineering the perfect (bacterial) cancer therapy. Nat Rev Cancer 2010;10(11):785–94.
[12] Vendrell A, et al. A novel *Salmonella* Typhi-based immunotherapy promotes tumor killing via an antitumor Th1-type cellular immune response and neutrophil activation in a mouse model of breast cancer. Vaccine 2011;29(4):728–36.
[13] Crull K, Bumann D, Weiss S. Influence of infection route and virulence factors on colonization of solid tumors by *Salmonella enterica* serovar Typhimurium. FEMS Immunol Med Microbiol 2011;62(1):75–83.
[14] Roider E, et al. Invasion and destruction of a murine fibrosarcoma by *Salmonella*-induced effector CD8 T cells as a therapeutic intervention against cancer. Cancer Immunol Immunother 2011;60(3):371–80.
[15] Yu B, et al. Explicit hypoxia targeting with tumor suppression by creating an "obligate" anaerobic *Salmonella Typhimurium* strain. Sci Rep 2012;2:436.
[16] Cheng X, et al. A *Salmonella Typhimurium* mutant strain capable of RNAi delivery: higher tumor-targeting and lower toxicity. Cancer Biol Ther 2014;15(8):1068–76.
[17] Frahm M, et al. Efficiency of conditionally attenuated *Salmonella enterica* serovar Typhimurium in bacterium-mediated tumor therapy. MBio 2015;6(2), e00254-15.
[18] Chavez-Navarro H, et al. *Salmonella enterica*: an ally in the therapy of cancer. Bol Med Hosp Infant Mex 2015;72(1):15–25.
[19] Zhang Y, et al. Tumor-targeting *Salmonella typhimurium* A1-R arrests growth of breast-cancer brain metastasis. Oncotarget 2015;6(5):2615–22.
[20] Felgner S, et al. aroA-deficient *Salmonella enterica* Serovar Typhimurium is more than a metabolically attenuated mutant. MBio 2016;7(5), e1220-16.
[21] Kasinskas RW, Forbes NS. *Salmonella typhimurium* lacking ribose chemoreceptors localize in tumor quiescence and induce apoptosis. Cancer Res 2007;67(7):3201–9.
[22] Wei MQ, et al. Facultative or obligate anaerobic bacteria have the potential for multimodality therapy of solid tumours. Eur J Cancer 2007;43(3):490–6.
[23] Crull K, et al. Biofilm formation by *Salmonella enterica* serovar Typhimurium colonizing solid tumours. Cell Microbiol 2011;13(8):1223–33.

[24] Kaimala S, et al. *Salmonella*-mediated tumor regression involves targeting of tumor myeloid suppressor cells causing a shift to M1-like phenotype and reduction in suppressive capacity. Cancer Immunol Immunother 2014;63(6):587–99.

[25] Tu DG, et al. *Salmonella* inhibits tumor angiogenesis by downregulation of vascular endothelial growth factor. Oncotarget 2016;7(25):37513–23.

[26] Lee CH, et al. *Salmonella* induce autophagy in melanoma by the downregulation of AKT/mTOR pathway. Gene Ther 2014;21(3):309–16.

[27] Brown JM, Wilson WR. Exploiting tumour hypoxia in cancer treatment. Nat Rev Cancer 2004;4(6):437–47.

[28] Barak Y, et al. Role of nitric oxide in *Salmonella typhimurium*-mediated cancer cell killing. BMC Cancer 2010;10:146.

[29] Coenon L, et al. Antitumoral microorganisms: the Swiss army knife of immunotherapy. Med Sci (Paris) 2021; 37(1):47–52.

[30] Faraj JA, et al. Reprogramming the tumor microenvironment to improve the efficacy of cancer immunotherapies. Med Oncol 2022;39(12):239.

[31] Lawson DA, Bhakta NR, Kessenbrock K, Prummel KD, Yu Y, Takai K, Zhou A, Eyob H, Balakrishnan S, Wang C-Y, Yaswen P, Goga A, Werb Z. Single-cell analysis reveals a stem-cell program in human metastatic breast cancer cells. Nature 2015;526(7571):131–5.

[32] Pangilinan CR, Lee CH. Highlights of immunomodulation in *Salmonella*-based cancer therapy. Biomedicine 2021;9(11):1566.

[33] Tsao YT, et al. Downregulations of AKT/mTOR signaling pathway for *Salmonella*-mediated suppression of matrix metalloproteinases-9 expression in mouse tumor models. Int J Mol Sci 2018;19(6):1630.

[34] Raman V, et al. The motility regulator flhDC drives intracellular accumulation and tumor colonization of *Salmonella*. J Immunother Cancer 2019;7(1):44.

[35] Yun M, et al. Effect of *Salmonella* treatment on an implanted tumor (CT26) in a mouse model. J Microbiol 2012;50 (3):502–10.

[36] Zhao M, et al. Targeted therapy with a *Salmonella typhimurium* leucine-arginine auxotroph cures orthotopic human breast tumors in nude mice. Cancer Res 2006;66(15):7647–52.

[37] Zhao M, et al. Tumor-targeting bacterial therapy with amino acid auxotrophs of GFP-expressing *Salmonella typhimurium*. Proc Natl Acad Sci USA 2005;102(3):755–60.

[38] Choe E, Kazmierczak RA, Eisenstark A. Phenotypic evolution of therapeutic *Salmonella enterica* serovar Typhimurium after invasion of TRAMP mouse prostate tumor. MBio 2014;5(4), e01182-14.

[39] Hoffman RM. Patient-derived Orthotopic xenograft (PDOX) models of melanoma. Int J Mol Sci 2017;18(9):1875.

[40] Kawaguchi K, et al. Tumor-targeting *Salmonella typhimurium* A1-R combined with temozolomide regresses malignant melanoma with a BRAF-V600E mutation in a patient-derived orthotopic xenograft (PDOX) model. Oncotarget 2016;7(52):85929–36.

[41] Vendrell A, et al. Therapeutic effects of *Salmonella typhi* in a mouse model of T-cell lymphoma. J Immunother 2013;36(3):171–80.

[42] Grille S, et al. A therapeutic vaccine using *Salmonella*-modified tumor cells combined with interleukin-2 induces enhanced antitumor immunity in B-cell lymphoma. Leuk Res 2013;37(3):341–8.

[43] Avogadri F, et al. Cancer immunotherapy based on killing of *Salmonella*-infected tumor cells. Cancer Res 2005;65(9):3920–7.

[44] Badie F, et al. Use of *Salmonella* bacteria in cancer therapy: direct, drug delivery and combination approaches. Front Oncol 2021;11, 624759.

[45] Al-Saafeen BH, Fernandez-Cabezudo MJ, Al-Ramadi BK. Integration of *Salmonella* into combination cancer therapy. Cancers (Basel) 2021;13(13):3228.

[46] Sawant SS, et al. Microbes as medicines: harnessing the power of bacteria in advancing cancer treatment. Int J Mol Sci 2020;21(20).

[47] Leigh ND, et al. A flagellin-derived toll-like receptor 5 agonist stimulates cytotoxic lymphocyte-mediated tumor immunity. PLoS One 2014;9(1), e85587.

[48] Cai Z, et al. Activation of toll-like receptor 5 on breast cancer cells by flagellin suppresses cell proliferation and tumor growth. Cancer Res 2011;71(7):2466–75.

[49] Flentie K, et al. Nucleoside diphosphate Kinase-3 (NME3) enhances TLR5-induced NFkappaB activation. Mol Cancer Res 2018;16(6):986–99.

[50] Hancz D, et al. Flagellin increases death receptor-mediated cell death in a RIP1-dependent manner. Immunol Lett 2018;193:42–50.
[51] Lee CH, Wu CL, Shiau AL. Toll-like receptor 4 mediates an antitumor host response induced by *Salmonella* choleraesuis. Clin Cancer Res 2008;14(6):1905–12.
[52] Leschner S, et al. Tumor invasion of *Salmonella enterica* serovar Typhimurium is accompanied by strong hemorrhage promoted by TNF-alpha. PLoS One 2009;4(8), e6692.
[53] Grille S, et al. *Salmonella enterica* serovar typhimurium immunotherapy for B-cell lymphoma induces broad anti-tumour immunity with therapeutic effect. Immunology 2014;143(3):428–37.
[54] Kim JE, et al. *Salmonella typhimurium* suppresses tumor growth via the pro-inflammatory cytokine interleukin-1beta. Theranostics 2015;5(12):1328–42.
[55] Chang SY, Kim YJ, Ko HJ. Potential therapeutic anti-tumor effect of a *Salmonella*-based vaccine. Hum Vaccin Immunother 2013;9(8):1654–60.
[56] Liu T, Chopra AK. An enteric pathogen *Salmonella enterica* serovar Typhimurium suppresses tumor growth by downregulating CD44high and CD4T regulatory (Treg) cell expression in mice: the critical role of lipopolysaccharide and Braun lipoprotein in modulating tumor growth. Cancer Gene Ther 2010;17(2):97–108.
[57] Hurley D, et al. *Salmonella*-host interactions - modulation of the host innate immune system. Front Immunol 2014;5:481.
[58] Yu YA, et al. Visualization of tumors and metastases in live animals with bacteria and vaccinia virus encoding light-emitting proteins. Nat Biotechnol 2004;22(3):313–20.
[59] O'Donnell H, McSorley SJ. *Salmonella* as a model for non-cognate Th1 cell stimulation. Front Immunol 2014;5:621.
[60] Kocijancic D, et al. Therapeutic benefit of *Salmonella* attributed to LPS and TNF-alpha is exhaustible and dictated by tumor susceptibility. Oncotarget 2017;8(22):36492–508.
[61] Saccheri F, et al. Bacteria-induced gap junctions in tumors favor antigen cross-presentation and antitumor immunity. Sci Transl Med 2010;2(44), 44ra57.
[62] Oviedo-Orta E, Howard Evans W. Gap junctions and connexin-mediated communication in the immune system. Biochim Biophys Acta 2004;1662(1–2):102–12.
[63] Shilling DA, et al. *Salmonella typhimurium* stimulation combined with tumour-derived heat shock proteins induces potent dendritic cell anti-tumour responses in a murine model. Clin Exp Immunol 2007;149(1):109–16.
[64] Zheng JH, Min JJ. Targeted cancer therapy using engineered *Salmonella typhimurium*. Chonnam Med J 2016; 52(3):173–84.
[65] Tang Q, et al. Current status and future directions of bacteria-based immunotherapy. Front Immunol 2022;13, 911783.
[66] Huang X, et al. Bacteria-based cancer immunotherapy. Adv Sci 2021;8(7):2003572.
[67] Bermudes D, et al. Tumour-selective *Salmonella*-based cancer therapy. Biotechnol Genet Eng Rev 2001;18: 219–33.
[68] Wang D, et al. Perspectives on oncolytic *Salmonella* in cancer immunotherapy-a promising strategy. Front Immunol 2021;12, 615930.
[69] Al-Saafeen BH, et al. Attenuated *Salmonella* potentiate PD-L1 blockade immunotherapy in a preclinical model of colorectal cancer. Front Immunol 2022;13:1017780.
[70] Xu W, et al. Bacteria-mediated tumor immunotherapy via photothermally-programmed PD1 expression. Nanoscale Adv 2022;4(6):1577–86.
[71] Newsome RC, et al. Interaction of bacterial genera associated with therapeutic response to immune checkpoint PD-1 blockade in a United States cohort. Genome Med 2022;14(1):35.
[72] Liu CC, et al. Tumour-associated antigens and their anti-cancer applications. Eur J Cancer Care 2017;26 (5). https://doi.org/10.1111/ecc.12446.
[73] Panthel K, et al. Prophylactic anti-tumor immunity against a murine fibrosarcoma triggered by the *Salmonella* type III secretion system. Microbes Infect 2006;8(9–10):2539–46.
[74] Xu X, et al. Effective cancer vaccine platform based on attenuated *Salmonella* and a type III secretion system. Cancer Res 2014;74(21):6260–70.
[75] Nishikawa H, et al. In vivo antigen delivery by a *Salmonella typhimurium* type III secretion system for therapeutic cancer vaccines. J Clin Invest 2006;116(7):1946–54.

[76] Jellbauer S, et al. CD8 T-cell induction against vascular endothelial growth factor receptor 2 by *Salmonella* for vaccination purposes against a murine melanoma. PLoS One 2012;7(4), e34214.
[77] Yoon W, et al. Engineered *Salmonella typhimurium* expressing E7 fusion protein, derived from human papillomavirus, inhibits tumor growth in cervical tumor-bearing mice. Biotechnol Lett 2014;36(2):349–56.
[78] Park SH, et al. RGD peptide cell-surface display enhances the targeting and therapeutic efficacy of attenuated *Salmonella*-mediated cancer therapy. Theranostics 2016;6(10):1672–82.
[79] Zhu X, et al. Tumor antigen delivered by *Salmonella* III secretion protein fused with heat shock protein 70 induces protection and eradication against murine melanoma. Cancer Sci 2010;101(12):2621–8.
[80] Gniadek TJ, et al. A phase I, dose escalation, single dose trial of oral attenuated *Salmonella typhimurium* containing human IL-2 in patients with metastatic gastrointestinal cancers. J Immunother 2020;43(7):217–21.
[81] Kung YJ, et al. Localization of *Salmonella* and albumin-IL-2 to the tumor microenvironment augments anticancer T cell immunity. J Biomed Sci 2022;29(1):57.
[82] Mai PT, et al. Constitutive expression of a cytotoxic anticancer protein in tumor-colonizing bacteria. Cancers (Basel) 2023;15(5):1486.
[83] Wang WK, et al. *Salmonella*-mediated cytolethal distending toxin transfer inhibits tumor growth. Hum Gene Ther 2018;29(11):1327–35.
[84] Jiang SN, et al. Engineering of bacteria for the visualization of targeted delivery of a cytolytic anticancer agent. Mol Ther 2013;21(11):1985–95.
[85] Friedlos F, et al. Attenuated *Salmonella* targets prodrug activating enzyme carboxypeptidase G2 to mouse melanoma and human breast and colon carcinomas for effective suicide gene therapy. Clin Cancer Res 2008; 14(13):4259–66.
[86] Blache CA, et al. Systemic delivery of *Salmonella typhimurium* transformed with IDO shRNA enhances intratumoral vector colonization and suppresses tumor growth. Cancer Res 2012;72(24):6447–56.
[87] Lalsiamthara J, Kim JH, Lee JH. Engineering of a rough auxotrophic mutant *Salmonella Typhimurium* for effective delivery. Oncotarget 2018;9(39):25441–57.
[88] Liu X, Hoft DF, Peng G. Tumor microenvironment metabolites directing T cell differentiation and function. Trends Immunol 2022;43(2):132–47.
[89] Garcia-Canaveras JC, Lahoz A. Tumor microenvironment-derived metabolites: a guide to find new metabolic therapeutic targets and biomarkers. Cancers 2021;13(13):3230.
[90] Li Y, Wu Y, Hu Y. Metabolites in the tumor microenvironment reprogram functions of immune effector cells through epigenetic modifications. Front Immunol 2021;12, 641883.
[91] Kong W, Wanda SY, Zhang X, Bollen W, Tinge SA, Roland KL, Curtiss III R. Regulated programmed lysis of recombinant *Salmonella* in host tissues to release protective antigens and confer biological containment. Proc Natl Acad Sci USA 2008;105(27):9361–6.
[92] Wang S, Li Y, Scarpellini G, Kong W, Shi HY, Baek CH, Gunn B, Wanda SY, Roland KL, Senechal-Willis P, Curtiss III R. *Salmonella* vaccine vectors displaying delayed antigen synthesis in vivo to enhance immunogenicity. Infect Immun 2010;78(9):3969–80.
[93] Kong W, et al. Turning self-destructing *Salmonella* into a universal DNA vaccine delivery platform. Proc Natl Acad Sci USA 2012;109(47):19414–9.
[94] Raetz CR, Whitfield C. Lipopolysaccharide endotoxins. Annu Rev Biochem 2002;71:635–700.
[95] Raetz CR, et al. Lipid A modification systems in gram-negative bacteria. Annu Rev Biochem 2007;76:295–329.
[96] Wang X, et al. MsbA transporter-dependent lipid A 1-dephosphorylation on the periplasmic surface of the inner membrane: topography of francisella novicida LpxE expressed in Escherichia coli. J Biol Chem 2004;279 (47):49470–8.
[97] Baldridge JR, Crane RT. Monophosphoryl lipid A (MPL) formulations for the next generation of vaccines. Methods 1999;19(1):103–7.
[98] Wheeler AW, Marshall JS, Ulrich JT. A Th1-inducing adjuvant, MPL, enhances antibody profiles in experimental animals suggesting it has the potential to improve the efficacy of allergy vaccines. Int Arch Allergy Immunol 2001;126(2):135–9.
[99] Vernacchio L, et al. Effect of monophosphoryl lipid A (MPL) on T-helper cells when administered as an adjuvant with pneumocococcal-CRM197 conjugate vaccine in healthy toddlers. Vaccine 2002;20(31–32):3658–67.
[100] Kong W. Tumor-navigating, self-eradicating, trail-armed *Salmonella* for precision cancer therapy; 2022.
[101] Kong W, Fu L, Shi Y, Ning B. Using tumor-navigating to modulate tumor metabolism; 2023.

[102] Kong W, Fu L, Shi Y. Tumor-delivered multi-target therapeutics for colon cancer; 2023.

[103] Kasinskas RW, Forbes NS. *Salmonella typhimurium* specifically chemotax and proliferate in heterogeneous tumor tissue in vitro. Biotechnol Bioeng 2006;94(4):710–21.

[104] Adler J. Chemotaxis in bacteria. Science 1966;153(3737):708–16.

[105] Wiley SR, et al. Identification and characterization of a new member of the TNF family that induces apoptosis. Immunity 1995;3(6):673–82.

[106] Pitti RM, et al. Induction of apoptosis by Apo-2 ligand, a new member of the tumor necrosis factor cytokine family. J Biol Chem 1996;271(22):12687–90.

[107] Walczak H, et al. Tumoricidal activity of tumor necrosis factor-related apoptosis-inducing ligand in vivo. Nat Med 1999;5(2):157–63.

[108] Ashkenazi A, et al. Safety and antitumor activity of recombinant soluble Apo2 ligand. J Clin Invest 1999; 104(2):155–62.

[109] Yin S, et al. Strategies targeting PD-L1 expression and associated opportunities for cancer combination therapy. Theranostics 2023;13(5):1520–44.

[110] Su Y, et al. Regulation of PD-L1 posttranslational modification and its application progress in tumor immunotherapy. Xi Bao Yu Fen Zi Mian Yi Xue Za Zhi 2022;38(11):1036–43.

[111] Wei R, et al. Targeting PD-L1 protein: translation, modification and transport. Curr Protein Pept Sci 2019;20 (1):82–91.

[112] Dermani FK, et al. PD-1/PD-L1 immune checkpoint: potential target for cancer therapy. J Cell Physiol 2019;234 (2):1313–25.

[113] Jia H, et al. Attenuated *Salmonella* carrying siRNA-PD-L1 and radiation combinatorial therapy induces tumor regression on HCC through T cell-mediated immuno-enhancement. Cell Death Discov 2023;9(1):318.

[114] Lu S, et al. PD-1-siRNA delivered by attenuated *Salmonella* enhances the antitumor effect of chloroquine in colon cancer. Front Immunol 2021;12, 707991.

[115] Chen MC, Pangilinan CR, Lee CH. *Salmonella* breaks tumor immune tolerance by downregulating tumor programmed death-ligand 1 expression. Cancers (Basel) 2019;12(1):57.

[116] Barker N, et al. Crypt stem cells as the cells-of-origin of intestinal cancer. Nature 2009;457(7229):608–11.

[117] Singh SK, et al. Identification of a cancer stem cell in human brain tumors. Cancer Res 2003;63(18):5821–8.

[118] Todaro M, et al. Colon cancer stem cells dictate tumor growth and resist cell death by production of interleukin-4. Cell Stem Cell 2007;1(4):389–402.

[119] Litviakov N, Ibragimova M, Tsyganov M, Kazantseva P, Deryusheva I, Pevzner A, Doroshenko A, Garbukov E, Tarabanovskaya N, Slonimskaya E. Amplifications of stemness genes and the capacity of breast tumors for metastasis. Oncotarget 2020;11(21):1988–2001.

[120] Vermeulen L, et al. Single-cell cloning of colon cancer stem cells reveals a multi-lineage differentiation capacity. Proc Natl Acad Sci USA 2008;105(36):13427–32.

[121] Hashimoto Y, et al. Malignant lymphoma originating from the earliest T-lineage precursor cell. Am J Hematol 1993;43(4):319–21.

[122] Pearce DJ, et al. Characterization of cells with a high aldehyde dehydrogenase activity from cord blood and acute myeloid leukemia samples. Stem Cells 2005;23(6):752–60.

[123] Hemmati HD, et al. Cancerous stem cells can arise from pediatric brain tumors. Proc Natl Acad Sci USA 2003;100(25):15178–83.

[124] Kawai T, et al. SOX9 is a novel cancer stem cell marker surrogated by osteopontin in human hepatocellular carcinoma. Sci Rep 2016;6:30489.

[125] Beyaz S, et al. High-fat diet enhances stemness and tumorigenicity of intestinal progenitors. Nature 2016;531 (7592):53–8.

[126] Barker N, et al. Identification of stem cells in small intestine and colon by marker gene Lgr5. Nature 2007;449 (7165):1003–7.

[127] Wielenga VJ, et al. Expression of CD44 in Apc and Tcf mutant mice implies regulation by the WNT pathway. Am J Pathol 1999;154(2):515–23.

[128] Shenoy AK, et al. Transition from colitis to cancer: high Wnt activity sustains the tumor-initiating potential of colon cancer stem cell precursors. Cancer Res 2012;72(19):5091–100.

[129] He TC, et al. PPARdelta is an APC-regulated target of nonsteroidal anti-inflammatory drugs. Cell 1999; 99(3):335–45.

[130] Van Raay TJ, et al. Frizzled 5 signaling governs the neural potential of progenitors in the developing Xenopus retina. Neuron 2005;46(1):23–36.
[131] Day TF, et al. Wnt/beta-catenin signaling in mesenchymal progenitors controls osteoblast and chondrocyte differentiation during vertebrate skeletogenesis. Dev Cell 2005;8(5):739–50.
[132] Kanwar SS, et al. The Wnt/beta-catenin pathway regulates growth and maintenance of colonospheres. Mol Cancer 2010;9:212.
[133] Sato T, et al. Single Lgr5 stem cells build crypt-villus structures in vitro without a mesenchymal niche. Nature 2009;459(7244):262–5.
[134] Fevr T, et al. Wnt/beta-catenin is essential for intestinal homeostasis and maintenance of intestinal stem cells. Mol Cell Biol 2007;27(21):7551–9.
[135] The Human Protein Atlas. FZD6; 2021. https://www.proteinatlas.org/ENSG00000164930-FZD6.
[136] Luga V, et al. Exosomes mediate stromal mobilization of autocrine Wnt-PCP signaling in breast cancer cell migration. Cell 2012;151(7):1542–56.
[137] Corda G, Sala A. Non-canonical WNT/PCP signalling in cancer: Fzd6 takes centre stage. Oncogenesis 2017;6(7), e364.
[138] Yu W, et al. PD-L1 promotes tumor growth and progression by activating WIP and beta-catenin signaling pathways and predicts poor prognosis in lung cancer. Cell Death Dis 2020;11(7):506.
[139] Lee Y, et al. CD44+ cells in head and neck squamous cell carcinoma suppress T-cell-mediated immunity by selective constitutive and inducible expression of PD-L1. Clin Cancer Res 2016;22(14):3571–81.
[140] Wu Y, Chen M, Wu P, Chen C, Xu ZP, Gu W. Increased PD-L1 expression in breast and colon cancer stem cells. Clin Exp Pharmacol Physiol 2017;44(5):3.
[141] Castagnoli L, et al. WNT signaling modulates PD-L1 expression in the stem cell compartment of triple-negative breast cancer. Oncogene 2019;38(21):4047–60.
[142] Wei F, et al. PD-L1 promotes colorectal cancer stem cell expansion by activating HMGA1-dependent signaling pathways. Cancer Lett 2019;450:1–13.
[143] Mansour FA, et al. PD-L1 is overexpressed on breast cancer stem cells through notch3/mTOR axis. Onco Targets Ther 2020;9(1):1729299.
[144] Dong P, et al. Tumor-intrinsic PD-L1 signaling in cancer initiation, development and treatment: beyond immune evasion. Front Oncol 2018;8:386.
[145] Fu L, et al. PD-L1 interacts with frizzled 6 to activate beta-catenin and form a positive feedback loop to promote cancer stem cell expansion. Oncogene 2022;41(8):1100–13.
[146] King I, et al. Tumor-targeted *Salmonella* expressing cytosine deaminase as an anticancer agent. Hum Gene Ther 2002;13(10):1225–33.
[147] Nemunaitis J, et al. Pilot trial of genetically modified, attenuated *Salmonella* expressing the *E. coli* cytosine deaminase gene in refractory cancer patients. Cancer Gene Ther 2003;10(10):737–44.
[148] Cunningham C, Nemunaitis J. A phase I trial of genetically modified *Salmonella typhimurium* expressing cytosine deaminase (TAPET-CD, VNP20029) administered by intratumoral injection in combination with 5-fluorocytosine for patients with advanced or metastatic cancer. Protocol no: CL-017. Version: April 9, 2001. Hum Gene Ther 2001;12(12):1594–6.
[149] Saltikva for metastatic pancreatic cancer. Available from: https://clinicaltrials.gov/study/NCT04589234.

CHAPTER

12

Oncolytic virus for cancer immunotherapy

Penghui Yang[a] *and Hao Yang*[a,b]

[a]Faculty of Hepatopancreatobiliary Surgery, The First Medical Center of Chinese PLA General Hospital, Beijing, China [b]Department of General Surgery, Affiliated Taian City Central Hospital of Qingdao University, Taian, China

Abstract

Oncolytic viruses (OVs) have been developed for a century and have become one of the most advanced types of tumor immunotherapy. It is primarily a natural or genetically modified virus, including DNA and RNA viruses. In recent years, with the rapid development of genetic engineering technology, the gene-modified oncolytic viruses have made great progress in the field of tumor treatment. Several types of viruses (including HSV, adenovirus, poxvirus, measles virus, and reovirus) are currently in preclinical studies, clinical trials, or have been approved in clinics, showing good safety and clinical efficacy. It is generally believed that oncolytic viruses target and kill tumor cells by selectively replicating themselves in tumor cells and finally lysing and killing tumor cells. At the same time, they can stimulate the immune responses of the body, thus enhancing the antitumor immunity; the tumor cells can be targeted and killed without obvious side effects. The combination of oncolytic virus and immune checkpoint by gene recombination and the breakthrough of tumor immunotherapy have made the application of oncolytic virus more extensive; however, there are still some bottleneck problems, such as virus targeting, security, and administration route. This chapter of the book provides a comprehensive and detailed overview of the development of OVs. We list some viruses that have been used as candidates for lysis of cancer cells and the clinical trials in the field of oncolytic virotherapy. And we discuss the immunological mechanism of oncolytic virus targeting to kill tumors and the challenges and prospects in the future.

Abbreviations

APC antigen-presenting cell
CAR-T chimeric-antigen receptor-T
CTLA-4 cytotoxic T-lymphocyte antigen-4
DC dendritic cell
ER endoplasmic reticulum
GM-CSF granulocyte-macrophage colony-stimulating factor
HSV-1 herpes simplex virus type 1

https://doi.org/10.1016/B978-0-443-13827-0.00002-0

IARC	International Agency for Research on Cancer
IC	intracranial
IFN	type I interferon
IM	intramuscular, intranasal
IP	intraperitoneal
IVE	intravesical
MHC	major histocompatibility complex
MHC	major histocompatibility complex
NK cells	natural killer cells
OVs	oncolytic viruses
PAMP	pathogen-related molecular pattern
ROS	reactive oxygen species
SC	subcutaneous
TAAs	tumor-related antigens
TAP	transport protein
TK	thymidine kinase
TLRs	toll-like receptors
TME	tumor microenvironment
T-VEC	talimogene laherparepvec

Conflict statement

We declare that we have no conflict of interest.

Introduction

There were close to 20 million new cases of cancer in the year 2022, alongside 9.7 million deaths from cancer (International Agency for Research on Cancer, IARC, 2022) [1]. It was the second leading cause of death worldwide [2,3]. The clinical treatment of tumors needs to improve the cure rate according to the physical condition, pathological type, invasion scope, and other conditions of the patient, including surgery, chemotherapy, radiotherapy, immunotherapy, interventional therapy, and other means, hoping to improve the cure rate and improve the quality of life of the patient. However, the clinical efficacy of most tumor patients is not ideal, and new treatment strategies and means are urgently needed to improve the prognosis of patients.

In recent years, the rapid development of oncolytic viruses (OVs) research has made it one of the most promising tumor immunotherapies. OVs can be natural or genetically modified viruses whose antitumor activity involves multiple mechanisms, including interactions between the virus, tumor cells, and the immune system. In addition, injection of OVs into the tumor can produce enrichment, and peripheral uninfected tumors can also produce an immune response. This enrichment effect occurs through the phased activity of OVs, starting with the selective replication of tumor cells and then inducing immune cell death, which in turn promotes the antitumor immune response. Recently, a few genetically engineered OVs have been developed to further improve the effect of tumor elimination by expressing specific immune regulatory factors and thus enhance the body's antitumor immunity.

OVs can selectively kill tumor cells without affecting normal cells, as well as activate the innate and adaptive immune systems in patients. To date, four OVs and one nononcolytic virus have been approved for clinical treatment; more than 150 OVs have entered clinical trials; and some products have entered clinical phase III (https://ClinicalTrials.gov). Although oncolytic virotherapy has become an important tool of tumor immunotherapy therapy. However, the immunosuppressive nature of tumor microenvironment (TME) poses a substantial hurdle to the development of OVs as effective immunotherapeutic agents, as it restricts the activation and recruitment of immune cells [4]. In order to make the OVs target the tumor cells more effectively and specifically, the combination of the virus and the immune checkpoint genes to improve the specific targeting of the tumors has become a new perspective of tumor immunotherapy. The development of molecular biology and genetic engineering technology has made the viral genome modification technology gradually mature, and the OVs have made significant progress in the targeting, safety, and oncolytic effect.

The development course of the OVs

In 1904, the Lancet reported a woman with chronic leukemia, inadvertently infected with flu virus and leukemia symptoms. This strange phenomenon made people begin to realize that viruses and tumors are two difficult-to-conquer diseases that medicine may influence each other and pioneered the viral treatment of tumors. In 1912, Italian doctor Depace found a cervical cancer patient; the tumor spontaneously subsided after inoculation with an attenuated rabies vaccine. Since then, he has opened the history of viral treatment of tumors, OVs on the stage of tumor immunotherapy. Since then, OVs as a promising tumor treatment has a lot of research. Between 1950 and 1980, many scientists directly used a variety of wild-type viruses (such as hepatitis virus, west Nile virus, dengue virus, and yellow fever virus) for infection treatment of hundreds of cases of different tumors, but due to poor medical conditions, low technical level, and insufficient scientific understanding of virus and tumor mechanism factors, curative effect is not ideal, and OVs development is very slow. In 1991, with the first genetic modification of herpes simplex virus type 1 (HSV-1) knockout of thymidine kinase (TK), the study of OVs began to soar. In 1996, it was demonstrated that adenovirus with E1B55k gene mutations could selectively replicate in p53-deficient tumor cells and thus function as tumor-dissolving viruses, and subsequently, the genetically modified adenovirus ONYX-015 was entered into phase I clinical trials. In 2004, Latvia approved a nonpathogenic human intestinal cytopathic orphan virus (enteric cytopathic human orphan virus), RIGVIR, for the clinical treatment of melanoma, becoming the first OVs with regulatory approval for tumor treatment. In 2005, the modified adenovirus H101 (Oncorine, Ancore) was approved for marketing in China. In 2015, the FDA-approved T-VEC (Talimogene laherparepvec, Imlygic), T-VEC is a genetically modified HSV-1 for the treatment of advanced melanoma, which can replicate in tumor cells and express granulocyte-macrophage colony-stimulating factor (GM-CSF), which is the first FDA-approved oncolytic viral therapy. In 2016, it was approved in Europe and Canada, respectively, marking the maturity of OVs technology and the formal recognition of the efficacy of OVs. In 2017, Cell reported that the combination of T-VEC and Keytruda was used in the treatment of melanoma. The tumor remission rate was as high as 62%, of which 33% were in complete remission, setting off a wave of research on OV combination therapy. In 2021, Japan San Pharmaceutical Co., Ltd. announced that the third-generation HSV-1 OVs product Delytact (Teserpaturev/G47) was approved for marketing, which is the first OVs product approved for the treatment of malignant glioma. This product has stronger replication capacity and higher antitumor activity, and it is effective in the treatment of glioma, breast cancer, prostate cancer, schwannoma, nasopharyngeal carcinoma, hepatocellular carcinoma, colorectal cancer, malignant peripheral schwannoma, and thyroid cancer. In December 2022, ADSTILADRIN, an adenovirus vector-based nonreplicating gene therapy developed by Ferring Pharmaceutical, was approved by the FDA for the treatment of nonmuscle-invasive bladder without response

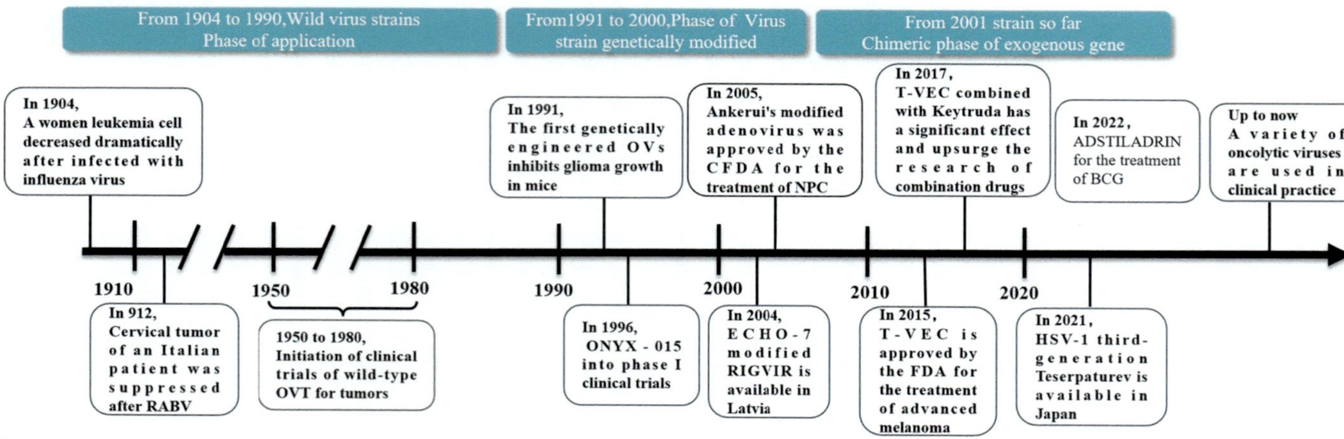

FIG. 1 Development of oncolytic virus.

to high-risk BCG (BCG) (Fig. 1) [5]. OV therapy has become the most promising clinical cancer immunotherapy and has attracted more attention worldwide.

Currently, four OVs and one nononcolytic virus have been approved for the treatment of cancer globally (Table 1), including RIGVIR of Sia Latima Latvia for melanoma and other malignancies, Anori H101 for the treatment of nasopharyngeal cancer, T-VEC, Biovex of Amgen, Delytact (Teserpaturev), and ADSTILADRIN from Ferring Pharmaceuticals for bladder cancer.

Classification of OVs

Ideal OVs must meet the following requirements: (1) high targeting of tumor cells; (2) rapid replication and expansion; (3) low immunogenicity with attenuated antiviral responses; (4) strong activation of the immune response. Previous studies have shown that OVs are small particles, and most of them are distributed between 20 and 200 nm. Selective replication in tumor cells is a very important feature for tumor immunotherapy, and replication capacity depends on the ability of virus infection, the characteristics of tumor cells, and host resistance. Viromes are composed of genetic material and proteins, consisting of single-stranded or double-stranded RNA or DNA (Table 2); a regular arrangement of protein subunits is called the capsid. DNA viruses have multiple advantages in that they have a more stable genome and lend themselves to genetic engineering more easily than RNA viruses. DNA viruses have a lower immunogenicity than RNA viruses and have more repair mechanisms, reducing the likelihood of errors during replication. The selection of OVs for tumor immunotherapy depends on many properties, including potential pathogenicity, immunogenicity, genes encoding therapeutic capacity, viral titers, and viral stability achievable during the manufacturing process. The viral life cycle has five different stages—adsorption, entry, and uncoating, replication, assembly, and release. At each stage, genetic engineering technology can be used to regulate the viral genome, and it can be designed to target tumor cell endocrine and replication. In addition, the pathogenicity and immunogenicity of the virus vary according to the virus type, dose, inoculation route, the original host immunity, and other factors, which are the characteristics that can promote effective antitumor immunity. At present, the OVs registered on the Clinical Trials website mainly include poxvirus, adenovirus, HSV-1, reovirus, measles virus, Newcastle disease virus, coxsackie virus, vesicular stomatitis virus, and poliovirus, among which vaccinia virus, adenovirus, and herpes simplex virus are the most common (Table 2) [6]. Due to the easy preparation of adenovirus gene recombination, wide clinical development, and most experiments, HSV-1 and vaccinia viruses can add any foreign gene because of their large genomes, which have great advantages in genetic modification, and their application is favored [7].

In our group, using influenza virus reverse genetics (RG) technology, we established a recombinant influenza virus vectored OVs rescue system with owned intellectual property rights and successfully generated several chimeric oncolytic flu viruses carrying the foreign gene GM-CSF and immune checkpoint inhibitors (PD-1, PD-L1, CTLA4, etc.) antibodies. In vitro and in vivo experiments displayed that these flu OVs exhibited good safety, oncolytic efficacy, and stability and have potential as a novel strategy for clinical tumor immunotherapy.

TABLE 1 OVs products on the market.

Name	Firm	Virus and genetic modification processes	Indication	Approval agency	Market year
RIGVIR	Sia Latima	ECHO-7	Melanoma, colorectal cancer, pancreatic cancer, bladder cancer, kidney cancer, prostate cancer, lung cancer, etc.	Latvia	2004
Shanghai Acrel Limited H101	Three-dimensional creatures	Recombinant human adenovirus (Gene fragment for deletion of E1B-55 KD and E2 region)	Primary, clinically advanced, recurrent head and neck tumors	China Food and Drug Administration	2005
T-VEC (Imlygic)	Biovex	Herpes simplex virus type 1, the ICP 34.5 and ICP 47 genes were removed, and the human GM-CSF integration was inserted into the ICP 34.5 locus and US11	Advanced melanoma	FDA	2015
Delytact (Teserpaturev)	The first three	Additional deletion mutations were introduced in the G207 genome of second-generation HSV-1 to enhance the antitumor effect	Glioma, breast cancer, prostate cancer, schwannoma, nasopharyngeal carcinoma, hepatocellular carcinoma, colorectal cancer, malignant peripheral schwannoma, and thyroid cancer	Japan	2021
ADSTILADRIN	Ferring	Copies of the gene encoding human interferon- α 2b (IFNα2b) are delivered to the bladder urothelium based on gene therapy with nonreplicating adenoviral vectors	For the treatment of adults with carcinoma in situ (CIS) of nonmuscle-invasive bladder carcinoma (NMIBC) with or without papillary neoplasms	FDA	2022

TABLE 2 Registered clinical trial of OVs.

Virus type	Virus name	Test number	Research name	Intervention study	Clinical research phase	Indication	Combination therapy	State
DNA virus	**Poxvirus**	NCT00794131	Safety Study of GL-ONC1, an Oncolytic Virus, in Patients with Advanced Solid Tumors	GL-ONC1	I	Advanced solid tumor		Accomplish
		NCT01443260	A Study of GL-ONC1, an Oncolytic Vaccinia Virus, in Patients with Advanced Peritoneal Carcinomatosis	GL-ONC1	I/II	Peritoneal cancer		Accomplish
		NCT01584284	Safety Study of Attenuated Vaccinia Virus (GL-ONC1) with Combination Therapy in Head & Neck Cancer	GL-ONC1	I	Head and neck cancers		Accomplish
		NCT02714374	Safety and Effect of GL-ONC1 Administered IV Prior to Surgery to Patients with Solid Organ Cancers Undergoing Surgery	GL-ONC1	I	Solid tumor		Stop
		NCT00429312	A Study of Recombinant Vaccinia Virus to Treat Malignant Melanoma	JX-594	I/II	Melanoma		Accomplish
		NCT02017678	Phase 2 Study of JX-594 in Patients with Peritoneal Carcinomatosis of Ovarian Cancer Origin	JX-594	II	Oophoroma		Recall
		NCT00625456	Safety Study of Recombinant Vaccinia Virus to Treat Refractory Solid Tumors	JX-594	I	Melanoma, lung cancer, renal cell carcinoma, and head and neck squamous cell carcinoma		Accomplish
		NCT01169584	Safety Study of Recombinant Vaccinia Virus to Treat Refractory Solid Tumors in Pediatric Patients	JX-594	I	Lymphoma, neuroblastoma, etc.		Accomplish

Continued

TABLE 2 Registered clinical trial of OVs—cont'd

Virus type	Virus name	Test number	Research name	Intervention study	Clinical research phase	Indication	Combination therapy	State
		NCT01394939	Recombinant Vaccinia Virus Administered Intravenously in Patients with Metastatic, Refractory Colorectal Carcinoma	JX-594	I/II	Colorectal cancer	Irinotecan	Complete results
		NCT01387555	A Phase 2b Study of Modified Vaccinia Virus to Treat Patients with Advanced Liver Cancer Who Failed Sorafenib	JX-594	II	Cancer of the liver		Accomplish
		NCT03294486	Safety and Efficacy of the Oncolytic Virus Armed for Local Chemotherapy, TG6002/5-FC, in Recurrent Glioblastoma Patients	TG 6002	I/II	Glioblastoma and brain cancer	5-FC	Recruit
		NCT03954067	A Study of an Intratumoral Oncolytic Virus in Patients with Advanced Metastatic Solid Tumors	ASP 9801	I/II	Metastatic tumor, solid tumor, and advanced cancer		Recruit
		NCT02562755	Hepatocellular Carcinoma Study Comparing Vaccinia Virus-Based Immunotherapy Plus Sorafenib vs. Sorafenib Alone	Pexa-V ec	III	Cancer of the liver	Sorafenib	Complete results
		NCT03206073	A Phase I/II Study of Pexa-Vec Oncolytic Virus in Combination with Immune Checkpoint Inhibition in Refractory Colorectal Cancer	Pexa-V ec	I/II	Colorectal cancer	Immune-based checkpoint inhibitors	Active, not recruited
		NCT04226066	Safety, Tolerability, and Pharmacokinetics Characteristics of Recombinant Oncolytic Vaccinia Virus Injection T601 as a Single Drug or in Combination with Oral Flucytosine (5-FC) in Patients with Advanced Malignant Solid Tumors	T601	I/II	Advanced malignant solid tumor	5-FC	Recruit

	NCT04301011	Study of TBio-6517, Given Intratumorally, Alone, or in Combination with Pembrolizumab, in Solid Tumors	TBio-6517	I/II	Solid tumors, triple-negative breast cancer, and colorectal cancer	Keytruda	Recruit
Adenovirus	NCT02143804	Safety and Efficacy of CG0070 Oncolytic Virus Regimen in Patients with High-Grade Nonmuscle-Invasive Bladder Cancer	CG 0070	II	Carcinoma of urinary bladder		Recall
	NCT01438112	Efficacy Study of Recombinant Adenovirus for Nonmuscle-Invasive Bladder Cancer	CG 0070	II/III	In situ carcinoma, and urinary bladder cancer		Stop
		Safety and Efficacy of CG0070 Oncolytic Virus Regimen in Patients with High-Grade Nonmuscle-Invasive Bladder Cancer			carcinoma of urinary bladder		Recall
	NCT01598129	ONCOS-102 (Previously CGTG-102) for Therapy of Advanced Cancers	ONCOS-102	I	Malignant solid tumor		Complete, have results
	NCT03190824	Evaluate Efficacy and Immunological Response of Intratumoral/Intralesional Oncolytic Virus (OBP-301) in Metastatic Melanoma	OBP-301	II	Melanoma		Active, not recruited
	NCT02705196	LOAd703 Oncolytic Virus Therapy for Pancreatic Cancer	LOAD 703	I/II	Pancreatic cancer/stage	Gemcitabine, paclitaxel, and no	Recruit
	NCT03916510	Chemoradiation with Enadenotucirev as a Radiosensitizer in Locally Advanced Rectal Cancer	Enadenotucirev	I	Locally advanced rectal cancer	Gemcitabine, and radiation therapy	Recruit
	NCT04217473	TNFalpha and Interleukin 2 Coding Oncolytic Adenovirus TILT-123 During TIL Treatment of Advanced Melanoma	TILT-123	I	Metastatic melanoma		Recruit

Continued

TABLE 2 Registered clinical trial of OVs—cont'd

Virus type	Virus name	Test number	Research name	Intervention study	Clinical research phase	Indication	Combination therapy	State
		NCT03714334	DNX-2440 Oncolytic Adenovirus for Recurrent Glioblastoma	DNX-2440	I	Glioblastoma		Recruit
		NCT03004183	SBRT and Oncolytic Virus Therapy Before Pembrolizumab for Metastatic TNBC and NSCLC	ADV/HSV-tk	II	Metastatic nonsmall cell lung cancer, metastatic triple-negative breast cancer	Vacylovir, pembrolizumab as well as radiotherapy	Active, not recruited
		NCT02045602	Phase I Dose Escalation Study of Intravenous VCN-01 with or without Gemcitabine and Abraxane in Patients with Advanced Solid Tumors	VCN-01	I	Advanced solid tumors, metastatic solid tumors, and pancreatic cancer	Gemcitabine, and Abraxane	Accomplish
		NCT02879760	Oncolytic MG1-MAGEA3 With Ad-MAGEA3 Vaccine in Combination With Pembrolizumab for Nonsmall Cell Lung Cancer Patients	Ad-MAGEA3	I/II	Nonsmall cell lung cancer	Keytruda 、 MG1-MA -GEA3	Accomplish
	Herpes simplex virus	NCT00289016	A Study of Talimogene Laherparepvec in Stage IIIc and Stage IV Malignant Melanoma	Talimogene	II	melanoma		Complete, have results
			An Extension Protocol for the Extended Use of Talimogene Laherparepvec for Eligible Patients Who Participated in Study 002/03 (NCT00289016)	Laherparepvec		melanoma		Complete, have results
		NCT01721018	Intrapleural Administration of HSV1716 to Treat Patients with Malignant Pleural Mesothelioma	HSV 1716	I/II	malignant pleural mesothelioma		Accomplish
		NCT02428036	A Study of TBI-1401(HF10) in Patients with Solid Tumors with Superficial Lesions	TBI-1401 (HF10)	I	Solid tumor		Accomplish
		NCT03153085	A Study of Combination With TBI-1401(HF10) and Ipilimumab in Japanese Patients with Unresectable or Metastatic Melanoma	TBI-1401(HF10)	II	Melanoma/stage	Ipilimumab	Accomplish

		NCT03259425	Neoadjuvant Trial of Nivolumab in Combination with HF10 Oncolytic Viral Therapy in Resectable Stage IIIB, IIIC, IVM1a Melanoma	HF10	II	Melanoma	Navutuzumab	Termination, results
		NCT03866525	OH2 Oncolytic Viral Therapy in Solid Tumors	OH2 injection	I	Solid tumors, gastrointestinal cancer	HX 008 Injection solution	Recruit
		NCT02062827	Genetically Engineered HSV-1 Phase 1 Study for the Treatment of Recurrent Malignant Glioma	M032	I	Glioblastoma		Recruit
		NCT04050436	Study Evaluating Cemiplimab Alone and Combined with RP1 in Treating Advanced Squamous Skin Cancer	RP1	II	Cutaneous squamous cell carcinoma	Cemiplimab	Recruit
		NCT03252808	Phase I Study of TBI-1401(HF10) Plus Chemotherapy in Patients with Unresectable Pancreatic Cancer	TBI-1401(HF10)	I	Pancreatic cancer/stage	Gemcitabine, paclitaxel, and TS-1	Active, not recruited
		NCT03657576	Trial of C134 in Patients with Recurrent GBM	C134	I	Glioblastoma		Recruit
		NCT03911388	HSV G207 in Children with Recurrent or Refractory Cerebellar Brain Tumors	G207	I	Glioblastoma, brain tumors, astrocytomas, nervous system tumors, etc.		Recruit
RNA virus	**Reovirus**	NCT00528684	Safety and Efficacy Study of REOLYSIN in the Treatment of Recurrent Malignant Gliomas	REOL YSIN	I	Glioblastoma		Accomplish
		NCT00503295	Safety and Efficacy Study of REOLYSIN in the Treatment of Bone and Soft Tissue Sarcomas Metastatic to the Lung	REOL YSIN	II	Fibroid sarcoma, leiomyosarcoma, etc.		Accomplish
		NCT02723838	Study of Intratumoral REOLYSIN in Combination with Gemcitabine and Cisplatin as Neoadjuvant Therapy in Muscle-invasive Transitional Cell Carcinoma of the Bladder	REOL YSIN	I	Invasive transitional cell carcinoma of the urinary bladder muscle	And gemcitabine and cisplatin	Cancel

Continued

TABLE 2 Registered clinical trial of OVs—cont'd

Virus type	Virus name	Test number	Research name	Intervention study	Clinical research phase	Indication	Combination therapy	State
		NCT00861627	Phase 2 Study of REOLYSIN in Combination with Paclitaxel and Carboplatin for Nonsmall Cell Lung Cancer with KRAS or EGFR Activation	REOL YSIN	II	Nonsmall cell lung cancer	Carboplatin, and paclitaxel	Accomplish
		NCT01274624	Study of REOLYSIN in Combination with FOLFIRI and Bevacizumab in FOLFIRI Naive Patients with KRAS Mutant Metastatic Colorectal Cancer	REOL YSIN	I	KRAS mutant metastatic colorectal carcinoma	5-FC, leucine, irinotecan, and Avastin	Accomplish
		NCT03605719	Dexamethasone, Carfilzomib, & Nivolumab with Pelareorep for Relapsed/Refractory Multiple Myeloma	Wild-type reovirus	I	Recurrent plasma cell myeloma	Carfilzomib, dexamethasone, et al	Recruit
		NCT04102618	A Window-of-opportunity Study of Pelareorep in Early Breast Cancer	PeLareorEp	I	Mammary cancer	Letrozole, trastuzumab, and Atezolizumab	Recruit
		NCT04215146	A Study to Assess Overall Response Rate by Inducing an Inflammatory Phenotype in Metastatic Breast Cancer with the Oncolytic Reovirus Pelareorep in Combination with Anti-PD-L1 Avelumab and Paclitaxel - BRACELET-1 Study	PeLareorEp	II	Metastatic breast cancer	Paclitaxel, and Avelumab	Recruit
	Measles virus	NCT02192775	UARK 2014-21 A Phase II Trial of Oncolytic Virotherapy by Systemic Administration of Edmonston Strain of Measles Virus	MV-NIS	II	Bozzolo's disease		Complete, have results
		NCT04195373	A Safety Study of TMV-018 in Patients with Tumors of the Gastrointestinal Tract	TMV-018	I/II	Gastrointestinal cancer	5-FC, and the anti-PD-1 antibody	Recall

Other viruses	NCT01174537	New Castle Disease Virus (NDV) in Glioblastoma Multiforme (GBM), Sarcoma, and Neuroblastoma	Avian pneumo-encephalitis virus	I/II	Glioblastoma, neuroblastoma		Recall
	NCT00438009	A Safety Study of Two Intratumoral Doses of Coxsackievirus Type A21 in Melanoma Patients (PSX-X03)	Coxsackie virus	I	Melanoma stage		Accomplish
	NCT01301430	Parvovirus H-1 (ParvOryx) in Patients with Progressive Primary or Recurrent Glioblastoma Multiforme	H-1PV	I/II	Pleomorphic glioblastoma		Accomplish
	NCT00314925	Safety Study of Seneca Valley Virus in Patients with Solid Tumors With Neuroendocrine Features	Seneca virus	I	Carcinoid type, neuroendocrine type		Unknown
	NCT02288897	PV-10 vs Chemotherapy or Oncolytic Viral Therapy for Treatment of Locally Advanced Cutaneous Melanoma	PV-10	III	Melanoma		Stop
	NCT03889275	A Study of MEDI5395 in Combination with Durvalumab in Subjects With Select Advanced Solid Tumors	MEDI5395	I	Advanced solid tumor	Drumumab	Recruit

Delivery of OVs

One of the greatest challenges for effective oncolytic virotherapy is adequate drug delivery. Local injection of OV into a single tumor site, such as IT administration, can directly target the location of the tumor. However, its use in metastatic tumors is limited. Systemic injection (e.g., intravenous) is logically simple and effective against metastatic disease, but OV is rapidly cleared from the circulation before it reaches the tumor. Carrier cells, including the transformation of tumor cells, immune cells, and stem cells, have been used in OV systemic delivery carriers. Intraperitoneal (IP), subcutaneous (SC), intracranial (IC), intramuscular (IM), intranasal, intravesical (IVE), aerosol, and inhalation administration are all used in preclinical and clinical practice [8]. In the preclinical setting, the choice of delivery method depends largely on the goals of the research program. Safety must be a top priority in the clinical setting. The bioavailability of systemically administered OVs is extremely poor. Furthermore, even in the presence of intravenous administration, the host immune system rapidly sequestrates and degrades the attenuated virus through the reticuloendothelial system led by red pulp macrophages in Kupffer cells of the spleen and liver. Viral particles are opsonized by antibodies, complement, and other factors to enhance binding and phagocytosis by endothelial cells and macrophages. Of note, there are no reports of poor dose tolerance to cancer virus therapy or recovery of virulence by inactivated particles. Balancing the degree of local immune suppression is a complex challenge in oncologic virotherapy. On the one hand, immunosuppression can increase the intratumoral distribution of therapy. In contrast, the enhanced host immune system will enhance the targeting of transfected tumor cells, but the spread of the virus within the tumor will be pruned. Thus, to date, the only way in which tumor virotherapy has been provided in sufficient quantities to achieve clinical efficacy is through locoregional or direct inoculation.

Intratumor injection

Intratumoral delivery is the most common route of delivery of OVs. The concentration of oncolytic virus in the target site can be precisely controlled, and the side effects caused by the accidental injection of virus into other organs can be prevented. Intratumoral delivery is more suitable for surface tumors such as melanoma, while deep tumors are difficult to operate and achieve. Maintaining tumor site soluble tumor virus concentration is a key advantage of tumor delivery. Intratumor delivery allows investigators to control the precise concentration of oncolytic virus at the tumor site and compare the results of in vitro experiments with those of in vivo studies. In addition, inoculation of the virus directly into the lesion will result in favorable pharmacokinetics. Intratumoral administration allows a lower dose to maximize drug concentration at the target lesion, maximize its effect, and reduce metabolic reduction during circulation in vivo. Because only delivery to the target site is more targeted, the dose to individual lesions is adjusted as precisely as possible by direct inoculation. Similarly, optimization of the timing of administration can be achieved as neoadjuvant, adjuvant, or primary therapy. In the T-VEC OPTIM phase III clinical trial, the clinical benefit of intratumoral injection of drugs to cancer viruses has been demonstrated for local and potentially systemic antitumor responses.

Direct injection can also be treated with novel or unique carriers such as polymeric micelles, nanoparticles, or implants. Image-guided localization and treatment would be by far the most resource-saving method, and the wasted or lost treatment would be negligible. By HSV thymidine kinase phosphorylation and positron emission substrate cell isolation, we have achieved the herpes virus direct imaging of tumors within the distribution. In animal models with iodinated and technetium-based media, gene splicing of the thyroid sodium-iodide sympotransporter has also been performed to monitor the distribution of cancer virus transcripts within the host, a concept validated in adenoviruses by pertechnetate-based SPECT imaging. Like all surgery, image-guided cancer virus treatment, of course there are related risks. However, the overall risk was quite low and comparable to the associated standard of care procedures. Risks can be categorized as those related to technology and treatment. From a technical point of view, bleeding and unexpected organ damage are the main potential adverse events and are considered highly unlikely. These risks are the same as those accepted when a mass is biopsied, which is sometimes the standard of care for the target lesion. Unlike other locoregional treatments such as thermal injury and nerve electrical conduction, direct cancer virotherapy does not require additional preventive measures. Furthermore, regarding the risks of the treatment itself, no cases of viral virulence recovery have been reported. Local inflammatory responses are certainly possible and to some extent desirable, with a theoretical risk of dysregulated inflammatory responses, although this has not been reported.

Another risk unique to oncolytic virotherapy is leakage of material through the needle tract, although the probability of this scenario is low, and local reactions may in principle occur even if OVs do not have systemic effects. If the risk is clinically meaningful, ways to minimize this may include the use of autologous blood for tracking repair, such as for certain lung and liver biopsies or the use of needle systems for tract ablation. Furthermore, even with direct vaccination, there is the possibility of neutralizing antibodies, and the tumor niche may be immunosuppressive, both of which would suppress the therapeutic response. In addition, if tumor cells are suspended in the growth phase due to hypoxia or acidosis or nearby necrosis, calcification, or high interstitial pressure, the efficacy of oncoviruses may be limited by the tumor niche. In addition, the rapid induction of apoptosis of cancer viruses may also be a disadvantage because there may not be optimal number of viruses reproduced. Acquired resistance or tumor adaptation to an oncolytic virus or associated tumor immune pressure is also a possibility.

Intravenous administration

At the stage of clinical trials, many investigators prefer to use intravenous injection rather than intratumoral injection, which may be related to the complexity of the surgery and the barriers to distant metastasis for intratumoral administration. Intravenous injection of OVs is a more simplified route of administration for physicians. Samson and others show that OV (reovirus) in patients after intravenous injection into the brain tumors. Reovirus RNA was widely detected in tumor cells of different histological types. Therefore, the intravenous route is expected to be an effective means of delivering OV to brain tumors, enabling regular scheduled treatment while avoiding the need for a neurosurgical access method. Miriam et al.

demonstrated that oncolytic adenovirus VCN-01 was effective by both intravenous and IT administration, significantly prolonged tumor control, and synergized with chemotherapy. The antitumor activity and replication potential of VCN-01 observed after intravenous administration will be important to support the clinical development of this agent through this simpler route of administration. In addition, its limited replication potential in normal liver also suggests that the expected liver injury caused by this class of drugs may be transient. Therefore, VCN-01 has also been administered intravenously in combination with chemotherapy in patients with advanced PDAC. The oncolytic virus HSV G207 has shown marked therapeutic efficacy in phase I clinical trials with intravenous administration in children with progressive or recurrent malignant supratentorial brain tumors [9].

Many clinical trials are currently underway, including intravenous poxvirus, soluble tumor of poison, M1 virus, blister stomatitis, and gonads soluble tumor viruses. In phase 1 b study, 67% of patients after intravenous solution zebu poxvirus Pexa - imaging stable after Vec. A phase I trial involving 23 patients with advanced solid tumors who received a single intravenous infusion of Pexa-Vec observed dose-related antitumor activity without clinical normal tissue involvement. In a phase I trial, Pelareorep, an intravenous product of the oncolytic Ehrlichiosis virus, induced a T cell inflammatory phenotype in pancreatic ductal adenocarcinoma; Ehrlichivirus replication, T cell infiltration, and PD-L1 upregulation were detected in tumor tissue from patients treated with Pelareorep, with cancer control in 3 of 10 patients. Intravenous HSV-1 soluble tumor virus clinical research Co., LTD. [10]. We suspect that this may be associated with the characteristics of HSV-1, such as HSV-1 has a greater viral particle, more surface protein, and is easier to clear the immune system. However, there are still many studies actively investigating the possibility of intravenous HSV-1 administration.

Immunological mechanisms of OVs targeting to kill tumor cells

OVs are highly promising immunotherapeutic agents in current tumor therapy, and their antitumor activity is mediated through multiple mechanisms that affect the tumor immune cycle. The mechanism of OVs killing tumor cells has not been fully explored. At present, the recognized mechanism of OVs includes: OVs proliferate in tumor cells through self-replication and eventually cleave tumor cells to release more OVs and cause tumor lysis; OVs can induce systemic and local antitumor immunity and activate T cells to kill tumor cells and make tumor subside.

(a) Targeted tumor cell lysis: As promising tumor gene therapy agents, OVs have the unique ability to replicate in tumor cells and lead to the decline or even death of tumor cells. They do not replicate in normal tissues, further leading to tumor-related antigen exposure to cause the body and immune response. OVs, a tumor addiction (the ability of an OVs to be selectively replicated in tumor cells), is an important property of an OVs as a tumor immunotherapy. Although some viruses exhibit an innate antitumor tropism, others must be processed by genetic engineering techniques to selectively infect tumor cells. After infection, the virus will use the protein factory of the tumor cells so that the tumor cells cannot produce enough protein to meet the growth needs, thus destroying the normal physiological process of the tumor cells.

(b) Antitumor immune response: The OVs induce local and systemic-specific antitumor immunity. Induction of tumor-specific immune responses in the body is a key factor for tumor eradication of OVs. The therapeutic efficacy of OVs depends on a combination of direct tumor cytolysis and indirect activation of an antitumor immune response. Upon infection with an OV, tumor cells initiate an antiviral response consisting of the endoplasmic reticulum (ER) and genotoxic stress. This response leads to the upregulation of reactive oxygen species (ROS) and the production of antiviral cytokines. Reactive oxygen species and cytokines, especially type I interferon (IFN), release and stimulate immune cells from infected tumor cells (antigen-presenting cells, CD8+ T cells, and natural killer (NK) cells) [11]. OVs infection tumor cells death, cause tumor lysis, then release of tumor-related antigens, offspring virus, pathogen-related molecular pattern (PAMP), risk-related molecular pattern signal (DAMP), and tumor-related antigens (TAAs), can promote adaptive immune response by activating the Toll-like receptors (TLRs) to stimulate the immune system.

How OVs as an endogenous antigen enter the ER, how they bind to MHC I molecules, and how they are presented to CD8+ T cells in APC cells is a complex process. The human major histocompatibility complex (MHC) is expressed on different cell surfaces. MHC genes regulate the optimal immune response to various pathogens to adapt to the variable internal and external environment. It is essential in T cell differentiation and development and plays an important role in the initiation and regulation of immune responses. Antigen-presenting cells (antigen-presenting cell, APC) are a class of cells with uptake, process, present extracellular (exogenous) antigen, activate CD4+ T cells, and can express MHC I molecules [12]. APC cells include macrophages (macrophage, M ɸ), dendritic cells (dendritic cell, DC), and B lymphocytes, of which dendritic cells are small but in the body and are widely distributed [13]. It has been shown that DC is activated after peripheral uptake of viral antigen, tumor-associated antigen/nascent antigen to enter the draining lymph nodes and differentiate into mature DC. Losing the ability to actively uptake antigen, upregulate and highly expressed MHC molecules, costimulatory molecules, and adhesion molecules. DC in lymph nodes are mainly concentrated in the T cell region, and they are no longer a phagocytotic function. When the OVs enter the cell, DC processes and presents them. Viral antigen presentation by MHC I and class molecules, activation of CD4+ and CD8+ T cells. The processed and presented antigens are divided into endogenous antigens and exogenous antigens. OVs entry into the cytoplasm after infection is an endogenous antigen, recognized by host cells and proteolysis, and the resulting peptide is selectively transported from the cytoplasm to the ER by the antigen processing-related transport protein (TAP), and here it binds to newly synthesized MHC I molecules [14]. It is then expressed together on the cell membrane, where it is recognized by the TCR on CD8+ T cells. The process of endogenous antigen peptide processing and binding of class I molecules to the antigenic peptide ends. The class I molecular structure after binding the peptide is stabilized, modified from the ER to the Golgi with glycosylation, and transported to the cell surface by exocytosis for recognition by CD8+ T cells. The processed antigen is expressed as MHC/peptide complexes on the target cell or APC surface and is selectively identified by CD4+ T cells or CD8+ T cells expressing specific TCR [15]. Unsensitized T cells meet the APC in the peripheral lymphoid organs, while the sensitized T cells leave the peripheral lymphoid organs and reach the distal tumor site to play effector functions.

In the resulting immunostimulatory environment, antigen-presenting cells release and absorb TAAs (tumor-associated antigens) and NEO (de novo antigens) to mediate tumor regression at distant tumor sites that are not exposed to the virus. These would activate the antigen-specific CD4+ and CD8+ T cell responses [16]. Once activated, CD8+ T cells can be expanded into cytotoxic effector cells that can access established tumor growth sites and mediate antitumor immunity after antigen recognition (Fig. 1) [17].

However, the ability of the virus to induce an antiviral immune response may clear the virus by neutralizing antiviral antibodies and cytotoxic T cell-mediated immune responses. The extent to which viral neutralization affects tumor immune induction is complex and may be influenced by many variables, most notably the characteristics of the viral and tumor microenvironments. The release of tumor-associated antigens, particularly coupled with local cytokine and DAMP release, facilitates the induction of innate and adaptive immune responses against tumor cells (Fig. 2). This effect may be extremely important for regulating tumor regression at distant tumor sites not injected or not exposed to the virus. Preclinical studies have demonstrated the importance of tumor-specific CD8+ T cells in the OV-mediated tumor immune response. Type I interferons and DAMP can also directly activate NK cells, which is an important part of the innate immune response. NK cells can kill target cells with downregulation of the major histocompatibility complex (MHC) class I expression, which is common in tumor cells [18]. The effect of NK cells may depend on the host species

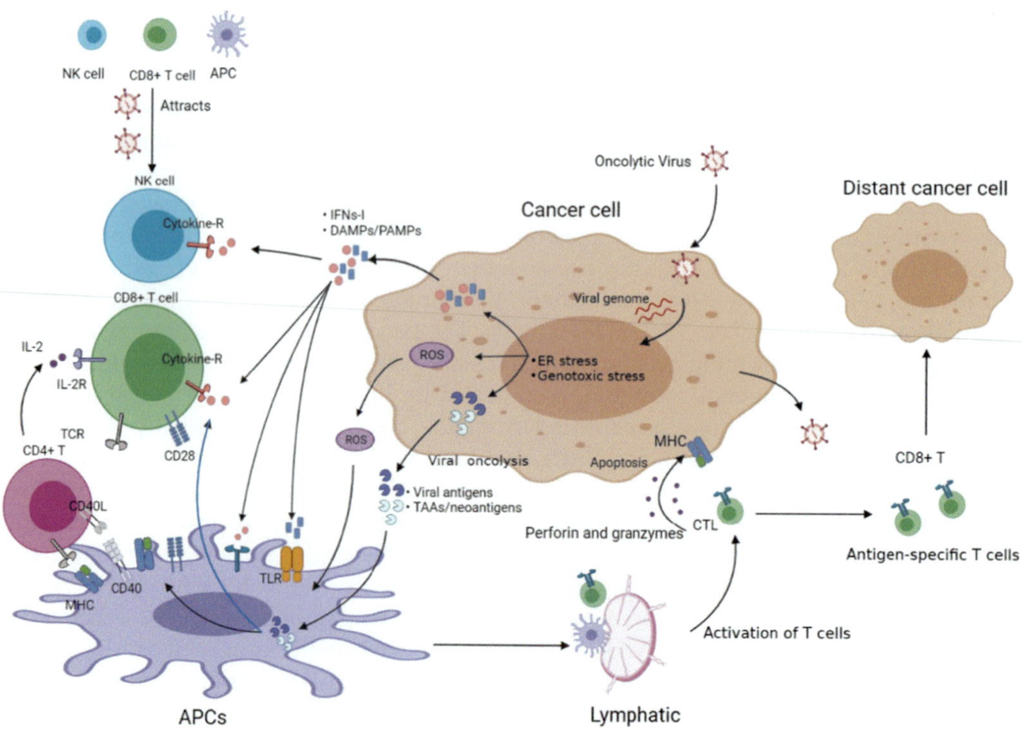

FIG. 2 Mechanism of action of OVs.

and viral characteristics. Furthermore, NK cells impair the efficacy of the OVs by clearing the virus-infected cells. Multiple factors collectively produce an immune response against virus-infected tumor cells and a de novo immune response against cells expressing TAAs/NEO antigens on uninfected tumor cells. But the factors affecting the balance between immune-mediated viral clearance and the induction of antitumor immunity are not fully understood.

In conclusion, OVs destroy virus-infected tumor cells, release cytokines, reduce tumor growth, and acquire virus and tumor antigens from dying tumor cells to present and activate T cells and activate local and systemic immunity in the body, with an antitumor effect.

Combination therapies

The mechanisms by which OVs eliminate tumor cells are distinctly different from those of other anticancer therapies; therefore, OVs are a rational candidate for combination with most other treatment modalities, including systemic chemotherapies, immunotherapies, targeted therapies, and/or radiotherapy. Furthermore, the toxicity profiles of OVs are limited and generally do not overlap with those associated with other therapeutic approaches. Indeed, data from established preclinical models demonstrate the improved antitumor activity of OVs when combined with other systemic therapies, including chemotherapies and cellular therapies. These promising data are supported by early clinical studies suggesting improved therapeutic activity with OV combination strategies.

Combination with chemotherapy

Chemotherapy remains a conventional anticancer therapy. OVs act synergistically with chemotherapy at various levels, including the inhibition of immune-mediated degradation and neutralization of the virus. Chemotherapy complements virotherapy through numerous mechanisms, such as the enhancement of tumor cell immunogenicity, direct killing of cancer cells, and blockage of antiviral immune responses (NCT00006106, NCT00634595, NCT01584284). Cyclophosphamide is an alkylating agent that increases the efficacy of virotherapy using different viruses. The immunomodulation triggered by this chemotherapeutic agent is thought to improve the replication, spread, and cytolytic activity of the virus. Considering the promising results in preclinical and phase I and II studies (NCT00006106), an engineered Ad vector, ONYX-015, was administered in association with 5-fluorouracil and cisplatin to treat squamous cell carcinoma of the head and neck or esophagus in a phase III trial. The response rate was 79% in patients treated with combination therapy, compared with 40% in those treated with chemotherapy alone.

In addition, the antitumor effects were intensified when the Ad was administered in combination with different drugs, such as doxorubicin, gemcitabine, paclitaxel, irinotecan, and cyclophosphamide. Furthermore, in preclinical studies, a combination of temozolomide with OV was tested in patients with glioma and showed a good toxicity profile and improved survival.

However, the efficacy of OVs in combination with chemotherapy remains controversial and requires further investigation. Although some clinical trials have shown promising results, others involving large cohorts have reported the inefficacy of this combination regimen. For example, Eigl et al. demonstrated a median survival time of 19.1 months in patients diagnosed with metastatic castration-resistant prostate cancer who were treated with a combination of docetaxel and pelareorep, compared with a median survival time of 21.1 months in patients who received docetaxel (NCT01619813).

Combination with radiotherapy

Several studies have demonstrated that a combination of virotherapy and radiotherapy leads to promising therapeutic results. Although robust preclinical data exist, this effect was due to the synergistic activity between these two types of therapies despite the lack of a well-established mechanism. This combination resulted in improved anticancer activity. Radiotherapy-induced cellular modifications can increase the capacity of viruses to replicate and disseminate through tumors. Pokrovska et al. proved that when OV was combined with radiotherapy, tumor-specific cytotoxicity was induced through increased virus production. In a phase I trial, a mutated HSV (G207), combined with radiotherapy for recurrent malignant glioma, showed the potential for a clinical response. In this study, six out of nine patients showed a partial response or stable disease, which proved to have a synergistic effect.

In another approach, virotherapy could increase tumor sensitivity to radiotherapy. A p53-expressing Ad induced greater sensitivity in infected p53-defective tumor cells to apoptosis induced by irradiation, followed by greater tumor inhibition, compared with the reduced apoptosis of tumor cells that did not benefit from radiotherapy.

Combination with immunotherapy

Despite the few reports in the literature on the combination of immunotherapy and virotherapy, the association of immune checkpoint inhibitors targeting PD-1 and/or cytotoxic T-lymphocyte antigen-4 (CTLA-4) with OV offers a new perspective in cancer treatment, particularly in cancers with an immunosuppressive microenvironment in which monotherapy with PD-1/PD-L1 inhibitors can become ineffective. The oncolytic activity of this therapy determines a release of cytokines and other molecules, which have the effect of converting a "cold" tumor to a "hot" one, a modification that could improve the effect of immunotherapy on the susceptible target. In addition to PD-1, a number of immune checkpoint molecules that are also expressed on TILS have been shown to induce an immunosuppressive microenvironment and to reduce activated CD8 cells. This category includes CTLA-4, TIGIT, TIM-3, and LAG-3. OVH-aMPD-1 acts synergistically with anti-TIGIT and substantially improves immune responses in MC38 and Hepa 1-6 in tumor models. Nevertheless, in refractory lung cancer, vvDD monotherapy in combination with anti-PD-1 and anti-TIM-3 mAbs failed to show a therapeutic advantage. Importantly, combining more than two ICIs as a therapeutic strategy can yield better antitumor results.

T-VEC is an intralesional therapy that is a leading combination of immunotherapies. The role of the GM-CSF in this association is to enhance cellular immunity. The action of T-VEC results from a combination of a direct oncolytic effect of the viral vector and an immune response trigger. Promising results of objective and enduring response rates were observed in a phase I trial combining T-VEC with anti-CTLA4, ipilimumab, for the treatment of stage IIIB-IV melanoma. Furthermore, in an open-label phase II trial in both primary and metastatic tumors, the objective response rate substantially increased with T-VEC in combination with ipilimumab as opposed to that in ipilimumab monotherapy. In a phase II trial of T-VEC with ipilimumab vs. ipilimumab alone for advanced melanoma, in a 5-year analysis, the median PFS was 13.5 months with the combination and 6.4 months with ipilimumab, and estimated 5-year OS was 54.7% in the combination arm compared with 48.4% in the ipilimumab arm (NCT01740297).

Pexastimogene Devacirepvec (Pexa-VEC) is an oncolytic vaccinia virus that has demonstrated encouraging results in a phase Ib trial in combination with an anti-PD-1 inhibitor, cepilimumab, in patients with renal neoplasms. An overall response of 37.5% was reported for this combination therapy [19]. Adenovirus DNX-2401 is an OV that was investigated in a dose-escalated study, which included patients diagnosed with malignant glioma. In that study, a single dose of DNX-2401, administered intratumorally, obtained variable tumor responses. Currently, this OV is being investigated in a phase II trial in combination with pembrolizumab in patients with glioblastoma. An interim analysis revealed a median overall survival of 12 months and a response rate of 47% in 48 patients treated with the combination therapy.

The idea of triple therapy combining anti-PD1/PD-L1 with anti-CTLA4 and an OV has recently become very attractive, with a study in which an Ad in combination with anti-PD-L1 and anti-CTLA4 significantly inhibited tumor growth and prolonged survival in a TNBC orthotopic mouse model. A synergistic increase in the antitumor effect through the enhancement of the proportion of CD8 T and T memory cells and the depletion of Tregs and tumor-associated macrophages polarized to the M1 phenotype, which modulated the TME, was reported. This combination therapy has great potential and represents a promising strategy for improving cancer cell therapy that should be available shortly. Considering the safety, toxicity profile, and promising results, a growing number of ICIs and OVs are entering clinical trials.

Combination with cell therapy

Chimeric-antigen receptor (CAR)-T therapy represents a novel and promising treatment for cancer that produces effective and durable clinical responses. Synthetic receptors that redirect lymphocytes, principally T cells, eliminate cells expressing a specific target antigen. In 2017, anti-CD19 CAR-T cell therapy was approved by the FDA for the treatment of B-cell malignancies. There are some limitations of CAR-T cell therapy, including major toxicities, which limit its efficacy against solid tumors, antigen escape, limited persistence, and poor tumor infiltration. These have led to new approaches in which CAR-T therapy has been associated with other anticancer treatments to improve antitumor efficacy and diminish toxicities. Anti-CD19 CAR-T cell therapy has revolutionized the treatment of B-cell malignancies. Several studies have demonstrated that CD-19 is expressed on the surface of multiple OV-infected tumor cell types, harboring the CD19-T coding gene (OV19T). Consequently, CD-19 stimulates CAR-T cells and triggers tumor lysis, stimulating the release of OV19T from dead tumor cells that can activate endogenous antitumor immunity. Moreover, virotherapy can be concurrently combined with immunotherapy and CAR-T cells to increase the number of immune cells in the TME and improve the antitumor response. For example, Ads expressing IL12 can be administered simultaneously with systemic HER-2- and PD-L1-blocking antibodies. In preclinical tumor models, this association exhibited enhanced or synergistic effects. Moreover, the timing of the administration of anti-CTLA4, virotherapy, anti-PD-(L)1 antibodies, and emerging ICI antibodies represents another important area that requires further observation. To increase the efficacy and improve tumor responsiveness, systemic exploration of the effects of immunotherapy administered before, after, and concurrently with OVs and CAR-T cells is required.

Challenges and prospects

In recent years, with the development of genetic engineering technology, a better understanding of the function and structure of viral genes has been deepening, and optimizing the design and operation of viral genomes to produce nonpathogenic viruses has become the direction of OVs development, which has greatly promoted the progress of OVs therapy. Genetically modified OVs can target infected tumor cells, enhance the body's immune response, reduce toxic and side effects, and become a more stable expression vector. In 2015, T-VEC was approved for clinical treatment of patients with advanced melanoma, and in 2021, Delytact was approved for clinical treatment of malignant glioma. Oncolytic viral therapy has made a significant leap forward in cancer treatment, and oncolytic virotherapy has become an essential means of tumor immunotherapy.

Although several clinical trials exhibit the good safety, efficacy, and stability of OV therapy, the therapy still has some limitations to be addressed. There are also many difficulties in the process of saving the recombinant oncolytic virus. First, how to optimize the design of viruses from the source of the gene sequence to improve the targeting of OVs, avoid the impact on normal cells, and reduce the individual differences; second, how to avoid the OVs cleared by the human immune system before entering the tumor cells; third, the biological safety of OVs therapy should be further investigated, whether the OVs can gene rearrangement, gene mutation, and even produce super viruses; fourth, whether the type of OVs used in clinical treatment or trials is limited, and whether there are other more friendly and operable viruses to be further explored. To further solve the above problems, how to improve the treatment effect of OVs therapy, make patients benefit, still need more clinical trials to ensure the safety and effectiveness of OVs therapy, determine the best dose and route of administration to achieve the optimal treatment effect, and in addition, to build more types of OVs for clinical treatment were urgently needed.

The combination of OVs with chemotherapeutic or immunotherapy drugs, or even the combination of multiple OVs, may more effectively control tumor progression or cure tumors, providing promising clinical treatment options for patients with tumors. Accumulating clinical evidence indicates that OVs can enhance the response rates of tumors to immunoregulatory checkpoint inhibitors, CAR-T cells, and are an ideal object in combination treatment strategies. We believe that with the deepening of OV research, OV therapy will play a vital role in the treatment of human diseases.

References

[1] Bray F, et al. Global cancer statistics 2022: GLOBOCAN estimates of incidence and mortality worldwide for 36 cancers in 185 countries. CA Cancer J Clin 2024;74(3):229–63.

[2] Ferlay J, et al. Cancer statistics for the year 2020: an overview. Int J Cancer 2021. https://doi.org/10.1002/ijc.33588.

[3] Sung H, et al. Global cancer statistics 2020: GLOBOCAN estimates of incidence and mortality worldwide for 36 cancers in 185 countries. CA Cancer J Clin 2021;71(3):209–49.

[4] Lin C, Teng W, Tian Y, Li S, Xia N, Huang C. Immune landscape and response to oncolytic virus-based immunotherapy. Front Med 2024;18:411–29.

[5] Lee A. Nadofaragene firadenovec: first approval. Drugs 2023;83(4):353–7.

[6] de Graaf JF, Huberts M, Fouchier RAM, van den Hoogen BG. Determinants of the efficacy of viro-immunotherapy: a review. Cytokine Growth Factor Rev 2020;56:124–32.

[7] Buijs PR, Verhagen JH, van Eijck CH, van den Hoogen BG. Oncolytic viruses: from bench to bedside with a focus on safety. Hum Vaccin Immunother 2015;11(7):1573–84.

[8] Hu PY, et al. The limiting factors of oncolytic virus immunotherapy and the approaches to overcome them. Appl Microbiol Biotechnol 2020;104(19):8231–42.

[9] Tang G, Wang D, Zhao X, Feng Z, Chen Q, Shen Y. The dilemma of HSV-1 oncolytic virus delivery: the method choice and hurdles. Int J Mol Sci 2023;24(4).

[10] Guedan S, Rojas JJ, Gros A, Mercade E, Cascallo M, Alemany R. Hyaluronidase expression by an oncolytic adenovirus enhances its intratumoral spread and suppresses tumor growth. Mol Ther 2010;18(7):1275–83.

[11] Veyer DL, Carrara G, Maluquer de Motes C, Smith GL. Vaccinia virus evasion of regulated cell death. Immunol Lett 2017;186:68–80.

[12] Schirrmacher V. Cancer vaccines and oncolytic viruses exert profoundly lower side effects in cancer patients than other systemic therapies: a comparative analysis. Biomedicine 2020;8(3):61.

[13] Xu X, Sun Q, Mei Y, Liu Y, Zhao L. Newcastle disease virus co-expressing interleukin 7 and interleukin 15 modified tumor cells as a vaccine for cancer immunotherapy. Cancer Sci 2018;109(2):279–88.

[14] Heath WR, et al. Cross-presentation, dendritic cell subsets, and the generation of immunity to cellular antigens. Immunol Rev 2004;199:9–26.

[15] Clark GJ, Silveira PA, Hogarth PM, Hart DNJ. The cell surface phenotype of human dendritic cells. Semin Cell Dev Biol 2019;86:3–14.

[16] de Graaf JF, de Vor L, Fouchier RAM, van den Hoogen BG. Armed oncolytic viruses: a kick-start for anti-tumor immunity. Cytokine Growth Factor Rev 2018;41:28–39.

[17] Russell L, Peng KW, Russell SJ, Diaz RM. Oncolytic viruses: priming time for cancer immunotherapy. BioDrugs 2019;33(5):485–501.

[18] Abd-Aziz N, Poh CL. Development of oncolytic viruses for cancer therapy. Transl Res 2021;237:98–123.

[19] Lauer UM, Beil J. Oncolytic viruses: challenges and considerations in an evolving clinical landscape. Future Oncol 2022.

CHAPTER

13

Advances in immunotherapy for colorectal cancer

Shuai Lu[a], *Benqiang Rao*[b], *and Penghui Yang*[c]

[a]Department of General Surgery, Beijing Anzhen Hospital Affiliated to Capital Medical University, Beijing, China [b]Department of Gastrointestinal Surgery/Clinical Nutrition, Beijing Shijitan Hospital Affiliated to Capital Medical University, Beijing, China [c]Faculty of Hepatopancreatobiliary Surgery, The First Medical Center of Chinese PLA General Hospital, Beijing, China

Abstract

Colorectal cancer (CRC) is the third most common cancer globally and the second leading cause of cancer-related death. CRC has poor response to radiotherapy and chemotherapy and tends to develop resistance, necessitating further targeted therapies to improve patient quality of life and prognosis. In recent years, immune therapy has rapidly become an exciting and promising cancer treatment approach. Immune checkpoint blockade therapy has made significant progress in the treatment of advanced malignancies, and patients who benefit from this therapy may achieve long-term and sustained responses. Since the approval of the first anti-PD-1/PD-L1 drug pembrolizumab for second-line treatment of advanced melanoma on September 4, 2014, and subsequently for second-line treatment of mCRC in 2017, immune therapy has seen great popularity over the past decade. However, in the field of CRC, immune therapy is only effective in a limited number of patients with microsatellite instability-high (MSI-H), and initial responders may develop acquired resistance. With further research on Immune checkpoint blockade (ICB) treatment, the mechanisms of immune tolerance in CRC patients are gradually becoming clear, and new predictive biomarkers continue to emerge. This article reviews the immune therapy strategies for CRC patients, discusses the mechanisms and applications of ICB in CRC treatment, outlines potential biomarkers for ICB efficacy, and lists other immune therapy strategies, in order to provide a theoretical basis and practical direction for CRC immune therapy strategies.

Abbreviations

ACT adoptive cell therapy
APCs antigen-presenting cells
CAR-T chimeric antigen receptor T-cell
CRC colorectal cancer

https://doi.org/10.1016/B978-0-443-13827-0.00020-2

CTLA-4 cytotoxic T-lymphocyte-associated protein 4
DCR disease control rate
EGFR epidermal growth factor receptor
GM-CSF granulocyte-macrophage colony-stimulating factor
ICB immune checkpoint blockade
ICIs immune checkpoint inhibitors
IDO indoleamine 2,3-dioxygenase 1
LAG-3 lymphocyte activation gene 3
mCRC metastatic CRC
MDSCs myeloid-derived suppressor cells
MEK mitogen-activated protein kinase
MMR mismatch repair
MSI microsatellite instability
MSI-H microsatellite instability-high
MSI-L low-level microsatellite instability
MSS microsatellite stability
NSCLC non-small cell lung cancer
ORFV Orf virus
ORR response rate
OS overall survival
PC peritoneal carcinomatosis
PD-1 programmed cell death protein 1
PDL1 programmed cell death 1 ligand 1
PFS progression-free survival
POLE polymerase epsilon
TCR-T T-cell receptor-engineered T-cell
TILs tumor-infiltrating lymphocytes
TMB tumor mutational burden
TME tumor microenvironment
Tregs regulatory T cells
VEGFR vascular endothelial growth factor receptor
VV vaccine virus

Conflict of interest

All authors have declared that no conflict of interest exists.

Introduction

Colorectal cancer (CRC) is the third most common cancer globally and the second leading cause of cancer-related death. It is estimated that by 2030, CRC will account for over 2.2 million new cases and 1.1 million deaths [1]. Although screening has reduced the incidence and mortality of CRC, approximately 25% of CRC patients are diagnosed with advanced disease, and among early-stage disease patients, 25–50% will experience metastasis [2]. The 5-year survival rate for non-metastatic patients after tumor resection and systemic therapy is 40%, while the 5-year survival rate for patients with metastatic CRC (mCRC) is only around 20% [3]. Conventional treatments for CRC include surgery, radiation therapy, chemotherapy, and targeted therapy. For localized cancer, when the tumor is easily accessible, total mesorectal excision or laparoscopic or transanal surgery may be the best options [4]. In most cases, complete eradication of all cancer cells is usually not feasible. Therefore, adjuvant chemotherapy or radiation therapy is often used in approximately 66% of stage II colon cancer patients and 61% of stage III rectal cancer patients [5]. However, the nonspecificity and cytotoxicity of these treatments often lead to many side effects [6]. Despite receiving neoadjuvant therapy, the recurrence rate remains high at 54% [7]. Therefore, CRC patients are in urgent need of more effective treatment strategies.

Immunotherapy aims to harness the immune system to fight cancer. Immune checkpoint inhibitors (ICIs) can enhance the anti-tumor immune response by modulating the interaction between T cells, antigen-presenting cells (APCs), and tumor cells. Due to the success of immunotherapy in preventing previously incurable solid tumors such as melanoma and lung cancer, it has gained great popularity in recent decades. When many types of tumors have a high tumor mutational burden (TMB), they are responsive to immunotherapy [8]. The highly mutated or microsatellite instability (MSI)-high state of dMMR/MSI-H CRC was approved in 2017 for immune checkpoint blockade treatment. In contrast, ICIs are ineffective in tumors that have both microsatellite stability (MSS) and low-level microsatellite instability (MSI-L). These tumors have low mutation loads and lack immune cell infiltration, which is thought to be a mechanism of immune resistance [9]. Unlike chemotherapy and radiation therapy, cancer immunotherapy addresses the issue of specificity. By alerting the immune system to foreign substances found in malignant cells, cancer immunotherapy triggers a coordinated immune response that leads to disease eradication. Importantly, cancer antigens selectively target cancer cells without affecting normal cells.

This section discusses the current immune therapy approaches for dMMR/MSI-H CRC, including ICI monotherapy, adoptive cell therapy, oncolytic viruses, and cancer vaccines. Additionally, we discuss potential biomarkers from existing clinical data to predict the efficacy of immunotherapy.

ICIs in dMMR/MSI-H CRC

dMMR status is often caused by mutations in the mismatch repair (MMR) proteins (MLH1, MSH2, MSH6, and PMS2), which are responsible for recognizing and correcting mismatched nucleotides, or by methylation of the MLH1 gene promoter [10]. If the MMR system fails, mismatches can accumulate and lead to changes in microsatellite sequence length or composition, known as microsatellite instability (MSI), and result in a highly mutated phenotype. The tumor microenvironment (TME) of dMMR/MSI-H CRC exhibits strong upregulation of immune checkpoint proteins, such as PD-L1, CTLA-4, lymphocyte activation gene 3 (LAG-3), and indoleamine 2,3-dioxygenase 1 (IDO), and produces tumor-associated immunogenic antigens [11]. Due to the defective recognition and repair of spontaneous mutations, the number of somatic mutations in dMMR/MSI-H CRC is 10–100 times higher than that in pMMR/MSS CRC, and almost all significantly differentially mutated genes are more common in dMMR/MSI-H CRC [12]. These observations have drawn attention to the feasibility of checkpoint blockade in the treatment of dMMR/MSI-H CRC.

Several clinical studies have evaluated the therapeutic potential of single ICI therapy. A phase II trial (NCT01876511) tested the efficacy of the PD-1 inhibitor pembrolizumab in 3 groups of 41 patients: (i) dMMR/MSI-H mCRC patients, (ii) pMMR mCRC patients, and (iii) dMMR non-CRC patients. At 20 weeks, the researchers reported an objective response rate (ORR) of 40% and a progression-free survival (PFS) of 78% in dMMR/MSI-H mCRC patients, while no patients in the pMMR mCRC group achieved objective response, and the PFS was only 11% [13].

In 2018, the KEYNOTE-164 trial (NCT02460198) evaluated the efficacy of pembrolizumab in the treatment of MSI-H mCRC. Among the 63 enrolled patients, the objective response rate (ORR) was 32% and the median PFS was 4.1 months. The median overall survival (OS) has not been reached. The 1-year progression free survival rate and overall survival rate were 41% and 76%, respectively [14].

The subsequent update published in 2020 also showed that single ICI has durable antitumor activity in the treatment of dMMR/MSI-H or mCRC patients [15]. The KEYNOTE-177 trial (NCT02563002) is a phase III trial comparing pembrolizumab with chemotherapy as first-line treatment for dMMR/MSI-H mCRC. In the 307 enrolled patients, 153 received pembrolizumab monotherapy and 154 received chemotherapy. At a median follow-up of 32.4 months, the median PFS was 16.5 months in the pembrolizumab group and 8.2 months in the chemotherapy group ($HR = 0.60$, $P = 0.0002$). The 12-month and 24-month PFS rates were 55% and 48% in the pembrolizumab group, while the rates were 37% and 19% in the chemotherapy group, showing a significant difference [16]. Based on the convincing data from this study, pembrolizumab was approved by the FDA in June 2020 for the first-line treatment of dMMR/MSI-H advanced unresectable or mCRC patients.

Another PD-1 inhibitor, nivolumab, is also being evaluated in some clinical studies for dMMR/MSI-H CRC. The phase II trial CheckMate142 (NCT02060188) investigated the efficacy of nivolumab in 74 patients with dMMR/MSI-H mCRC. With a median follow-up of 12 months, the objective response rate was 31% (23/74), and the median progression-free survival was 14.3 months. The median overall survival rate was not reached. The 12-month progression-free survival rate and overall survival rate were 50% and 73%, respectively. In this cohort trial, the safety profile of nivolumab was consistent with previous reports in other solid tumor studies, and no new safety events were identified [3]. Based on these study data, nivolumab was approved by the FDA in August 2017 for the treatment of dMMR/MSI-H mCRC in adults and children over 12 years old.

Meanwhile, the CheckMate142 study also explored the combination of nivolumab with the CTLA-4 monoclonal antibody ipilimumab. A total of 119 patients with refractory dMMR/MSI-H mCRC received nivolumab 3 mg/kg in combination with ipilimumab 1 mg/kg every 3 weeks for four doses, followed by nivolumab 3 mg/kg every 2 weeks. With a median follow-up of 13.4 months, the objective response rate was 55%. Eighty percent of patients achieved disease control for $\geq$12 weeks, and the median progression-free survival was not reached. The 9-month and 12-month progression-free survival rates were 76% and 71%, respectively. The median overall survival was not reached, and the 9-month and 12-month overall survival rates were 87% and 85%, respectively [17].

These data indicate that combination ICIs can provide clinical benefits for patients with dMMR/MSI-H mCRC. It should be noted that the combination of nivolumab and ipilimumab is associated with relatively higher grade 3/4 toxicities (32%), with the most common being elevated aspartate aminotransferase and/or alanine aminotransferase (11%), elevated lipase (4%), anemia (3%), and colitis (3%) [17]. Based on the results of this trial, the FDA accelerated the approval of nivolumab in combination with ipilimumab for the treatment of dMMR/MSI-H mCRC patients in July 2018. The latest data update of the trial with 2 years of follow-up was presented at the ASCO 2020 virtual meeting. With a median follow-up of 29 months, the objective response rate and disease control rate were 69% (95% CI, 53–82%) and 84% (95% CI, 70.5–93.5%), respectively. The median progression-free survival and median overall survival were not reached, and the 2-year progression-free survival rate and overall survival rate were 74% and 79%, respectively [18].

In conclusion, these data suggest that single-agent and combination ICIs provide clinical benefits for dMMR/MSI-H mCRC patients.

Application of ICIs in pMMR/MSS CRC

pMMR/MSS CRC is considered a "cold tumor" and accounts for approximately 95% of all mCRC. Unlike dMMR/MSI-H inflammatory tumors, monotherapy with ICIs has shown poor efficacy in pMMR/MSS cancer treatment. Results from studies on combination ICIs therapy suggest that the immunotherapy response in this population can be improved. Several studies evaluating the combination therapy of ICIs with chemotherapy, targeted therapy, ICI therapy, and radiation therapy have been conducted to explore how to make cold tumors hot and improve sensitivity to immunotherapy.

ICIs in combination with chemotherapy

In patients with refractory cancer receiving conventional chemotherapy and immunotherapy, particularly non-small cell lung cancer (NSCLC), triple-negative breast cancer, and gastric/gastroesophageal junction carcinoma the combination of ICIs with systemic chemotherapy has been shown to significantly improve tumor response compared to single-agent ICI therapy. Evidence suggests that immunogenic chemotherapy can enhance the effects of ICIs by inducing the infiltration of CD4+ and CD8+ T cells into tumors and disrupting the activity of immune-suppressive cells such as regulatory T cells (Tregs) and myeloid-derived suppressor cells (MDSCs) [19]. There are few studies evaluating the safety and efficacy of combined chemotherapy and ICIs in the treatment of pMMR/MSS CRC.

One single-arm, multicenter phase 1b trial (NCT02375672) was designed to test the efficacy of pembrolizumab in combination with mFOLFOX6 as a first-line treatment for primary mCRC. Among the 30 patients enrolled in this trial, 87% (26/30) had pMMR/MSS mCRC. In the median follow-up of 19.9 months, the authors observed an objective response rate of 57% (17/30) in the selected patients. The average response duration was 37.57 weeks. The median progression-free survival was 8.8 months (80% CI 7.7-11.3 months), and the median overall survival was not reached (80% CI NR-NR) [20]. Another ongoing trial, the METIMMOX trial (NCT03388190), is evaluating the combination of oxaliplatin-based chemotherapy (FLOX) with nab-paclitaxel and FOLFOX or FLOX alone as first-line treatment for MSS mCRC. Preliminary results presented at the 2021 ASCO Annual Meeting showed a median progression-free survival of 6.6 months (0.5-20 months) and 5.6 months (0.5–15 months) in the chemotherapy plus nab-paclitaxel and chemotherapy alone groups, respectively. The complete response rate was 0% in the chemotherapy alone group and 16% in the chemotherapy plus nab-paclitaxel group, while the objective response rate was 23% and 32%, respectively. The incidence of grade $\geq$3 FLOX-related adverse events was comparable between the two groups. Twelve immune-related grade 3–4 adverse events were reported, with no new safety signals. The results of this study suggest that short-term oxaliplatin-based chemotherapy may stimulate the responsiveness of MSS mCRC patients to ICIs.

ICIs in combination with targeted therapy

Preclinical data from animal models have shown that targeted therapies such as anti-vascular endothelial growth factor receptor (VEGFR) and anti-epidermal growth factor receptor (EGFR) monoclonal antibodies, as well as mitogen-activated protein kinase (MEK) inhibitors, can enhance the sensitivity of tumors to checkpoint blockade [21]. Preclinical studies have also found that activation of the oncogenic EGFR pathway upregulates the expression of programmed death-ligand 1 (PD-L1), while EGFR inhibitors decrease the expression of PD-L1 in tumor cell lines with activated EGFR. Based on these compelling data, a phase 1b trial, REGONIVO (NCT03406871), was initiated to study the effects of the VEGFR inhibitor regorafenib in combination with nivolumab on PD-1 inhibition [22]. This study includes two cohorts of patients with refractory advanced GC and CRC. In the 11 GC patients (44%) and 9 CRC patients (36%). Only one out of 25 CRC patients had MSI-H status. After excluding this patient, the objective response rate for MSS CRC patients reached 33% (8/24), which was much higher than the reported incidence in previous studies on MSS CRC. Updated survival data in 2020 showed a median progression-free survival of 7.9 months (95% CI, 2.9-NR) and a median overall survival that was not reached (95% CI, 9.8-NR) for CRC. The 12-month progression-free survival and overall survival rates for CRC were 42% and 68%, respectively. Overall, the combination of regorafenib and nivolumab showed significant efficacy in the salvage therapy of CRC.

The ongoing Phase 3 study, REGONIVO (NCT03866269), is planned to further investigate the long-term follow-up of MSS CRC. The results of future studies may change clinical practice. Another Phase 2 trial, REGOMUNE (NCT034759553), is evaluating the effectiveness of regorafenib in combination with the PD-L1 inhibitory antibody avelumab. In the 43 patients with metastatic or unresectable MSS CRC enrolled in the study, the median follow-up time was 7.2 months. None of the patients achieved objective remission, 23 (53.5%) patients had stable disease, and 17 (39.5%) patients had disease progression. The median progression-free survival and overall survival were 3.6 months (95% CI, 1.8-5.4) and 10.8 months (95% CI, 5.9-NA), respectively [23]. The IMblaze370 trial (NCT02788279) investigated the combination of the PD-L1 inhibitor atezolizumab and the MEK inhibitor cobimetinib with standard third-line therapy regorafenib. In the phase 1b study, the combination of atezolizumab and cobimetinib showed a higher objective response rate (8%) and disease control rate (31%) compared to atezolizumab monotherapy or regorafenib. The results of the phase 3 study published in 2019 showed a median overall survival of 8.9 months (95% CI 7.0–10.6) for atezolizumab plus cobimetinib, 7.1 months (95% CI 6.1–10.1) for atezolizumab monotherapy, and 8.5 months (95% CI 6.4–10.7) for regorafenib. There was no significant difference in overall survival among the three groups. However, the incidence of grade ≥3 adverse events was higher in the combination therapy group. Several ongoing trials are evaluating the combination of ICIs and targeted therapy in pMMR/MSS CRC patients.

ICIs in combination with radiation therapy

Preclinical animal models have demonstrated that radiation therapy in combination with PD-L1 antibodies enhances the infiltration of CD8+ T cells and reduces the local accumulation of myeloid-derived suppressor cells and regulatory T cells [24]. In the tumor microenvironment, radiation-induced cell damage enhances the efficacy of immunotherapy by inducing the "abscopal effect," which upregulates immune responses. Therefore, radiation therapy is considered to have immunogenic potential in solid tumors and increases sensitivity to ICI responses [25]. One phase 2 trial, NCT03104439, recently investigated the combination of radiation therapy with dual ICI (ipilimumab and nivolumab) in 40 MSS mCRC patients. In this study, the disease control rate and response rate in the intention-to-treat population were 17.5% (7/40) and 7.5% (3/40), respectively. The median time to disease control in the intention-to-treat population was 77 days. At 3 months, the time to achieving disease control ($n = 7$) was 252 days, while the time to disease progression in patients ($n = 17$) or those who did not receive radiation therapy due to clinical progression ($n = 13$) was 70.5 days. Treatment-related adverse events (TRAEs) were reported in 22/40 (55%) patients, 20/40 (50%) of whom experienced grade $\geq$3 toxicities, with fatigue, nausea, vomiting, diarrhea, infusion-related reaction, and dyspnea being the most common. However, it is noteworthy that 13/40 patients did not receive radiation therapy due to clinical disease progression. The latest results of this study showed a disease control rate of 25% (10/40) and an objective response rate of 10% (4/10) in 40 MSS mCRC patients, with a median progression-free survival and median overall survival of 2.4 months and 7.1 months, respectively. Among the 27 patients who received radiation therapy, disease control and objective response rates were better, at 37% (10/27) and 15% (4/27), respectively, with a median progression-free survival of 2.5 months and a median overall survival of 10.9 months.

Adoptive cell therapy

As an important approach for tumor immunotherapy, adoptive cell therapy (ACT) involves the administration of immunologically active cells that have been expanded and modified ex vivo with direct anti-cancer activity for anti-tumor purposes [26]. Currently, three forms of ACT are being developed for cancer treatment: chimeric antigen receptor T-cell (CAR-T) therapy, tumor-infiltrating lymphocytes (TILs), and T-cell receptor-engineered T-cell (TCR-T) therapy.

CAR-T cell therapy involves ex vivo modification of T cells in a MHC-independent manner, enabling them to target tumor antigens and generate corresponding anti-cancer immune responses. CAR-T cell therapy has shown significant efficacy in leukemia, multiple myeloma, and certain types of lymphoma, demonstrating great potential for application and development. Regarding the treatment prospects of CAR-T cell therapy in colorectal cancer, a large number of clinical studies are currently being conducted to evaluate its safety and efficacy. Cyad-01 is a CAR-T therapy targeting the NKG2D protein, which is a target that is either not expressed or expressed at low levels in normal cells; however, the expression of these ligands is significantly increased when cells are infected or undergo malignancy. The ongoing Phase II trial THINK (NCT03018405) is evaluating the use of the multi-dose regimen of CYAD-01 alone in seven types of refractory cancer, including five solid tumors and two hematologic tumors. The study consists of two sequential parts, including a dose-escalation part and an expansion part. The dose-escalation part focuses on evaluating three different doses of CYAD-01, namely 3 × 10^8, 1 × 10^9, and 3 × 10^9 CYAD-01 injected every two weeks. The expansion part evaluates the recommended dose for the seven different tumor types defined in the first part [27]. CYAD-101 is a non-gene encoded homozygous form of CYAD-01. The Phase I AlloSHRINK trial (NCT03692429) is evaluating the safety and clinical efficacy of three doses of CYAD-101 in combination with chemotherapy in refractory mCRC patients. Among the 15 patients, 2 patients (13.3%) achieved clinical partial response, 9 patients (60.0%) achieved stable disease, and the median overall survival was 10.6 months. The treatment was well tolerated, with no grade 3 or higher TRAEs. A range of ongoing clinical trials targeting CD133 (NCT02541370), CEA-positive tumors (NCT02349724), HER2 (NCT03740256), and other targets are further evaluating the efficacy and safety of CAR-T therapy in mCRC.

TIL therapy involves targeting T cells against multiple antigens in cancer cells, thus stimulating cytotoxic responses against cancer cells through multiple targets. It is generally believed that the increase in CD8+ T cells plays a key role in enhancing the clinical activity of TIL therapy. Following the demonstrated durable efficacy of TIL therapy in metastatic melanoma, encouraging results have been observed in the mCRC population. Tran et al. reported a case of an mCRC patient who received TIL therapy targeting the KRAS G12D mutation. After 9 months of treatment, 6 out of 7 tumors in the patient's lung metastases regressed significantly, with 3 tumors completely disappearing. The researchers further identified CD8+ T cells that recognize the mutant KRAS G12D specifically. In addition, immune checkpoint inhibitors and concurrent radiation therapy may enhance the anti-tumor response of TIL therapy by eliminating potential inhibitory immune cells [28].

Oncolytic viruses

Using the vaccine virus (VV) as a smallpox vaccine in humans is safe. Recombinant VVs have been used to treat patients with advanced solid tumors and tumor models as expression vectors, enhancing tumor lysis. The VG9-IL-24 strain was created to target vaccinia virus. VG9-IL-24 induces apoptosis in colorectal cancer cells through multiple apoptotic signaling pathways, inducing a specific and persistent immune response against colorectal cancer. Peritoneal carcinomatosis (PC) usually develops after the spread of colorectal cancer (CRC) to the peritoneal cavity, and approximately one-fourth of CRC patients experience peritoneal metastasis, which is considered as an end-stage of the disease. It has a poor prognosis and is a major cause of patient death. In one study, researchers found that infection with the tumor lytic vaccinia virus JX-594 expressing GM-CSF effectively inhibited the metastasis of peritoneal carcinomatosis cells by infecting and lysing the peritoneal carcinomatosis cells and activating CD8+ T cells in the peritoneum, restoring the anti-cancer immune function [29].

Reovirus

The oncolytic reovirus pelareorep selectively lyses colorectal tumor cells after infection with mutated kras. Researchers found that pelareorep upregulates several autophagy proteins and induces apoptosis in colorectal cancer cells, leading to increased cell apoptosis and death. In addition, pelareorep can stimulate the activation of CD4+ T cells and CD8+ T cells in patients with KRAS-mutant colorectal cancer, enhancing immune efficacy [30].

HSV

There are two types of herpes simplex viruses (HSV-1 and HSV-2). Researchers have enhanced the anti-tumor immune response by engineering the HSV-1 virus to express the IL-12 gene and infect and kill colorectal cancer cells. The expression of IL-12 in HSV-1 infected and killed colorectal cancer cell lines enhanced the cytolytic activity of natural killer cells and cytotoxic T lymphocytes. oHSV-2 can effectively eliminate primary tumors, generate tumor-specific immunity, and induce immune memory to suppress tumor recurrence and metastasis. Viral therapy can reshape the immune status of the spleen and tumor microenvironment in mice, further improving the therapeutic anti-tumor effect. In addition, oHSV-2 has shown significant anti-tumor activity when combined with immune checkpoint inhibitors in patients with metastatic colorectal cancer.

Adenovirus

The efficacy of oncolytic adenoviruses in the treatment of colorectal cancer has been improved by genetically modifying the adenovirus genome. SPDD-UG carries the SNORD44 (a C/D box snoRNA) and GAS5 genes, which inhibit the growth and induce apoptosis of colorectal cancer cells. Adenoviruses expressing granulocyte-macrophage colony-stimulating factor (GM-CSF) and core protein polysaccharide have also been found to inhibit the growth and distant metastasis of colorectal cancer cells. Additionally, the combination of CD55-TRAIL and luteolin can inhibit the proliferation of colorectal cancer cells while minimizing cytotoxicity to normal cells. According to a related experimental study, when adenoviruses encoding multiple novel antigens are combined with ICIs, tumor growth can be controlled in mice, and a new type of GAd with high tumor clearance potential is generated. The combination of adenoviruses and ICIs in the treatment of colorectal cancer is still in the early stage of exploration, and it is believed that modifying adenoviruses and combining them with ICIs remains a promising treatment strategy.

Other viruses

Colorectal cancer can be effectively induced to undergo apoptosis by oncolytic measles viruses encoding the IL-12 gene and activating the IL-12/interferon-c/TNF-α pathway. Orf virus (ORFV) can potentially serve as an oncolytic virus vector because it can stimulate both innate and acquired immune responses in vivo. In colon cancer cells, the ORFV strain NA1/11 induces apoptosis and inhibits the growth or metastasis of colorectal cancer. Researchers have also created the vesicular stomatitis virus (VSV) clone M51R with engineered M protein to inhibit the peritoneal surface metastasis of colorectal cancer in vivo. By reversing the immune-suppressed microenvironment of the peritoneum, M51R VSV enhances its oncolytic potential. In a mouse model of colorectal cancer, infection and lysis of CRC cells by coxsackievirus B3 strain PD were effective and showed good safety, which was obtained from the recombinant CVB3 variant PD-H. PD-H significantly reduced the malignancy of tumors in immune-active mice and inhibited the progression of colorectal cancer.

Tumor vaccines

Tumor vaccines are one of the newest treatment methods in cancer immunotherapy. Unlike immune checkpoint inhibitors and cell penetration therapies, cancer vaccines enhance immune responses against cancer cells by targeting transgenic genes or neoantigens, using the body's own immune function to kill tumor cells, provide anti-tumor effects, and can be continually monitored to prevent their regrowth. Over the past few decades, cancer vaccines have been extensively studied in animal models and humans. While there are currently no approved vaccines for advanced colorectal cancer, related research is ongoing.

In 1973, Professors Zanvil Cohn and Ralph Steinman discovered dendritic cells in the spleen of mice, a novel type of immune cell that plays a critical role in connecting innate and acquired immunity. In 2010, the vaccine Sipuleucel-T based on autologous immunotherapy became the first vaccine approved by the FDA for the treatment of castration-resistant prostate cancer. DC vaccine therapy targets and focuses on cancer cells through the recognition of tumor cell-specific antigens, effectively inhibiting the proliferation of residual lesions and enhancing the function of the patient's immune system. Multiple prospective trials have studied the effects of autologous tumor lysate DC vaccines against colorectal cancer. For example, in two early-phase Phase II trials, favorable efficacy and beneficial survival outcomes were observed for autologous tumor lysate DC vaccines in mCRC [31].

In addition, a phase II clinical trial compared the efficacy of autologous tumor lysate dendritic cell (DC) vaccine with placebo in mCRC patients. In this trial, a tumor-specific immune response was observed with the vaccine, but the study was terminated early as the results showed no benefit in median progression-free survival or median overall survival in the group of 28 patients who received DC vaccine [32]. In the immune microenvironment, the maturity of DC, tumor tolerance, and the expression of co-stimulatory ligands are the major factors that influence the immune response to DC vaccination. This may explain the conflicting results observed in the above studies. Combining DC vaccination with other cancer treatment modalities such as chemotherapy, radiation therapy, and ICIs may activate antigen-specific effector T cells and may be an ideal approach to overcome tumor immune suppression.

Biomarkers for immune therapy response

The response of mCRC patients to ICIs is not ideal, and there is a significant variability in patient sensitivity to immune therapy. Therefore, its application in clinical practice is inevitably limited. It is urgent to identify biomarkers that can effectively distinguish between sensitive and resistant populations to select appropriate treatment strategies. Recent studies predicting the efficacy of immune checkpoint inhibitors have mainly focused on factors such as dMMR/MSI, TMB, PD-L1 expression, tumor-infiltrating lymphocyte (TIL) density, and POLE.

Microsatellite instability

The MMR genes encode corresponding mismatch repair proteins. Four major genes associated with genomic instability are MLH1, MSH2, MSH6, and PMS2. Deficiencies in the DNA mismatch repair system can lead to MSI. Generally, MSI detection evaluates five selected microsatellite loci: Bat-25, Bat-26, d5s346, d2s123, and d17s250. MSI status is classified into high instability (MSI-H), low instability (MSI-L), and stability (MSS). MMR is divided into deficient mismatch repair and proficient mismatch repair. Clinically, dMMR is equivalent to MSI-H, and pMMR is equivalent to MSI-L or MSS. The accumulation of mutations and unrepaired genes can lead to MSI-H and carcinogenesis. Approximately 15% of CRC is caused by the activation of MSI pathway. MSI-H is associated with a large number of neoantigens, which can lead to higher immunogenicity and effective immune responses. In colorectal cancer, MSI is an excellent predictive biomarker, and the FDA has approved Pembrolizumab for the treatment of MSI-H/dMMR mCRC patients.

Tumor mutation burden

TMB is a quantifiable biomarker, and its exact definition depends on the size and location of the test region and the types of mutations included. Tumors with high TMB are considered to have a high clonal neoantigen load, making them immunogenic and responsive to immune therapy. TMB is currently used to predict the immunotherapy efficacy in various types of solid tumors. However, there is limited clinical research data on using TMB as a biomarker to predict the immune therapy response in mCRC. One study emphasized the utility of TMB as a predictive biomarker for anti-PD-1/PD-L1 immune therapy in 22 patients with MSI-H mCRC. The cohort study found that responders had a median TMB of 54 mut/Mb, while non-responders had a median TMB of 29 mut/Mb. Subsequently, the enrolled patients were divided into high TMB and low TMB groups, and all 13 patients in the high TMB group achieved remission, while 6 out of 9 (66.7%) patients in the low TMB group experienced disease progression. Based on these data, the optimal threshold estimate for TMB was determined to be 37–41 mut/Mb. In the CCTG CO.26 trial (NCT02870920) of ICI combination therapy in advanced refractory CRC, the authors observed that patients with high TMB ($\geq$28 mt/Mb) had a greater overall survival benefit compared to patients with low TMB ($<$28 mt/Mb). Similarly, consistent findings were reported in another trial, the REGONIVO study. This study divided colorectal cancer patients into high and low expression groups based on the optimal threshold value of TMB 22.6. Among the 25 patients included in the study, it was found that the high TMB group had a median overall survival of 12.5 months compared to 7.9 months in the low TMB group. Overall, TMB seems to be a predictive indicator of the efficacy of immunotherapy in advanced colorectal cancer. Further exploration of this conclusion requires more large-scale prospective phase III clinical studies. However, TMB also has limitations as some high TMB patients do not respond to immune therapy, or low TMB patients can achieve optimal treatment responses. This is because high TMB does not necessarily correspond to high tumor neoantigen levels. Therefore, tumor neoantigen burden (TNB), which reflects the total amount of neoantigens in tumor cells, can be used as a complementary indicator to the TMB index [33].

Expression of PD-L1

The detection of PD-L1 expression on tumor cells or tumor-infiltrating immune cells has become the most widely used biomarker in ICI therapy. For example, pembrolizumab has specific indications for PD-L1-expressing NSCLC and GC. In 2017, the National Comprehensive Cancer Network guidelines stated that if the PD-L1 expression rate is $\geq$50%, the PD-1 monoclonal antibody pembrolizumab can be used as a first-line treatment for patients with primary NSCLC without significant driving mutations (EGFR, ALK, or ROS1). Based on the phase II study KEYNOTE-059 (NCT02335411), the FDA accelerated the approval of pembrolizumab for the treatment of PD-L1 (+) advanced GC patients resistant to chemotherapy. Among the enrolled patients, 143/259 (55.2%) expressed PD-L1. The objective response rate in PD-L1 (+) patients was 16%, significantly higher than the 6% in PD-L1 (−) patients.

However, there is still controversy over the relationship between PD-L1 expression and the efficacy of ICIs in mCRC based on existing clinical research data. In the REGONIVO trial, the median progression-free survival in CRC patients with PD-L1 CPS ≤ 1 ($n = 8$) was 6.0 months, while it was not reached in patients with PD-L1 CPS > 1 ($n = 16$). Although the sample size of this study was small, it was observed that patients with higher PD-L1 CPS scores were more likely to benefit in terms of survival outcomes. In contrast, the results from the KEYNOTE-016 trial for advanced CRC showed no significant correlation between PD-L1 expression and patients' progression-free survival or overall survival. Similarly, the Checkmate142 study also obtained negative results, indicating that the effectiveness in MSI-H patients and ineffectiveness in MSS patients were not related to PD-L1 expression levels.

The biological characteristics of PD-L1 may lead to inconsistent utility of PD-L1 as a biomarker and predictor of immune checkpoint blockade response. For example, tumors are heterogeneous, and PD-L1 expression varies between different regions within the same lesion and different lesions. In addition, PD-L1 expression is inducible and dynamic, and different treatment modalities can affect PD-L1 expression at different stages of treatment. Therefore, PD-L1 is not a perfect biomarker for predicting the efficacy of immune therapy.

Tumor-infiltrating lymphocytes (TILs)

TILs are an important component of the tumor microenvironment (TME), and their abundance is associated with PD-L1 upregulation and various clinical benefits. The anti-tumor effects of ICB require the involvement of lymphocytes in the vicinity of the tumor. Thus, the abundance of TILs can also serve as a marker for predicting the efficacy of ICB. In general, immunohistochemical analysis is used to assess the infiltration of CD8+ T cells in tumor tissue. Among the characteristics of TILs, a higher proportion of CD28+ in TIL cells usually indicates a higher response to treatment. It has been reported that only a small fraction of CD8+ TILs are sensitive to tumor antigens, while the majority of TILs are not. Further research has shown that the differential expression of CD39 is a key factor in distinguishing the sensitivity of CD8+ TILs [34]. CD39 is a molecule with chronic antigen persistent stimulation, indicating that the activity of this type of TIL is suppressed. They retain anti-tumor capabilities but require additional stimuli to overcome the inhibition, which ICB can provide. Therefore, the higher the proportion of CD39+ in CD8+ TILs, the more likely the PD-1/PD-L1 signaling axis is involved. Thus, CD39+ is also a potential marker. In addition, TMIT (tumor microenvironment immune types) has been constructed based on PD-L1 expression and TILs, and it has been divided into four groups to describe different TMEs. The characteristic of TILs is the expression of CD8A mRNA and cytolytic activity score (CYT, GZMA) based on the mRNA expression levels of GZMA and PRF1. This stratification emphasizes the importance of PD-L1 expression and TIL recruitment. PD-L1-positive TILs are categorized as TMIT I type. ICB therapy can benefit patients with PD-L1-positive TILs. Furthermore, in CRC, the mutation or neoantigen quantity of cancers of TIMT I type (high PD-L1 and CD8A/CYT) is significantly higher than that of cancers of TIMT II type (low PD-L1 and CD8A/CYT). The stratification of TMIT serves as a supplementary method for distinguishing "cold" tumors from "hot" tumors and developing optimal immune therapy strategies. Examining TMIT can quantitatively predict immune infiltration based on mRNA-seq analysis.

Polymerase epsilon (POLE) mutations

POLE is a key enzyme in DNA synthesis and repair, and the proofreading function of POLE is crucial for ensuring the accuracy of replication in the cell cycle and the proper replication of the genome. After POLE mutations, DNA repair defects and genetic material errors cannot be repaired. Over time, a large number of mutations accumulate, even up to 10 times that of MSI-H CRC [35]. Retrospective analysis has shown that among 6517 CRC patients, 66 (1.0%) had somatic POLE mutations. Patients with POLE mutations tend to be relatively young and male, and their lesions are mainly on the right side. At the initial diagnosis, they are also in the early stage of the disease (stage II-III). Approximately 1–2% of MSS CRCs exhibit POLE mutations, and the frequency can reach between 5% and 7% in patients under 50 years old. Previous studies have indicated that colorectal cancer patients with POLE mutations often have elevated levels of TILs, upregulated PD-L1 expression, increased expression of cytotoxic T cell markers and effector cytokines, suggesting enhanced tumor immunogenicity. POLE mutations appear to be a promising new predictor for the response to ICB therapy. Wang et al. demonstrated that POLE mutation is an independent biomarker for predicting the efficacy of immunotherapy in multiple cancer types [36]. It is worth noting that MSS CRC patients with POLE mutations exhibit durable clinical responses in ICB therapy. This suggests that POLE mutations are a promising indicator for improving the benefits of immune therapy in MSS mCRC patients.

Conclusion

Immunotherapy for CRC has made great progress in recent years. Based on compelling data from two Phase 2 clinical trials, pembrolizumab and nivolumab (alone or in combination with Ipilimumab) have been approved by the FDA for second-line treatment of dMMR-MSI-H mCRC patients. Based on the convincing data from the KEYNOTE-177 trial, pembrolizumab was approved by the FDA in 2020 for first-line treatment of this subset. Several ongoing trials are exploring the potential benefits of ICIs as first-line treatment for dMMR-MSI-H CRC patients and early-stage dMMR-MSI-H CRC patients, and are highly anticipated. However, the key challenge remains to overcome primary resistance to immunotherapy in the majority of pMMR-MSS CRC patients. Various strategies based on ICIs have been tested in the subgroup to modulate immune cells and improve treatment efficacy, such as dual blockade with anti-PD-1 and anti-CTLA-4 antibodies, ICIs in combination with radiation therapy, ICIs in combination with small molecule TKIs (such as MEK inhibitors and anti-angiogenic drugs), and ICIs in combination with bispecific antibodies. While some early trials have provided promising data, further research is needed to validate safety and efficacy. Alternative attempts to stimulate tumor immunity and enhance anti-tumor activity, such as ACT, oncolytic viruses, and cancer vaccines, are promising areas of active research. Additionally, treatment based on biomarkers is an inevitable trend in immunotherapy. The selection criteria are essential for identifying patients who may benefit from these drugs. While some promising biomarkers have been mentioned above, the discovery and validation of sensitive and specific biomarkers still require further exploration. With insights gained from additional trials aimed at developing effective treatment strategies, novel combinations and biomarkers may help guide clinicians in more personalized treatment of CRC patients, and immunotherapy for CRC is expected to make significant breakthroughs in the future.

References

[1] Ghahremanloo A, Javid H, Afshari AR, et al. Investigation of the role of neurokinin-1 receptor inhibition using aprepitant in the apoptotic cell death through PI3K/Akt/NF-κB signal transduction pathways in colon cancer cells. Biomed Res Int 2021;2021:1383878.

[2] Marabelle A, Le DT APA, et al. Efficacy of pembrolizumab in patients with noncolorectal high microsatellite instability/mismatch repair-deficient cancer: results from the phase II KEYNOTE-158 study. J Clin Oncol 2020;38(1):1–10.

[3] Overman MJ, McDermott R, Leach JL, et al. Nivolumab in patients with metastatic DNA mismatch repair-deficient or microsatellite instability-high colorectal cancer (CheckMate142): an open-label, multicentre, phase 2 study. Lancet Oncol 2017;18(9):1182–91.

[4] Bonjer HJ, Deijen CL, Abis GA, et al. A randomized trial of laparoscopic versus open surgery for rectal cancer. New Engl J Med 2015;372(14):1324–32.

[5] Miller KD, Siegel RL, Lin CC, et al. Cancer treatment and survivorship statistics, 2016. CA Cancer J Clin 2016; 66(4):271–89.

[6] Javid H, Karimi-Shahri M, Khorramdel M, et al. Probiotics as an adjuvant for management of gastrointestinal cancers through their anti-inflammatory effects: a mechanistic review. Curr Med Chem 2023;30(4):390–406.

[7] Ogura A, Konishi T, Cunningham C, et al. Neoadjuvant (chemo)radiotherapy with total mesorectal excision only is not sufficient to prevent lateral local recurrence in enlarged nodes: results of the multicenter lateral node study of patients with low cT3/4 rectal cancer. J Clin Oncol Off J Am Soc Clin Oncol 2019;37(1):33–43.

[8] Samstein RM, Lee CH, Shoushtari AN, et al. Tumor mutational load predicts survival after immunotherapy across multiple cancer types. Nat Genet 2019;51(2):202–6.

[9] Le DT, Durham JN, Smith KN, et al. Mismatch repair deficiency predicts response of solid tumors to PD-1 blockade. Sci (New York, NY) 2017;357(6349):409–13.

[10] Salvador MU, Truelson MRF, Mason C, et al. Comprehensive paired tumor/germline testing for lynch syndrome: bringing resolution to the diagnostic process. J Clin Oncol Off J Am Soc Clin Oncol 2019;37(8):647–57.

[11] Toh JWT, De Souza P, Lim SH, et al. The potential value of immunotherapy in colorectal cancers: review of the evidence for programmed death-1 inhibitor therapy. Clin Colorectal Cancer 2016;15(4):285–91.

[12] The Cancer Genome Atlas Network. Comprehensive molecular characterization of human colon and rectal cancer. Nature 2012;487(7407):330–7.

[13] Le DT, Uram JN, Wang H, et al. PD-1 blockade in tumors with mismatch-repair deficiency. N Engl J Med 2015;372(26):2509–20.

[14] Le DT, Diaz Jr LA, Kim TW, et al. Pembrolizumab for previously treated, microsatellite instability-high/mismatch repair-deficient advanced colorectal cancer: final analysis of KEYNOTE-164. Eur J Cancer (Oxford, Engl) 1990;2023(186):185–95.

[15] Le DT, Kim TW, Van Cutsem E, et al. Phase II open-label study of pembrolizumab in treatment-refractory, microsatellite instability-high/mismatch repair-deficient metastatic colorectal cancer: KEYNOTE-164. J Clin Oncol Off J Am Soc Clin Oncol 2020;38(1):11–9.

[16] Diaz LA, Jr SKK, Kim TW, et al. Pembrolizumab versus chemotherapy for microsatellite instability-high or mismatch repair-deficient metastatic colorectal cancer (KEYNOTE-177): final analysis of a randomised, open-label, phase 3 study. Lancet Oncol 2022;23(5):659–70.

[17] Overman MJ, Lonardi S, Wong KYM, et al. Durable clinical benefit with nivolumab plus ipilimumab in DNA mismatch repair-deficient/microsatellite instability-high metastatic colorectal cancer. J Clin Oncol Off J Am Soc Clin Oncol 2018;36(8):773–9.

[18] Lenz HJ, Van Cutsem E, Luisa Limon M, et al. First-line nivolumab plus low-dose ipilimumab for microsatellite instability-high/mismatch repair-deficient metastatic colorectal cancer: the phase II CheckMate142 study. J Clin Oncol 2022;40(2):161–70.

[19] Pfirschke C, Engblom C, Rickelt S, et al. Immunogenic chemotherapy sensitizes tumors to checkpoint blockade therapy. Immunity 2016;44(2):343–54.

[20] Herting CJ, Farren MR, Tong Y, et al. A multi-center, single-arm, phase Ib study of pembrolizumab (MK-3475) in combination with chemotherapy for patients with advanced colorectal cancer: HCRN GI14-186. Cancer Immunol Immunother 2021;70(11):3337–48.

[21] Yang H, Lee WS, Kong SJ, et al. STING activation reprograms tumor vasculatures and synergizes with VEGFR2 blockade. J Clin Invest 2019;129(10):4350–64.

[22] Fukuoka S, Hara H, Takahashi N, et al. Regorafenib plus nivolumab in patients with advanced gastric or colorectal cancer: an open-label, dose-escalation, and dose-expansion phase Ib trial (REGONIVO, EPOC1603). J Clin Oncol 2020;38(18):2053–61.

[23] Cousin S, Cantarel C, Guegan JP, et al. Regorafenib-Avelumab combination in patients with microsatellite stable colorectal cancer (REGOMUNE): a single-arm, open-label, phase II trial. Clin Cancer Res 2021;27(8):2139–47.

[24] Gong X, Li X, Jiang T, et al. Combined radiotherapy and anti-PD-L1 antibody synergistically enhances antitumor effect in non-small cell lung cancer. J Thorac Oncol 2017;12(7):1085–97.

[25] Sharabi AB, Nirschl CJ, Kochel CM, et al. Stereotactic radiation therapy augments antigen-specific PD-1-mediated antitumor immune responses via cross-presentation of tumor antigen. Cancer Immunol Res 2015;3(4):345–55.

[26] Rosenberg SA, Restifo NP. Adoptive cell transfer as personalized immunotherapy for human cancer. Sci (New York, NY) 2015;348(6230):62–8.

[27] Lonez C, Verma B, Hendlisz A, et al. Study protocol for THINK: a multinational open-label phase I study to assess the safety and clinical activity of multiple administrations of NKR-2 in patients with different metastatic tumour types. BMJ Open 2017;7(11), e017075.

[28] Li J, Chen QY, He J, et al. Phase I trial of adoptively transferred tumor-infiltrating lymphocyte immunotherapy following concurrent chemoradiotherapy in patients with locoregionally advanced nasopharyngeal carcinoma. Onco Targets Ther 2015;4(2), e976507.

[29] Lee YS, Lee WS, Kim CW, et al. Oncolytic vaccinia virus reinvigorates peritoneal immunity and cooperates with immune checkpoint inhibitor to suppress peritoneal carcinomatosis in colon cancer. J Immunother Cancer 2020;8(2).

[30] Parakrama R, Fogel E, Chandy C, et al. Immune characterization of metastatic colorectal cancer patients post reovirus administration. BMC Cancer 2020;20(1):569.

[31] Toh HC, Wang WW, Chia WK, et al. Clinical benefit of allogeneic melanoma cell lysate-pulsed autologous dendritic cell vaccine in MAGE-positive colorectal cancer patients. Clin Cancer Res 2009;15(24):7726–36.

[32] Caballero-Baños M, Benitez-Ribas D, Tabera J, et al. Phase II randomised trial of autologous tumour lysate dendritic cell plus best supportive care compared with best supportive care in pre-treated advanced colorectal cancer patients. Eur J Cancer (Oxford, Engl) 1990;2016(64):167–74.

[33] Wang P, Chen Y, Wang C. Beyond tumor mutation burden: tumor neoantigen burden as a biomarker for immunotherapy and other types of therapy. Front Oncol 2021;11, 672677.

[34] Yeong J, Suteja L, Simoni Y, et al. Intratumoral CD39(+)CD8(+) T cells predict response to programmed cell death protein-1 or programmed death ligand-1 blockade in patients with NSCLC. J Thorac Oncol 2021;16(8):1349–58.

[35] Heitzer E, Tomlinson I. Replicative DNA polymerase mutations in cancer. Curr Opin Genet Dev 2014;24 (100):107–13.

[36] Wang F, Zhao Q, Wang YN, et al. Evaluation of POLE and POLD1 mutations as biomarkers for immunotherapy outcomes across multiple cancer types. JAMA Oncol 2019;5(10):1504–6.

Index

Note: Page numbers followed by *f* indicate figures and *t* indicate tables.

U

V

W

Z

Printed in the United States
by Baker & Taylor Publisher Services